SPHEROMAKS

A Practical Application of Magnetohydrodynamic Dynamos and Plasma Self-Organization

SPHEROMAKS

A Practical Application of
Magnetohydrodynamic Dynamos
and Plasma Self-Organization

Paul M. Bellan
California Institute of Technology

Imperial College Press

Published by

Imperial College Press
57 Shelton Street
Covent Garden
London WC2H 9HE

Distributed by

World Scientific Publishing Co. Pte. Ltd.
P O Box 128, Farrer Road, Singapore 912805
USA office: Suite 1B, 1060 Main Street, River Edge, NJ 07661
UK office: 57 Shelton Street, Covent Garden, London WC2H 9HE

British Library Cataloguing-in-Publication Data
A catalogue record for this book is available from the British Library.

SPHEROMAKS: A PRACTICAL APPLICATION OF MAGNETOHYDRODYNAMIC DYNAMOS AND PLASMA SELF-ORGANIZATION

Copyright © 2000 by Imperial College Press

All rights reserved. This book, or parts thereof, may not be reproduced in any form or by any means, electronic or mechanical, including photocopying, recording or any information storage and retrieval system now known or to be invented, without written permission from the Publisher.

For photocopying of material in this volume, please pay a copying fee through the Copyright Clearance Center, Inc., 222 Rosewood Drive, Danvers, MA 01923, USA. In this case permission to photocopy is not required from the publisher.

ISBN 1-86094-141-9

Printed in Singapore.

To Josette

CONTENTS

Preface   xi

1 Introduction   1
    1.1   Brief description of spheromaks   1
    1.2   History and time-line   6

2 Basic Concepts   17
    2.1   Vacuum magnetic fields   17
    2.2   Poloidal and toroidal fields   18
    2.3   Magnetic stress tensor   20
    2.4   Beta   24
    2.5   Magnetic flux and symmetry   24
    2.6   Poloidal flux   24
    2.7   Poloidal flux and particle confinement   25
    2.8   Relation between field, field lines, and flux   26
    2.9   Safety factor   28
    2.10   The plasma as a magnetic flux conserver   32
    2.11   The condition for frozen-in flux   33
    2.12   Tendency of the plasma to maximize its inductance   35
    2.13   Cowling's theorem   35

3 Magnetic Helicity   37
    3.1   The issue of analyticity in Gauss's and Stokes's theorems   37
    3.2   Definition of magnetic helicity   39
    3.3   Helicity, safety factor, and twist of an isolated flux tube   42
    3.4   Gauge invariance   45
    3.5   Relative helicity   45
    3.6   Simply connected volumes v. doubly connected volumes   49
    3.7   Helicity conservation equation   50
    3.8   Single species helicity   58
    3.9   Magnetic reconnection   60
    3.10   Geometric interpretation of magnetic helicity   61
    3.11   Magnetic reconnection and helicity conservation   63
    3.12   Reconnection and dissipation   66

4 Relaxation of an Isolated Configuration to the Taylor State   71
    4.1   Introduction   71

	4.2	Helicity decay rate v. magnetic energy decay rate	72
	4.3	Derivation of the isolated Taylor state	73
	4.4	Relationship between helicity, energy, eigenvalue	75
	4.5	Cylindrical force-free states	77
	4.6	Comparison of minimum energy states in a long cylinder	79
	4.7	Spheromaks in spherical geometry	80

5 Relaxation in Driven Configurations   87
    5.1 Taylor relaxation in systems with open field lines   87
    5.2 Helicity injection   91
    5.3 Impedance of the driven force-free configuration   92

6 The MHD Energy Principle, Helicity, and Taylor States   95
    6.1 Derivation of the MHD Energy Principle   95
    6.2 Relationship of the energy principle to Taylor states   101
    6.3 Beta limit   103

7 Survey of Spheromak Formation Schemes   109
    7.1 Magnetized coaxial gun   110
    7.2 Non-axisymmetric gun method   117
    7.3 The inductive method   117
    7.4 Z-Theta pinch method   120

8 Classification of Regimes: an Imperfect Analogy to Thermodynamics   123
    8.1 Analogy to thermodynamics   123
    8.2 Classification of thermodynamic problems   123
    8.3 Analogy between lambda and temperature   126
    8.4 Strong and weak coupling   128
    8.5 Overview of next five chapters   128

9 Analysis of Isolated Cylindrical Spheromaks   129
    9.1 Flux, current, magnetic field, helicity and energy   129
    9.2 Experimental measurements   135
    9.3 Safety factor   136

10 The Role of the Wall   143
    10.1 Helicity insulation   143
    10.2 Equilibrium   143
    10.3 Tilt stability   146

11 Analysis of Driven Spheromaks: Strong Coupling — 155
   11.1 Force-free equilibria with open field lines — 156
   11.2 Flux surfaces — 162
   11.3 Safety factor variation with lambda — 167
   11.4 Flux amplification — 170
   11.5 Relative helicity — 171
   11.6 Relative energy — 174
   11.7 Gun efficiency — 175
   11.8 Gun impedance and load line — 177

12 Helicity Flow and Dynamos — 183
   12.1 Downhill flow of helicity — 183
   12.2 Dynamos and relaxation mechanisms — 185
   12.3 Observations of dynamo behavior — 189
   12.4 Deviation from the Taylor state — 196
   12.5 MHD dynamo, helicity flux, and lambda gradient — 199

13 Confinement and Transport in Spheromaks — 207
   13.1 Overview — 207
   13.2 Confinement times — 209
   13.3 Survey of transport mechanisms — 209
   13.4 Experiments on transport in spheromaks — 218
   13.5 Anomalous ion heating — 222

14 Some Important Practical Issues — 227
   14.1 Breakdown and Paschen curves — 227
   14.2 Gas puff valves — 232
   14.3 Wall desorption and contamination — 234
   14.4 Impurity line radiation — 236
   14.5 Refractory electrode materials — 239
   14.6 Skin effect and the wall as a flux conserver — 239
   14.7 Inductance budget — 240
   14.8 Mechanical forces — 241
   14.9 Noise radiation from pulsed power supplies — 241
   14.10 Ground loops — 242

15 Basic Diagnostics for Spheromaks — 243
   15.1 Magnetic field and electric current measurement — 243
   15.2 Equilibrium reconstruction using measurements at the wall — 248
   15.3 Voltage measurements — 248

15.4	Density measurement	249
15.5	Ion temperature measurement	255
15.6	Electron temperature measurement	257
15.7	Impurity radiation measurements	260

16 Applications of Spheromaks — 263

16.1	The spheromak as a fusion reactor	263
16.2	Accelerated spheromaks	269
16.3	Tokamak Fuel injection	274
16.4	Helicity injection current drive in tokamaks	275
16.5	Colliding spheromaks to investigate magnetic reconnection	277
16.6	Proposed additional spheromak applications	281

17 Solar and Space Phenomena Related to Spheromaks — 283

17.1	Sun-Earth connection viewed as helicity flux/relaxation	291
17.2	A spheromak-like laboratory model of solar prominences	293
17.3	S-shapes	297
17.4	Flux tube bifurcation and breakup	298
17.5	Comparison of magnetic field, field lines, flux tubes	299
17.6	Relaxation and line tying	300
17.7	Prominence simulation experiment	300

References		303
Appendix A:	*Vector Identities and Operators*	315
Appendix B:	*Bessel Orthogonality Relations*	317
Appendix C:	*Capacitor Banks*	321
Appendix D:	*Selected Formulae*	325
Index		333

PREFACE

Turning adversity into good fortune has always been an attractive proposition. Spheromaks present this sort of appeal because spheromaks offer the possibility of turning what was previously considered catastrophic instability into the basis of a low-cost plasma confinement scheme. The essential idea is to harness magnetohydrodynamic instability in such a way as to let the plasma relax or self-organize into a toroidal confinement configuration which otherwise would have to be created artificially at great expense. The physics of the spheromak has intrigued its devotees because this physics depends on intrinsic three dimensionality and complexity.

This book assumes the reader has a modest background in plasma physics but no prior knowledge of spheromaks. It has been written to be as self-contained as possible and whenever possible complete mathematical derivations are provided from first principles. However, to show that spheromaks are not just algebra, there is ample discussion of the practical experimental issues that must be addressed in order to make real spheromaks.

Spheromak research to date has been primarily motivated by the goal of developing a controlled thermonuclear fusion reactor, but spheromak physics is quite interesting in itself because it provides a conjunction for many of the most pressing and least understood issues in contemporary plasma physics. Thus, spheromak physics has much relevance beyond fusion. This book considers spheromaks from the related points of view of fusion relevance and fundamental plasma physics.

The book is organized as follows: Chapters 1-3 provide an introduction and present basic concepts. Chapter 1 places spheromaks in context relative to other toroidal confinement schemes and gives a brief history of spheromak research. Chapter 2 discusses several standard magnetohydrodynamic concepts used throughout the book. Chapter 3 introduces and explores magnetic helicity, a delightful and fascinating concept which quantifies magnetic field topology and is at the heart of spheromak physics.

Chapters 4-7 show how spheromaks are created and give examples. Chapters 4 and 5 explain the Taylor relaxation principle (Chapter 4 derives Taylor relaxation for isolated configurations and Chapter 5 derives Taylor relaxation for driven configurations). This powerful principle shows that plasmas left to their own devices spontaneously self-organize into spheromak equilibria, but the principle also has an enigmatic aspect because it does not explain how this self-organization takes place. Chapter 4 gives arguments supporting the essential premise underlying Taylor relaxation, namely that magnetic helicity is much better conserved than magnetic energy when there is fine-scale dissipation. Chapter 6 shows how Taylor relaxation is related to the well-known magnetic energy principle. Chapter 7 surveys the various schemes that have been used to create laboratory spheromaks.

Chapters 8-13 discuss more subtle aspects of spheromak formation and sustainment. Chapter 8 puts this discussion in context by arguing that there is a close (but imperfect) analogy between spheromak physics and thermodynamics and that this analogy can be used to classify regimes. Chapter 9 derives relationships between the main parameters characterizing an isolated spheromak. Chapter 10 explores the various roles of the wall surrounding the spheromak. Chapter 11 considers an idealized driven spheromak and derives the parametric dependence of the spheromak energy, helicity, and impedance. Chapter 12 takes up the more realistic viewpoint that magnetic helicity is not exactly conserved, but instead dissipates somewhat and must therefore be replenished by an external source; this leads to the issue of dynamos and sustained fluctuations. Chapter 13 surveys confinement and transport, deviations from ideal behavior which provide the main limits on spheromak performance.

Chapter 14 discusses several practical issues relevant to making laboratory spheromaks. Chapter 15 gives an overview of several of the standard diagnostics used on laboratory spheromaks. Chapter 16 discusses spheromak applications, beginning with fusion confinement and then going on to several fusion-related and non-fusion applications. The book concludes with Chapter 17, a survey of spheromak-related phenomena in solar and space physics; many of the ideas discussed in earlier chapters reappear in this much larger venue.

As in most fields, the notation in spheromak literature has developed haphazardly and is not always consistent. For clarity, this book always uses ψ to denote poloidal flux, Φ to denote toroidal flux, and χ to denote the Helmholtz function related to force-free states. Another consistency problem is that variables in different contexts are often represented by the same symbol; for example ϕ can denote electrostatic potential or toroidal angle depending on the context. Similarly V denotes voltage or volume, etc. Rather than define a set of completely consistent, non-overlapping, but unfamiliar symbols, I have elected to use standard notation as much as possible, and state that ϕ is an angle or a potential at the beginning of each separate discussion. Following this approach, cylindrical coordinates are (r, ϕ, z) while spherical coordinates are (r, θ, ϕ) and it will be explicitly stated whether an analysis is in cylindrical or in spherical coordinates.

Acknowledgments:

I would like to thank Michael Brown and Cris Barnes for their very thorough reading of the entire manuscript and their many useful comments. I would also like to thank Tom Jarboe and Masaaki Yamada for their comments about spheromak history and fusion issues, Bick Hooper for comments about dynamos and helicity flux, Ming Chu for reading and commenting on the modeling of externally driven spheromaks, and David Rust for commenting on the solar and space chapter. I also wish to express my appreciation to Torkil Jensen for a multitude of stimulating conversations over the years about magnetic helicity and relaxation.

I would like to thank the many authors who provided copies of artwork from their publications, to Alexei Pevtsov for providing the Yohkoh S-shape photo, and to Lennart Lindberg for graciously reconstructing the figures from his 1959 paper (Alfvén, Lindberg, and Mitlid).

Appreciation is also due to Wayne Waller and Carolyn Patterson of the Caltech Digital Media Center for help in reconstructing several figures for which original artwork was not available, to Greg Dunn for assistance in obtaining original figures and permission to use copyrighted material, and to the technical support group at MacKichan Software for assistance with Scientific WorkPlace, the Latex-based text editor used for writing this book.

I would also like to thank Cris Barnes for use of his very extensive bibliography of spheromak literature. I wish to express gratitude to all the researchers whose efforts have provided the material in this book.

Finally I would like to thank my family for putting up with what seemed to be an obsession about the behavior of twisted magnetic fields.

CHAPTER 1
Introduction

1.1 Brief description of spheromaks

Plasmas are gases composed of free electrons and ions. Typically, the electron and ion charge densities are nearly the same so that the plasma is an approximately neutral electrically conducting gas which is subject to electrical and magnetic forces in addition to the usual hydrodynamic forces. The process by which an ordinary gas is transformed into plasma is called ionization. For most plasmas, ionization take place when free electrons strike neutral atoms with sufficient force to eject bound electrons, thereby creating more free electrons and ions. In order for this process to occur, there must be some free electrons with kinetic energy exceeding the binding energy of the most weakly bound outer electron in a neutral atom. This means that plasmas typically have an electron temperature of at least a few electron volts (1 eV=11,604 K). Plasmas occur naturally in space environments (e.g., the solar corona, Earth's magnetosphere, the aurora) but must be created in the laboratory using artificial means.

If one wishes to trap a laboratory plasma, then some kind of confinement scheme is required, because otherwise the plasma will quickly convect to the surrounding walls and recombine. Substantial effort has been directed during the past half century towards developing devices which use magnetic fields to confine plasmas. These magnetic confinement schemes can be understood at many levels of sophistication, but ultimately are based on the magnetic force $\mathbf{F} = q\mathbf{v} \times \mathbf{B}$ acting on individual charged particles.

Spheromaks are a toroidal confinement configuration where the magnetic field is produced almost entirely by currents flowing in the plasma. The spheromak configuration is defined as an axisymmetric magnetohydrodynamic equilibrium with (i) a simply connected bounding surface, (ii) both toroidal and poloidal magnetic fields, and (iii) at least some closed poloidal flux surfaces. What distinguishes spheromaks from other toroidal configurations is that the toroidal magnetic field in spheromaks vanishes at the bounding surface (i.e., at the wall). Therefore no external coils link the spheromak and so the spheromak manages to have an internal toroidal field while still being simply connected. In contrast, tokamaks, reversed field pinches (RFP's), and stellarators all have finite toroidal magnetic field at the wall; this corresponds to having external coils linking the plasma. Field reversed configurations (FRC's) have zero toroidal magnetic field everywhere and so, like spheromaks, do not have coils linking the plasma. Thus, spheromaks manage to have a toroidal field without having toroidal field coils; FRC's do not have toroidal field coils but also do not have a toroidal field.

Figure 1.1 compares spheromak topology to the other toroidal confinement methods and Table 1.1 lists the similarities and differences. The device complexity increases going down the table; this is also obvious from Fig. 1.1. All devices except the stellarator use a toroidal current to produce the poloidal field required for confinement; the poloidal field in the stellarator is created by external helical coils so that current-free operation is obtained at the expense of loss of axisymmetry. The FRC is the simplest device but, having no toroidal field, is MHD-unstable and also has a field null on the magnetic axis.

device	Axi-symmetric	Poloidal field B_{pol}	Toroidal field B_ϕ	B_ϕ at wall	Chamber topology
FRC	yes	yes	no	no	spheroidal
spheromak	yes	yes	yes	no	spheroidal
RFP	yes	yes	yes	yes	toroidal
tokamak	yes	yes	yes	yes	toroidal
stellarator	no	yes	yes	yes	toroidal

Table 1.1   Comparison of topologies of various toroidal confinement devices

According to the magnetohydrodynamic (MHD) point of view, plasma is modeled as an electrically conducting fluid and confinement involves balancing the outward force of hydrodynamic pressure against the inward force due to interaction between magnetic fields and electric currents in the plasma. This balancing is most effective when the magnetic field lines in the plasma form nested surfaces called flux surfaces. The existence of flux surfaces means that any field line traces out a surface in three dimensional space and does not fill up a volume.

A point of view complementary to MHD and also more physically correct is provided by Hamiltonian-Lagrangian theory which shows that if there is symmetry about an axis, then confinement results from the conservation of canonical angular momentum for each particle. In this case, particle trajectories are restricted to surfaces on which the canonical angular momentum is a constant and confinement is akin to a spinning top standing upright because of conservation of angular momentum. Both the microscopic Hamiltonian-Lagrangian point of view and the macroscopic magnetohydrodynamic point of view arrive at the same conclusion because as particle mass goes to zero, invariance of canonical angular momentum becomes equivalent to the existence of flux surfaces. Thus, symmetry is important for confinement whether one uses the MHD point of view or the single particle point of view.

Flux surfaces are formed from the magnetic field produced by the combined effect of internal plasma currents and external coil currents. The various schemes for producing flux surfaces can be categorized according to the extent to which the flux surfaces are

1.1 Brief description of spheromaks

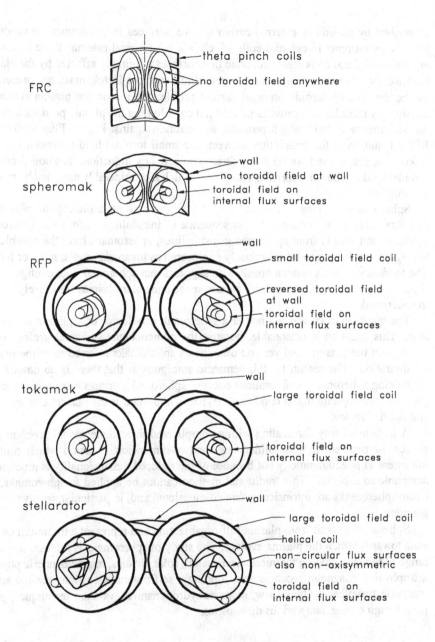

Fig. 1.1 Comparison between various toroidal confinement devices. FRC's and spheromaks have simply connected vacuum chambers, others have doubly connected vacuum chambers.

prescribed by plasma or external currents. Flux surfaces in stellarators are produced entirely by currents in external coils which link the toroidal plasma: these precision-engineered helical coils create accurate flux surfaces minimally affected by the plasma because the plasma is nearly current-free. Flux surfaces in tokamaks are prescribed by the dominantly toroidal internal current profile; the reason the plasma current is dominantly toroidal is because large external coils linking the plasma produce a strong toroidal magnetic field which provides stabilization against kinks. Flux surfaces in RFP's result from the interaction between the small toroidal field produced by coils linking the plasma and poloidal flux directly injected by induction. The coil-produced toroidal field can be considered as a seed field which is considerably modified by plasma instabilities.

Spheromaks are closely related to RFP's but have no coils linking the plasma so that flux surfaces are entirely the consequence of instabilities. Since the spheromak configuration results from spontaneous instabilities, spheromaks have the notable advantage of not having to be as precisely engineered as tokamaks, stellarators, or RFP's. The tendency to form spontaneously also suggests that spheromak-like configurations should occur in nature, and indeed, certain space and solar plasmas are closely related to spheromaks.

The question often arises whether a spheromak is a device or a plasma configuration. This question is reasonable, because the nomenclature 'tokamak' refers to the device, not the plasma, and yet one often hears spheromaks referred to as the plasma configuration. The reason for this semantic ambiguity is that there is no unique way for making spheromak configurations because spheromak plasmas form spontaneously given the appropriate initial conditions. What is important is the plasma configuration and not the device.

A traditional way for dealing with a complicated three dimensional problem is to reduce the problem to a simplified one- or two-dimensional version which contains the essential phenomenology but because of the reduced dimensionality is much more amenable to analysis. This traditional method cannot be applied to spheromaks, because spheromaks are intrinsically three dimensional and, in particular, involve helical geometry.

Spheromaks result from plasma self-organization and represent a minimum energy state towards which the plasma evolves. The study of spheromaks is relevant to a wide range of topics including thermonuclear fusion, solar physics, magnetospheric physics, astrophysics, magnetic reconnection, topology, self-organization, inaccessible states, magnetic turbulence, Ohm's law, magnetohydrodynamics, vacuum techniques, pulse power engineering, and various diagnostics.

1.2 History and time-line

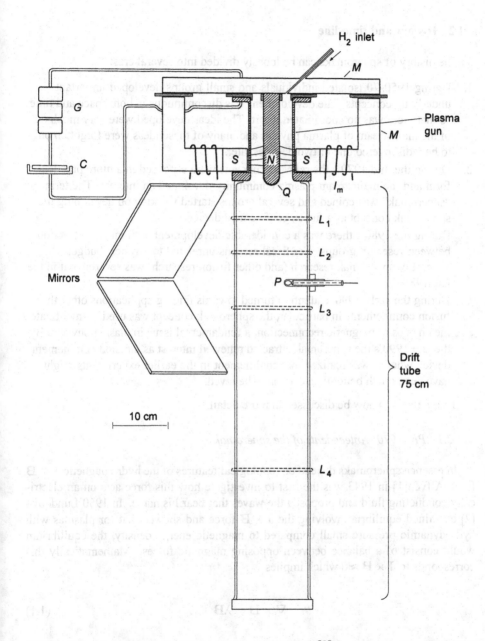

Fig. 1.2 Experimental setup of Alfvén, Lindberg, and Mitlid [8] (reconstructed drawing kindly provided by L. Lindberg).

1.2 History and time-line

The history of spheromaks can be loosely divided into several eras:

1. During 1950-70 isolated individuals and small groups developed important underlying concepts. The development was discontinuous in both space and time so that there was no coordinated effort. The ideas developed were very much out of the mainstream of plasma physics and many of these ideas were forgotten only to be rediscovered one or two decades later.
2. During the mid 1970's the relevant concepts were espoused in a more practical form and the mainstream plasma community developed an interest. The term "spheromak" was coined and several groups started working on developing the spheromak concept as a fusion confinement device.
3. During the 1980's there was a considerable development with much interaction between research groups. For fiscal reasons unrelated to physics, budgetary support of spheromak research (and other fusion research) was severely cut in the late 1980's.
4. During the early 1990's attention turned towards finding applications other than fusion confinement. In particular, the spheromak concept was used to investigate the physics of magnetic reconnection, a fundamental issue in plasma physics. By the late 1990's the spheromak attracted renewed interest as a fusion confinement device when it was realized that confinement in the earlier experiments might have been much better than originally believed.

These eras will now be discussed in more detail.

1.2.1 Pre-1970: Antecedents of the spheromak

In essence spheromaks depend on topological features of the hydromagnetic $\mathbf{J} \times \mathbf{B}$ force. Alfvén[1] in 1943 was the first to investigate how this force acts on an electrically conducting fluid and proposed the waves that bear his name. In 1950 Lundquist [2] examined equilibria involving the $\mathbf{J} \times \mathbf{B}$ force and showed that for plasmas with hydrodynamic pressure small compared to magnetic energy density, the equilibrium would consist of a balance between opposing magnetic forces. Mathematically this corresponds to $\mathbf{J} \times \mathbf{B} \approx 0$ which implies

$$\nabla \times \mathbf{B} = \lambda \mathbf{B} \qquad (1.1)$$

where λ is effectively an eigenvalue. Lundquist showed this balancing of magnetic forces provides the very simple equilibrium

$$\mathbf{B} = \bar{B}J_1(\lambda r)\hat{\theta} + \bar{B}J_0(\lambda r)\hat{z} \qquad (1.2)$$

where J_0 and J_1 are Bessel functions and $\bar{B}$ is a constant. This is called the Lundquist solution or the Bessel function model (BFM) and prescribes a helical magnetic field.

In the early 1950's researchers in the USA, the UK, the USSR, and several other countries started to work on the problem of controlled thermonuclear fusion. This required confinement of high temperature plasmas and considerable resources were devoted towards developing magnetic configurations exploiting the $\mathbf{J} \times \mathbf{B}$ force so as to provide confinement. Both Alfvén and Lundquist were thinking in terms of space plasmas (e.g., the magnetosphere, the solar corona, and astrophysical plasmas) while the fusion researchers were preoccupied with designing and constructing laboratory plasmas. Thus began a parallelism between space and laboratory MHD research which has continued and which has lead to many useful interchanges of ideas.

In 1957 Furth, Levine, and Waniek[3] considered the physical limits on coils producing large transient magnetic fields and showed that the ultimate limitation was coil rupture due to large magnetic forces. They proposed winding a coil in such a way that the $\mathbf{J} \times \mathbf{B}$ force would vanish within the coil and derived the required magnetic field profile to produce this force-free situation. This profile is precisely the same as the magnetic field profile of a spheromak confined by a cylindrical flux conserver.

In 1958 Woltjer[4, 5] considered the various constraints acting on a magnetohydrodynamic system, noted these constraints could be expressed in terms of integrals, and showed these constraints could be used to determine minimum energy states for a plasma. One constraint is the conservation of the magnetic helicity, a measure of the linkage of magnetic flux tubes with each other. Woltjer showed that, for a given magnetic helicity, the lowest energy state satisfied $\mathbf{J} = \lambda \mathbf{B}$ with λ spatially uniform. Chandrasekhar and Kendall [6] discussed this equation in cylindrical geometry and derived solutions now called Chandrasekhar - Kendall functions; these are generalizations of the Bessel function solution given by Lundquist. Chandrasekhar[7] also examined solutions in spherical geometry.

During the period 1959-1964 Alfvén, Lindberg, and Mitlid[8] and Lindberg and Jacobsen[9, 10] built and operated a device which produced rings of magnetized plasma. Figure 1.2 shows the setup of this experiment while Fig.1.3 shows the sequence of operation. While these experiments pre-date the modern spheromak concept, they can be considered as the first spheromak-related experiments because the essential features of modern coaxial spheromak guns were observed and identified. The original purpose [8] of these experiments was to determine whether RFP properties depended explicitly on how the configuration was formed; this issue was addressed by forming an RFP-like plasma using a coaxial magnetized plasma gun instead of the conventional method, transformer induction. During the course of the experiments several interesting and unexpected features were noted and these were investigated in some detail[9, 10]. Features relevant to spheromaks included: magnetic reconnection resulting in a detached plasma ring breaking off from the electrodes, conversion of toroidal flux into poloidal flux by helical instabilities, formation of closed poloidal flux surfaces, and amplification of the poloidal flux[11]. It is amazing that after these experiments were completed, the coaxial magnetized plasma gun concept lay dormant for nearly two decades.

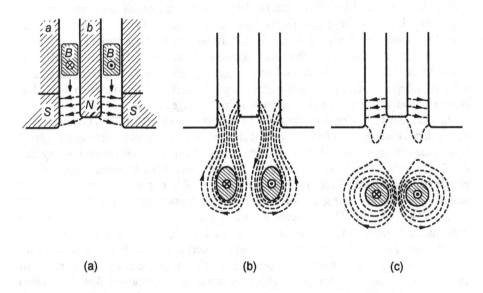

Fig.1.3 Sequence of operation as described by Alfvén, Lindberg, and Mitlid[8]: A poloidal field is established going between the N and S poles shown in (a). An accelerated plasma ring is initially captured by this poloidal field and distends it as shown in (b). Finally, as shown in (c) the poloidal field breaks and reconnects around the ring (reconstructed drawing kindly provided by L. Lindberg).

The 1950's and 1960's were a time of great diversity in fusion magnetic confinement concepts: stellarators were being developed in the USA, tokamaks in the USSR, and mirror machines in both the USA and USSR. In the UK development efforts focussed on the toroidal Z-pinch, now known as the RFP. The RFP looks similar to a tokamak, but differs in having a much weaker toroidal field and a much larger toroidal current. ZETA, a large RFP operated at Culham in the 1950's and 1960's displayed a mysterious behavior: after an initial period of violent instability, the plasma would settle into a quiescent state. If the ratio of toroidal current to toroidal field in this state exceeded a threshold, the toroidal field had a spontaneously reversed polarity near the surface[12], hence the name Reversed Field Pinch (RFP). The ZETA program was discontinued in 1968 and from 1969 onward attention in fusion research shifted to tokamaks (and to a lesser extent, magnetic mirrors).

In the 1960's Bostick and Wells[13] investigated the conical θ-pinch and found that this developed a spontaneous toroidal field; this unexpected effect was attributed to Hall terms. Wells[14] interpreted the dynamical evolution of the conical θ-pinch in terms of Woltjer's helicity-constrained energy minimization and proposed that these plasmas would develop a $\mathbf{J} = \lambda\mathbf{B}$ equilibrium. Wells and colleagues built a series of small conical θ-pinches first at the Princeton Plasma Physics Lab and later at the University of Florida. These devices involved extremely rapid dynamics which were difficult to follow with the diagnostics available at the time. However, it is interesting that magnetic probe measurements on one of these devices demonstrated[15] the toroidal magnetic field profile which is the hallmark of spheromaks, i.e., the toroidal field was zero on axis, rose to a maximum at some interior point, and then went to zero at the wall; Fig. 1.4 shows the data from Wells' measurement. In 1972 Nolting, Jindra, and Wells[16] discussed the magnetic field profile for force-free spherical MHD configurations produced by a conical theta pinch and presented measurements consistent with the theoretical profiles. Research activity on conical θ-pinches ceased in the 1970's except for one device [17] at the University of Washington.

1.2.2 Advances in theory: Taylor relaxation and development of the theoretical model for the spheromak

In 1974, long after ZETA had been shut down, Taylor[18] proposed an explanation for ZETA's mysterious tendency to develop reversed toroidal magnetic fields at the plasma edge. Taylor proposed that magnetic turbulence does not change the global helicity content of plasma but does dissipate magnetic energy. The turbulence would cause changes in magnetic topology such that every time a microscopic flux linkage was broken, another would be created; thus, global helicity, the measure of flux linkages, would be conserved. This point of view recast Woltjer's abstract variational principle into a practical prescription for how a real plasma would behave — a turbulent plasma would spontaneously relax (or self-organize) to a simple, well-defined state now called the Taylor state. The relaxation process would conserve helicity but dissipate energy until reaching a lowest energy state. The relaxed state (Taylor state) satisfies Eq.(1.1) and, for a large aspect ratio RFP, the solutions of this equation are just Lundquist's Bessel function equilibrium. The field reversal was simply a consequence of the $J_0(\lambda r)$ Bessel function passing through zero when λr became larger than the first root of J_0.

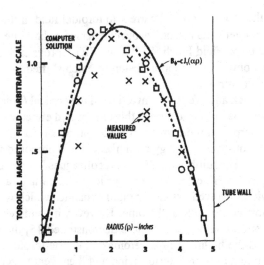

Fig.1.4 Wells' [15] 1964 conical theta pinch measurement showing that the toroidal field vanishes at both the geometric axis ($r = 0$) and the wall ($r = 5$ inches).

The ZETA observations were in good agreement with the Bessel function model, and this agreement persisted for other RFP's. The Taylor relaxation model explained the dominant features of RFP equilibrium and received much attention.

Rosenbluth and Bussac[19] extended Taylor's approach to spherical geometry and discussed minimum energy states having zero toroidal field on the bounding surface. This means that there is no externally driven current on the device axis and so there are no external toroidal field coils. The remarkable consequence of this freedom from external coils is that the magnetohydrodynamic equilibrium becomes simply connected. Thus the plasma container has the topology of a sphere in contrast to tokamaks, stellarators and RFP's all of which are doubly connected and require containers with the more complicated topology of a toroid (doughnut). Figure 1.5 shows the distinction between simply connected and doubly connected topologies.

The mathematical form of Rosenbluth and Bussac's spherical solutions was the same as what Chandrasekhar[7] had studied many years earlier. Rosenbluth and Bussac called this simply-connected, spherical, force-free equilibrium a "spheromak". Thus, a spheromak is a configuration with the topology of a sphere and with a magnetic field satisfying Eq.(1.1). Because of its minimalist design the spheromak immediately attracted widespread attention. The features of simply connected geometry, no external coils linking the plasma, and no toroidal magnetic field at the wall offered the possibility of a fusion confinement device much smaller and less costly than the more familiar doubly-connected devices. The interest in the spheromak was so great in the late 1970's that several different groups decided to attempt making spheromaks using a variety of methods.

1.2.3 The 1980.'s: The spheromak investigated as a fusion confinement scheme

It was not clear at the time which, if any, of the proposed methods for spheromak formation would work. This is because the spheromak was essentially a detached magnetic bubble and there seemed a possibility that the spheromak equilibrium might be mathematically self-consistent, but physically inaccessible. A rough analogy to this quandary would be demonstrating the concept of a soap bubble without knowing any technique for actually making bubbles.

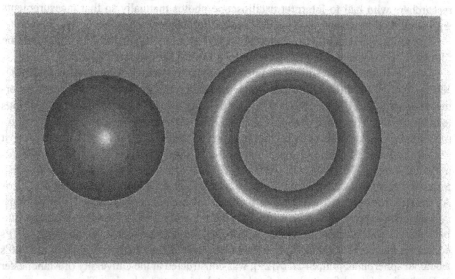

Fig.1.5 Left: Simply connected topology (spheroid); right: doubly connected topology (toroid).

The first spheromak experiments were at Nihon University [20], the PS-1 $Z - \theta$ experiment[21] at the University of Maryland, the Beta II experiment[22, 23] at Livermore, the CTX experiment[24] at Los Alamos, the proto-S1 experiment [25] at Princeton, and an experiment at Osaka University[26]. Even though these experiments used three different methods for making spheromaks ($z - \theta$ pinch, coaxial magnetized plasma guns, inductive source), they all succeeded in producing spheromak configurations. This broad success was remarkable and showed that detached spheromaks would form spontaneously given the right initial conditions. The demonstration that there is no unique way to make a spheromak showed convincingly that the spheromak is a lowest energy state towards which a magnetohydrodynamic system naturally evolves. The status of these early spheromak experiments was reviewed in 1983 by Furth[27].

The 1980's were a golden age for spheromak research; machines were built and upgraded, diagnostics were improved, and the Taylor relaxation concept was generalized. The advent of multi-channel transient digitizers (devices which collect temporal sequences of data in digital form and store this data in easily accessed computer memory) made it possible to measure internal magnetic fields with high spatial and temporal resolution and so verify the details of the relaxed state model. These transient digitizers gave the researchers of the 1980's a tremendous advantage compared to previous researchers who had to interpret oscilloscope photos manually so that measurements with combined space and time resolution were impractical.

The gross MHD stability of spheromaks was also investigated and, in particular, the tilt instability was identified experimentally and simulated numerically[28].

Despite these improvements in understanding spheromak formation, equilibrium, and gross MHD stability, little was known regarding the intrinsic confinement properties of spheromaks because the intrinsic confinement was typically masked by spurious loss mechanisms, especially atomic line radiation. The observed confinement properties of spheromaks were certainly not competitive with tokamaks and, in particular, it was not possible to obtain electron temperatures higher than about 100 eV in spheromaks for most of the 1980's. However, by the end of the 1980's achieved confinement improved significantly (electron temperatures as high as 400 eV were obtained at Los Alamos). Unfortunately, and despite these promising results, all major US spheromak confinement experiments were shut down by the early 1990's because of budget cutbacks in US fusion research. Smaller spheromak programs begun in Japan in the 1980's continued and some smaller experiments also continued in the USA. In the late 1980's a coaxial spheromak gun, SPHEX[29], was constructed at the University of Manchester in England and used for studying fundamental spheromak physics. There was also dialogue with the space physics community, and in particular, Heyvaerts and Priest [30] in 1984 applied the Taylor relaxation hypothesis to model the topology of solar coronal structures.

1.2.4 The 1990's: Search for other applications and renaissance in confinement efforts

Faced with the prospect that spheromaks might not be developed as fusion confinement devices, by the late 1980's and early 1990's spheromak researchers in the USA looked for other possible applications. Hammer, Hartman et al.[31] showed that spheromaks could be translated in space and accelerated to extremely high velocity using a coaxial rail gun and proposed several interesting applications for an accelerated spheromak. Perkins et al.[32] and Parks[33] examined how accelerated spheromaks could be used to refuel tokamaks; this was investigated experimentally by the Caltech group [34] and by Raman et al.[35] at INRS in Canada. Degnan, Peterkin et al. [36, 37] at Kirtland Airforce Base investigated the possibility of using accelerated spheromaks as a high power x-ray source (x-rays would be produced when a rapidly moving spheromak collided with a target); they also considered using the moving spheromak as the armature in a plasma opening switch[38].

Jensen and Chu[39] proposed that spheromak-like plasma guns could transfer magnetic helicity to a tokamak and act as an effective current drive. This was demonstrated in 1990 on a small scale by Brown and Bellan[40] at Caltech and in 1994 served as the basis of Jarboe and Nelson's Helicity Injection Tokamak[41, 42] at the University of Washington. Raman et al.[35] demonstrated non-disruptive refueling of the medium size Tokamak de Varennes. The SPHEX group[43, 44] continued to investigate many important aspects of spheromak physics, especially dynamo action. Spheromak research from 1979-1993 was summarized in a 1994 review article by Jarboe[45].

In the late 1990's spheromak concepts were applied towards the study of fundamental plasma physics, especially the problem of magnetic reconnection. Yamada et al.[46] and Ono et al.[47] at the University of Tokyo investigated the three dimensional magnetic reconnection associated with colliding, merging spheromaks. At Princeton, Yamada, Ji, et al.[48] built MRX, a spheromak-based device specifically designed to provide a well-defined reproducible magnetic reconnection layer. Geddes, Kornack and Brown[49] used an interacting double spheromak experiment at Swarthmore College to investigate magnetic reconnection.

Also during the 1990's interest increased among space and solar physicists in the spheromak-related concepts of magnetic helicity and relaxation. In particular, Rust and Kumar[50, 51] applied these concepts to solar prominences and modeled the dynamical evolution of these structures in terms of Taylor's relaxed states. Bellan and Hansen[52] exploited spheromak techniques in a laboratory experiment which simulated solar prominences.

In the mid 1990's Fowler et al.[53] and Mayo[54] re-evaluated the confinement performance of the Los Alamos CTX spheromak (which had ceased operating several years earlier) and postulated that core confinement was substantially better than previously believed. By the late 1990's, with energy prices low, the strategy of the US fusion program shifted towards developing speculative non-tokamak concepts which might ultimately prove more attractive than tokamaks. In 1998 construction began on a new spheromak at Livermore. This device, the Sustained Spheromak Physics Experiment (SSPX) is designed[55] to take into account the revised analysis of the 1980's experiments. Arguments in favor of the spheromak as a fusion reactor were summarized in Ref. [56].

Selected spheromak-related papers are listed in Tables 1.2-1.4; the time-line provided by these tables gives a sense of the pace and direction of spheromak development.

Year (type)	Description
1950 (Theory)	Lundquist[2] proposes force-free equilibrium $\mathbf{J} \times \mathbf{B} = 0$
1954 (Theory)	Lüst and Schlüter[57] discuss force free magnetic fields
1957 (Theory)	Furth et al.[3] investigate force-free coils and derive magnetic equilibria analogous to spheromaks
1958 (Theory)	Woltjer[4] shows $\mathbf{J} = \lambda \mathbf{B}$ is a minimum energy state, introduces concept of conservation of magnetic helicity
1959 (Expt.)	Alfvén et al.[8] produce magnetized plasma rings with a coaxial plasma gun, observe reconnection and flux amplification
1964 (Expt.)	Wells[15] observes toroidal field going to zero at wall in conical θ pinch, proposes $\mathbf{J} \times \mathbf{B} = 0$ equilibrium
1974 (Theory)	Taylor[18] proposes that magnetic turbulence causes relaxation to $\mathbf{J} = \lambda \mathbf{B}$ equilibrium, shows this provides good model for RFP
1979 (Theory)	Rosenbluth and Bussac[19] describe the "spheromak", a simply-connected force-free equilibrium, consider tilt stability

Table 1.2   Selected spheromak-relevant publications from 1950-1979

1.2 History and time-line

Year (type)	Description
1980 (Expt.)	Nogi et al.[20] form $Z - \theta$ spheromak at Nihon University
1980 (Expt.)	Goldenbaum et al.[21] form $Z - \theta$ spheromak at U. Maryland, start of PS program
1980 (Expt.)	Jarboe et al.[24] form coaxial gun spheromak at Los Alamos, start of CTX program
1981 (Expt.)	Yamada et al.[25] form flux-core spheromak at Princeton, start of S-1 program
1981 (Expt.)	Turner et al.[22] describe Beta II coaxial gun experiment at Lawrence Livermore
1981 (Expt.)	Watanabe et al.[26] form a detached spheromak in an experiment at Osaka University
1982 (Theory)	Katsurai and Yamada[58] discuss spheromak fusion reactor design
1983 (Expt.)	Katsurai et al.[59] initiate the TS series of spheromak experiments at Univ. of Tokyo
1983 (Theory)	Sato and Hayashi[28] create 3D numerical simulation of spheromak tilt instability
1983 (Expt.)	Jarboe et al.[60] demonstrate slow formation and sustainment of spheromak using coaxial gun
1984 (Theory)	Jensen and Chu[39] propose that helicity injection could be used for toroidal current drive in tokamaks
1985 (Expt.)	Hagenson and Krakowski[61] discuss spheromak fusion reactor design
1986 (Expt.)	Barnes et al.[62] at Los Alamos provide experimental verification of helicity conservation in the CTX spheromak
1987 (Expt.)	Honda et al.[63] at Osaka Univ. describe the CTCC-1 spheromak
1987 (Expt.)	Bruhns et al.[64] at Univ. of Heidelberg add vacuum toroidal field to a spheromak to make ultra low aspect ratio tokamak (ULART)
1988 (Expt.)	Hammer et al.[31] at LLNL demonstrate spheromak acceleration/compression on RACE
1988 (Expt.)	Wysocki et al.[65] find pressure-driven instability in CTX to be well above β limit predicted by MHD

Table 1.3   Selected spheromak-relevant publications from the 1980's

Year	Description
1990 (Expt.)	Wysocki et al.[66] demonstrate 0.18 ms energy confinement times in CTX
1990 (Expt.)	Jarboe et al.[67] report $T_e \sim 400$ eV in a spheromak with small flux conserver
1990 (Expt.)	Brown and Bellan[40] at Caltech demonstrate helicity injection current drive on the Encore tokamak
1990 (Expt.)	Yamada, Ono et al.[46] investigate magnetic reconnection of two colliding spheromaks
1990 (Expt.)	Wira and Pietrzyk[68] demonstrate spheromak formation by a conical θ pinch
1993 (Expt.)	al-Karkhy, Browning et al. [69] observed dynamo effect in SPHEX spheromak
1993 (Expt.)	Degnan et al.[36] describe the very large MARAUDER spheromak at Kirtland AFB
1994 (Theory)	Fowler et al.[53] proposed possibility of Ohmic ignition in a spheromak fusion reactor
1994 (Expt.)	Raman et al.[35] demonstrate central fueling in Tokamak de Varennes by spheromak injection
1994 (Expt.)	Nelson et al.[42] report formation and sustainment of Helicity Injection Tokamak at U. Wash.
1997 (Expt.)	Yamada et al.[48] use spheromak concepts to investigate magnetic reconnection on MRX at Princeton
1998 (Expt.)	Geddes, Kornack and Brown[49] investigate magnetic reconnection at Swarthmore
1998 (Expt.)	Hooper et al.[55] initiate new spheromak program at Lawrence Livermore National Lab

Table 1.4 Selected spheromak-relevant publications from the 1990's

CHAPTER 2
Basic Concepts

This chapter develops certain basic concepts which underlie the analysis presented in later chapters.

2.1 Vacuum magnetic fields

Magnetic fields are generated by currents but the current need not be at the location of the magnetic field. For example, suppose that a certain volume V has internal magnetic fields but no internal currents. The magnetic field inside V is called a vacuum magnetic field because this field must be generated by currents external to V. Vacuum magnetic fields are also called potential magnetic fields because they can be expressed as the gradient of a potential, since if $\mu_0 \mathbf{J} = \nabla \times \mathbf{B} = 0$ then $\mathbf{B} = \nabla \chi$. The function χ must satisfy Laplace's equation $\nabla^2 \chi = 0$ in order to have $\nabla \cdot \mathbf{B} = 0$ and so the vacuum magnetic field is uniquely determined by normal boundary conditions prescribed on the surface S bounding V.

Other magnetic fields (i.e., non-vacuum fields) may also satisfy the same normal boundary conditions on S, but the lowest energy magnetic field satisfying these prescribed boundary conditions is the unique vacuum field. It therefore always requires work to perturb the vacuum field while holding the boundary condition on S fixed. To prove this statement, consider a volume V with $\mathbf{B}$ prescribed on the bounding surface S and consider a perturbation that does not change this boundary condition. Let $\mathbf{B}_{\min}$ be the lowest energy magnetic field satisfying the prescribed boundary condition and let $\mathbf{B} = \mathbf{B}_{\min} + \delta \mathbf{B}$ be a field slightly different from this minimum-energy field, but satisfying the same boundary condition. The magnetic energy associated with $\mathbf{B}$ is

$$W = \int_V \frac{(\mathbf{B}_{\min} + \delta \mathbf{B})^2}{2\mu_0} d^3 r \tag{2.1}$$

so that the increment in energy above the energy associated with $\mathbf{B}_{\min}$ is

$$\delta W = \int_V \frac{(\mathbf{B}_{\min} + \delta \mathbf{B})^2 - B_{\min}^2}{2\mu_0} d^3 r$$

$$= \int_V \frac{2 \mathbf{B}_{\min} \cdot \delta \mathbf{B} + (\delta \mathbf{B})^2}{2\mu_0} d^3 r. \tag{2.2}$$

Since $\mathbf{B}_{min}$ was assumed to be the minimum-energy field, δW must be quadratic in $\delta \mathbf{B}$. Consequently the term linear in $\delta \mathbf{B}$ must vanish so

$$\int \mathbf{B}_{min} \cdot \delta \mathbf{B} \, d^3r = 0 \tag{2.3}$$

for arbitrary $\delta \mathbf{B}$. Writing $\delta \mathbf{B} = \nabla \times \delta \mathbf{A}$ this becomes

$$\begin{aligned} 0 &= \int [\nabla \cdot (\delta \mathbf{A} \times \mathbf{B}_{min}) + \delta \mathbf{A} \cdot \nabla \times \mathbf{B}_{min}] \, d^3r \\ &= \int_S d\mathbf{S} \cdot \delta \mathbf{A} \times \mathbf{B}_{min} + \int \delta \mathbf{A} \cdot \nabla \times \mathbf{B}_{min} \, d^3r. \end{aligned} \tag{2.4}$$

Because $\mathbf{B} = \mathbf{B}_{min}$ on S, the component of $\delta \mathbf{A}$ tangential to S must vanish on S for, if not, then finite $\nabla \times \delta \mathbf{A}_\parallel$ would imply a finite normal component $\delta \mathbf{B}$ on S. Using $\delta \mathbf{A}_\parallel = 0$ on S, the surface integral in Eq.(2.4) vanishes. Since $\delta \mathbf{A}$ is arbitrary within V, the volume integral term in Eq.(2.4) shows that $\nabla \times \mathbf{B}_{min}$ must vanish within V, proving that $\mathbf{B}_{min}$ is indeed a current-free, potential field.

2.2 Poloidal and toroidal fields

Magnetic fields are solenoidal, that is they have zero divergence. The nomenclature poloidal[1] and toroidal is most easily understood by considering solenoidal vector fields in the context of the Earth's geometry. Fields entering or leaving the poles (i.e., parallel to lines of longitude) are called poloidal; fields parallel to the equator (or lines of latitude) are called toroidal. If ϕ denotes the toroidal direction, then toroidal vectors are parallel to $\nabla \phi = \hat{\phi}/r$ and poloidal vectors are perpendicular to $\nabla \phi$. The most general axisymmetric form for poloidal fields will be $\mathbf{B}_{pol} = \nabla F \times \nabla \phi = \nabla \times F \nabla \phi$, while the most general form for toroidal fields will be $\mathbf{B}_{tor} = I \nabla \phi$. Thus the most general form of axisymmetric magnetic field will be a sum of poloidal and toroidal fields, $\mathbf{B} = \nabla F \times \nabla \phi + I \nabla \phi$. The curl of a toroidal field is poloidal and vice-versa. The labeling F and I has been selected because these are respectively the poloidal flux and the poloidal current except for a factor of 2π. The poloidal flux at a location r, z is the magnetic flux linked by a circle of radius r coaxial with the z axis and with center at axial location z; this geometry is illustrated in Fig. 2.1. The poloidal current is the current flowing through a similarly defined circle.

A feature distinguishing poloidal fields from toroidal fields is that in spherical geometry toroidal fields are solenoidal fields with no radial component whereas poloidal

---

[1] The word poloidal is derived from the word polar.

2.2 Poloidal and toroidal fields

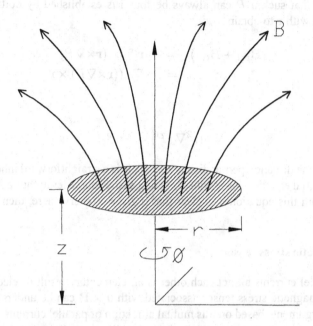

Fig.2.1 Poloidal flux is flux penetrating a circle of radius r at axial location z. The circle center is on the z axis and the normal to the surface is in the direction of $\hat{z}$.

fields are solenoidal fields that may have radial components[70]. Denoting **r** as the radius vector from the origin (in spherical coordinates), toroidal fields have the form

$$\mathbf{B}_{tor} = -\mathbf{r} \times \nabla T = \nabla \times (T\mathbf{r}) \qquad (2.5)$$

which is obviously solenoidal (i.e., $\nabla \cdot \mathbf{B} = 0$).

The poloidal field is defined as $\mathbf{B}_{pol} = \mathbf{B} - \mathbf{B}_{tor}$ and it is postulated that $\mathbf{B}_{pol}$ can always be expressed as

$$\mathbf{B}_{pol} = \nabla \times \nabla \times (P\mathbf{r}) = -\nabla \times (\mathbf{r} \times \nabla P). \qquad (2.6)$$

This postulate can be validated if it can be demonstrated that an appropriate scalar function P can be found for any **B**.

The proof that such a P can always be found is established by dotting the total magnetic field with $\mathbf{r}$ to obtain

$$\begin{aligned}\mathbf{r} \cdot (\mathbf{B}_{pol} + \mathbf{B}_{tor}) &= -\mathbf{r} \cdot \nabla \times (\mathbf{r} \times \nabla P) \\ &= -\nabla \cdot ((\mathbf{r} \times \nabla P) \times \mathbf{r})\end{aligned} \quad (2.7)$$

which gives

$$\nabla \cdot \left(r^2 \nabla_\perp P\right) = -r B_r; \quad (2.8)$$

here $\perp$ means the direction perpendicular to $\hat{r}$. This is a straightforward inhomogeneous partial differential equation for P with B_r acting as the source term; thus a P can always be found. From this equation it is clear that if $B_r = 0$ everywhere, then the poloidal field vanishes.

2.3 Magnetic stress tensor

That parallel currents attract each other is an elementary result of electromagnetic theory. The magnetic stress tensor associated with $\mathbf{J} \times \mathbf{B}$ can be understood qualitatively using arguments based on this mutual attraction of parallel currents.

A distributed current can be imagined as a bundle of parallel currents which mutually attract each other; this constitutes the well-known pinch force. The magnetic field associated with the bundle of parallel currents encircles the bundle, and so if one wants to imagine the pinch force as due to this magnetic field, one pretends that the magnetic field line has an elastic tension, just like an elastic band wrapped around a bundle of toothpicks. While this image is often invoked, it is strictly speaking incorrect because there is actually no force along the magnetic field line. The only force is $\mathbf{J} \times \mathbf{B}$ and clearly $\mathbf{B} \cdot (\mathbf{J} \times \mathbf{B}) = 0$. A better way for characterizing the magnetic force will be provided in the next few paragraphs.

Now consider a circular loop of current. Currents on opposite sides of this loop are antiparallel and hence repel each other, tending to expand the loop diameter; this effect is called the hoop force. Since the magnetic field line density interior to the loop exceeds the field line density exterior to the loop, magnetic field line density can be construed as exerting a pressure perpendicular to the magnetic field. From this point of view, excess magnetic pressure in the loop interior causes the hoop force.

These qualitative ideas can now be quantified. The magnetohydrodynamic equation of motion is

$$\rho \frac{d\mathbf{U}}{dt} = \mathbf{J} \times \mathbf{B} - \nabla \cdot \mathbf{P} \quad (2.9)$$

where **P** is the pressure tensor. Using Ampere's law the magnetic force term can be written as

$$\mathbf{J} \times \mathbf{B} = \frac{(\nabla \times \mathbf{B})}{\mu_0} \times \mathbf{B}. \tag{2.10}$$

Invoking the vector identity $\nabla(\mathbf{F} \cdot \mathbf{G}) = \mathbf{F} \cdot \nabla \mathbf{G} + \mathbf{G} \cdot \nabla \mathbf{F} + \mathbf{F} \times \nabla \times \mathbf{G} + \mathbf{G} \times \nabla \times \mathbf{F}$ this becomes

$$\mathbf{J} \times \mathbf{B} = -\frac{1}{\mu_0} \nabla \cdot \left[\frac{B^2}{2} \mathbf{I} - \mathbf{B}\mathbf{B} \right] \tag{2.11}$$

which can be expressed in matrix form as

$$\mathbf{J} \times \mathbf{B} = -\frac{1}{2\mu_0} \nabla \cdot \left\{ \begin{bmatrix} B^2 & 0 & 0 \\ 0 & B^2 & 0 \\ 0 & 0 & 0 \end{bmatrix} \mathbf{I}_\perp - \begin{bmatrix} 0 & 0 & 0 \\ 0 & 0 & 0 \\ 0 & 0 & B^2 \end{bmatrix} \hat{B}\hat{B} \right\} \tag{2.12}$$

where $\mathbf{I}_\perp = \mathbf{I} - \hat{B}\hat{B}$ is the unit tensor perpendicular to the magnetic field. If the pressure is anisotropic so that $\mathbf{P} = P_\perp \mathbf{I}_\perp + P_\parallel \hat{B}\hat{B}$ then the equation of motion can be written as

$$\rho \frac{d\mathbf{U}}{dt} = -\nabla \cdot \left\{ \begin{bmatrix} P_\perp + \frac{B^2}{2\mu_0} & 0 & 0 \\ 0 & P_\perp + \frac{B^2}{2\mu_0} & 0 \\ 0 & 0 & 0 \end{bmatrix} \mathbf{I}_\perp + \begin{bmatrix} 0 & 0 & 0 \\ 0 & 0 & 0 \\ 0 & 0 & P_\parallel - \frac{B^2}{2\mu_0} \end{bmatrix} \hat{B}\hat{B} \right\} \tag{2.13}$$

showing that the magnetic stress (i) acts as a pressure $B^2/2\mu_0$ in the direction perpendicular to the magnetic field since it adds to the thermal pressure $P_\perp$ and (ii) acts as a tension $B^2/2\mu_0$ along the magnetic field since it subtracts from the thermal pressure $P_\parallel$. As noted above, this explanation is somewhat glib, because there is actually no magnetic force parallel to the magnetic field. In fact, the existence of a $\hat{B}\hat{B}$ term in the magnetic stress tensor does *not* imply there is a magnetic force in the $\hat{B}$ direction.

A more useful decomposition of the magnetic stress tensor focuses on the importance of magnetic curvature and uses the relation

$$\hat{B} \cdot \nabla \hat{B} = -\frac{\hat{R}}{R} \tag{2.14}$$

where R is the radius of curvature of the magnetic field line and $\hat{R}$ is the unit vector pointing from the center of curvature to the field line. This relation may be easily validated by defining a local cylindrical coordinate system (R, ϕ, z) with z axis passing through the center of curvature as shown in Fig.2.2. The magnetic field unit vector is therefore in the azimuthal direction so that $\hat{\phi} = \hat{B}$ and

$$\hat{\phi} \cdot \nabla \hat{\phi} = \hat{\phi} \cdot \left(\hat{R}\frac{\partial}{\partial r} + \hat{\phi}\frac{1}{R}\frac{\partial}{\partial \phi} + \hat{z}\frac{\partial}{\partial z} \right) \hat{\phi} = \frac{1}{R}\frac{\partial \hat{\phi}}{\partial \phi} = -\frac{\hat{R}}{R}. \quad (2.15)$$

The magnetic force can be expressed as

$$\begin{aligned} \mathbf{J} \times \mathbf{B} &= \frac{1}{\mu_0} \left[\nabla \times \left(B\hat{B} \right) \right] \times B\hat{B} \\ &= \frac{B^2}{\mu_0} \left(\nabla \times \hat{B} \right) \times \hat{B} + \left[\nabla \left(\frac{B^2}{2\mu_0} \right) \times \hat{B} \right] \times \hat{B}. \end{aligned} \quad (2.16)$$

Using the identities $0 = \nabla \left(\hat{B} \cdot \hat{B} \right) = 2\hat{B} \cdot \nabla \hat{B} + 2\hat{B} \times \nabla \times \hat{B}$ and $\left(\hat{B} \times \mathbf{Q} \right) \times \hat{B} = \mathbf{Q}_\perp$ where $\mathbf{Q}_\perp$ is the component of $\mathbf{Q}$ perpendicular to $\hat{B}$, the magnetic force becomes

$$\mathbf{J} \times \mathbf{B} = -\frac{B^2}{\mu_0}\frac{\hat{R}}{R} - \nabla_\perp \left(\frac{B^2}{2\mu_0} \right). \quad (2.17)$$

The curvature term is sometimes defined as

$$\kappa = \hat{B} \cdot \nabla \hat{B} = -\hat{B} \times \left(\nabla \times \hat{B} \right) = -\frac{\hat{R}}{R} \quad (2.18)$$

so that the magnetic force is

$$\mathbf{J} \times \mathbf{B} = \frac{B^2}{\mu_0}\kappa - \nabla_\perp \left(\frac{B^2}{2\mu_0} \right). \quad (2.19)$$

The first term is a spring-like force that acts to *straighten out* magnetic field curvature while the second term is the pressure in the direction perpendicular to the magnetic field. This is a much better interpretation of the magnetic stress tensor because it does not invoke non-existent magnetic forces along the magnetic field. The spring-like force $B^2\kappa/\mu_0$ and the pressure force $-\nabla_\perp \left(B^2/2\mu_0 \right)$ are not necessarily orthogonal and can offset each other. In fact, for the case of the vacuum field (i.e., $\mathbf{J} = 0$) these two forces are finite but balance each other exactly.

Curved magnetic field lines can therefore be in force balance and furthermore may be the lowest energy state for prescribed boundary conditions. For example the magnetic field lines of a horseshoe magnet are the lowest energy state for the boundary conditions prescribed by the fields at the surface of the magnet. In order to determine the free energy in a system, one must compare to the vacuum state which may very well

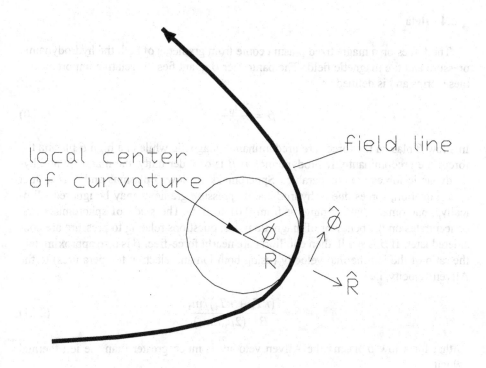

Fig.2.2 For any location on a curved magnetic field a local cylindrical coordinate system can be defined with origin at the center of curvature of the magnetic field. Thus, $\hat{\phi} = \hat{B}$.

have curved field lines. A state with straight field lines is not necessarily the lowest energy state for prescribed boundary conditions.

It is also clear that more complicated, non-vacuum states could exist which have zero net magnetic force. For these cases the system is not in the absolute lowest energy (i.e., vacuum field) state but is instead in a state corresponding to a *local* minimum of energy. This non-trivial situation can occur if **J** is finite but everywhere parallel to **B**. One can make a rough analogy to quantum energy levels. For prescribed boundary conditions, the vacuum (i.e., **J** =0) state is analogous to the ground state of a quantum system, whereas states with **J** finite and everywhere parallel to **B** are analogous to quantum states above the ground state. These higher energy states are relevant to spheromak equilibria and are called force-free equilibria.

2.4 Beta

The forces on a magnetized plasma come from gradients of both the hydrodynamic pressure and the magnetic field. The parameter β quantifies the relative importance of these terms and is defined as

$$\beta = \frac{2\mu_0 P}{B^2}. \quad (2.20)$$

In a low β plasma the forces are predominantly magnetic while in a high β plasma the forces are predominantly hydrodynamic. If β is of order unity, the magnetic and hydrodynamic forces are comparable. Spheromaks are low β plasmas and thus to first approximation, forces due to hydrodynamic pressure gradients may be ignored. Typically, spheromaks have β ranging from 0.01 to 0.2. The study of spheromaks first concentrates on the behavior of magnetic fields; questions relating to pressure are considered later. If β is small, then equilibria are nearly force-free. β is also approximately the ratio of the ion thermal velocity (using both ion and electron temperatures) to the Alfvén velocity, i.e.,

$$\beta = \frac{(\kappa T_e + \kappa T_i)/m_i}{B^2/(2\mu_0 n m_i)} \quad (2.21)$$

so that for a low β plasma the Alfvén velocity is much greater than the ion thermal velocity.

2.5 Magnetic flux and symmetry

The magnetic flux $\psi = \int_S \mathbf{B} \cdot d\mathbf{s}$ intercepting a surface S is very useful for characterizing fundamental behavior of magnetized plasma providing there is some sort of geometric symmetry. If symmetry exists then a flux surface $\psi = const.$ has geometrical meaning and is closely related to the vector potential $\mathbf{A}$ which in turn is closely related to the canonical momentum. In such situations, magnetic flux can be used as a generalized coordinate. However, for the general case of no geometric symmetry, the flux concept loses most of its utility because flux cannot be used as a coordinate. One must always bear in mind that any description of a plasma in terms of flux functions implicitly assumes some kind of symmetry. In reality there may or may not be symmetry.

2.6 Poloidal flux

Let us now assume the magnetic field is rotationally symmetric about the z axis.

The most general form for such a magnetic field is

$$\mathbf{B} = \frac{1}{2\pi}\nabla\psi(r,z)\times\nabla\phi + rB_\phi(r,z)\nabla\phi \qquad (2.22)$$

where $\nabla\phi = \hat{\phi}/r$. The toroidal magnetic field is the magnetic field in the ϕ direction and the poloidal magnetic field is the magnetic field orthogonal to the ϕ direction. The poloidal magnetic flux at the point r,z is

$$\text{poloidal flux} = \int_S \mathbf{B}\cdot d\mathbf{s} \qquad (2.23)$$

where S is a circle of radius r with center at location z on the z axis; this geometry is shown in Fig.2.1. Since $d\mathbf{s} = 2\pi r\, dr\, \hat{z}$ this becomes

$$\text{poloidal flux} = \int_S \frac{1}{2\pi r}\nabla\psi\times\hat{\phi}\cdot 2\pi r\, dr\, \hat{z} = \psi(r,z) \qquad (2.24)$$

so that $\psi(r,z)$ is indeed the poloidal flux function.

The curl of a toroidal field is in the poloidal direction and vice-versa, and in particular the poloidal magnetic field results from the curl of the toroidal vector potential. The poloidal flux is closely related to the toroidal component of the vector potential since

$$\mathbf{B}_{pol} = \frac{1}{2\pi}\nabla\psi(r,z)\times\nabla\phi = \nabla\times\left(\frac{\psi(r,z)}{2\pi r}\hat{\phi}\right). \qquad (2.25)$$

2.7 Poloidal flux and particle confinement

From Eq.(2.25) it is seen that the poloidal flux and toroidal vector potential are related by

$$\psi = 2\pi r A_\phi. \qquad (2.26)$$

This relation has important consequences for particle confinement. Since the plasma has been assumed to be symmetric in the ϕ direction, the canonical angular momentum $P_\phi = mrv_\phi + qrA_\phi$ is a constant of the motion for each particle. Thus, using Eq.(2.26) it is seen that

$$P_\phi = mrv_\phi + q\psi/2\pi \qquad (2.27)$$

is a constant of the motion. For strong magnetic fields, mrv_ϕ is small compared to $q\psi/2\pi$ by the ratio of the Larmor orbit radius in the poloidal magnetic field to the macroscopic scale length and so may be ignored. Particles are therefore constrained to move on surfaces of constant ψ and their excursion from these surfaces is of the order of the poloidal Larmor radius. *Hence, an effective way to confine a plasma is to create a system of nested poloidal flux surfaces.* Particles initially on a flux surface stay on that surface. This argument depends on the existence of axisymmetry; field errors (i.e., deviations from axisymmetry) tend to destroy flux surfaces and therefore adversely affect confinement.

2.8 Relation between field, field lines, and flux

There are subtle and sometimes important distinctions between the concepts of magnetic field, magnetic field lines, and magnetic flux. The magnetic field is a local variable and satisfies differential equations which describe local relationships. Magnetic flux is an integral of the magnetic field over a specified surface and in general is not a local variable. However, when there is axisymmetry, flux becomes a local variable since each point in three dimensional space has a well-defined associated surface obtained by rotation of the point about the symmetry axis. Field lines are also an integral quantity, but in a different sense, because one has to integrate the equation $d\mathbf{r}/ds = \hat{B}$ to find the trajectory of the field line. When there is symmetry, the planar projection of field lines assumes a simple form related to flux surfaces.

In transient situations, the meaning of the integral quantities can become ambiguous because information propagates at a finite velocity through the system. For magnetohydrodynamic systems this velocity is the Alfvén velocity, $v_A = B/\sqrt{\mu_0 \rho}$, and because this velocity is finite, one has to be careful not to assume simultaneity for events that are spatially separated. For example, consider two points, $\mathbf{x}_1$ and $\mathbf{x}_2$ initially on the same field line and spatially separated by the distance d_{12}. Suppose that at time $t = 0$ the magnetic field in the vicinity of $\mathbf{x}_1$ changes the field line topology so drastically that, when this change is completed, the points $\mathbf{x}_1$ and $\mathbf{x}_2$ are no longer on the same field line. The information about this change propagates to point $\mathbf{x}_2$, arriving at time $t = d_{12}/v_A$. Before this arrival time, an observer at $\mathbf{x}_2$ would be unaware of the change in field line topology. Thus, the definition of the field line is ambiguous for times $0 < t < d_{12}/v_A$. Similarly, integration along the field line would be ambiguous during this time interval and so the meaning of the flux linked by a path following the field line would be ambiguous.

An important consequence of these distinctions is that it is possible for the magnetic field to change continuously as parameters are varied, and yet at the same time the field lines might change discontinuously. Furthermore, an infinitesimal localized change in the magnetic field can completely change the overall topology of magnetic field lines.

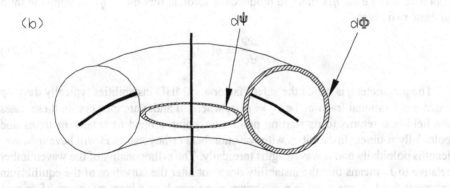

Fig. 2.3 (a) Definitions of poloidal flux ψ and toroidal flux Φ; (b) differentials of flux.

2.9 Safety factor

If the plasma magnetic field consists of closed magnetic flux surfaces, then each flux surface can be characterized by its poloidal flux ψ and its toroidal flux Φ. As shown in Fig.2.3(a) the toroidal flux is the flux through a cross-section of the flux tube while the poloidal flux is the flux linked by the flux tube. We consider the relationship between $d\psi$ and $d\Phi$. As shown in Fig.2.3(b) these two differential flux functions involve two orthogonal surfaces between adjacent nested flux tubes, namely

1. a ribbon-like surface that goes around toroidally; the magnetic flux through this surface is $d\psi$
2. an annular surface in the poloidal plane between the two concentric flux tubes; the magnetic flux through this surface is $d\Phi$.

We now follow the path of a typical field line located in the thin toroidal shell between two nested toroidal flux tubes with respective poloidal flux ψ and $\psi + d\psi$; because its path is helical, this field line will penetrate both surfaces #1 and #2. We assume this field line has a small but finite cross-section as shown in Fig.2.4 and, by choosing suitable units for measuring the magnetic field, assume this microscopic flux tube has unit flux. Let q be the number of times the field line goes around toroidally for each time it goes around poloidally. Each time the field line penetrates surface #1 it provides one unit of poloidal flux, while each time it penetrates surface #2 it provides one unit of toroidal flux. If the field line goes around poloidally some large number of times, say N, it will penetrate surface #1 N times to produce the poloidal flux $d\psi = N$ and it will penetrate surface #2 qN times to produce the toroidal flux $d\Phi = qN$. Taking the ratio of these two gives

$$\frac{d\Phi}{d\psi} = q(\psi). \qquad (2.28)$$

The parameter q is called the safety factor and MHD instabilities typically develop when q is a rational fraction, i.e., $q = m/n$ where m and n are integers. In these cases the field line returns to its starting point after going around toroidally m times and poloidally n times. Instabilities with wavenumbers orthogonal to $\mathbf{B}$ will have m wavelengths poloidally and n wavelengths toroidally. The orthogonality of the wavenumber relative to $\mathbf{B}$ means that the instability does not alter the curvature of the equilibrium magnetic field. Since a stable equilibrium corresponds to a local minimum of energy, changing the curvature of field lines from their equilibrium value would tend to increase the energy stored in the magnetic field. Instabilities which do not bend the equilibrium field are therefore the most dangerous; these instabilities can occur only at the rational surfaces where $q = m/n$. The stability properties of a toroidal equilibrium are largely

2.9 Safety factor

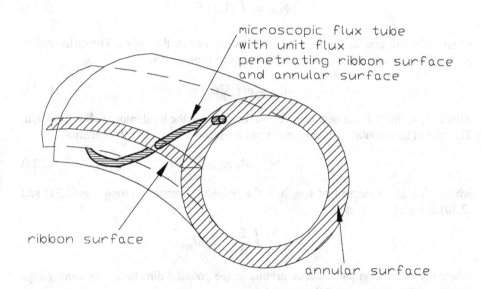

Fig. 2.4 A microscopic flux tube with unit flux penetrates both the annular surface thereby contributing to $d\Phi$ and penetrates the ribbon surface, thereby contributing to $d\psi$.

determined by its q profile. Tokamaks typically have $q \simeq 1$ near the magnetic axis and then have q increase going outwards from the magnetic axis; thus tokamak rational surfaces typically have $n = 1$ and $m \geq 1$. In contrast, spheromaks have $q \sim 1$ in the vicinity of the magnetic axis and have q decrease going outwards from the magnetic axis. Thus the rational surfaces of spheromaks typically have $m = 1$ and $n \geq 1$.

The safety factor for an axisymmetric configuration may be calculated[71] using Eq.(2.22). The toroidal flux between two adjacent nested flux tubes is

$$d\Phi_{tor} = \int B_{tor} dS \tag{2.29}$$

where dS is the annular surface element between the two flux tubes. The difference in poloidal flux between these two flux tubes can be expressed as

$$d\psi = dl_\perp |\nabla \psi| \tag{2.30}$$

where $dl_\perp$ is the distance between the two flux tubes in the local minor radius direction. The area of the annular surface element can be decomposed into sub-elements

$$dS = dl_\perp dl_{pol} \tag{2.31}$$

where dl_{pol} is an element of length in the poloidal direction. Using Eqs.(2.31) and (2.30) in Eq.(2.29) gives

$$d\Phi_{tor} = \oint \frac{B_{tor}}{|\nabla \psi|} d\psi dl_{pol} \tag{2.32}$$

where the integration path is once around in the poloidal direction. Combining Eqs. (2.28) and (2.32) shows that the safety factor is

$$q = \oint \frac{B_{tor}}{|\nabla \psi|} dl_{pol}. \tag{2.33}$$

From Eq. (2.22) the magnitude of the poloidal field is $B_{pol} = |\nabla \psi|/2\pi r$ so

$$q = \frac{1}{2\pi} \oint \frac{B_{tor}}{r B_{pol}} dl_{pol} = \frac{1}{2\pi} \oint \frac{\mathbf{B} \cdot \nabla \phi}{B_{pol}} dl_{pol}. \tag{2.34}$$

Equation (2.34) may be further simplified by noting that the components of the differential length along the poloidal field must satisfy

$$\frac{dr}{B_r} = \frac{dz}{B_z}. \tag{2.35}$$

2.9 Safety factor

Thus,

$$\frac{dl_{pol}}{B_{pol}} = \frac{\sqrt{dr^2 + dz^2}}{\sqrt{B_r^2 + B_z^2}} = \frac{\sqrt{dr^2 + (B_z dr/B_r)^2}}{\sqrt{B_r^2 + B_z^2}} = \frac{dr}{B_r} \qquad (2.36)$$

so that the safety factor may also be expressed as

$$q = \frac{1}{2\pi} \oint \frac{B_\phi}{rB_r} dr \qquad (2.37)$$

where again, the integration path is once around in the poloidal direction.

From Eq.(2.25) it is seen that the poloidal field follows contours of constant ψ and therefore encircles any local maximum or minimum of ψ. The extrema in ψ thus constitute magnetic axes for the poloidal flux.

In order to evaluate q at the magnetic axis $r = r_{axis}, z = 0$, the poloidal flux function is Taylor expanded[72] in the vicinity of the magnetic axis as

$$\psi(r,z) = \psi_0 + \frac{1}{2}(r - r_{axis})^2 \psi_{rr} + \frac{1}{2}z^2 \psi_{zz} \qquad (2.38)$$

where ψ_0 is the poloidal flux at the magnetic axis and the second derivatives are evaluated on the magnetic axis.

The poloidal magnetic field in the vicinity of the magnetic axis is therefore

$$\mathbf{B}_{pol} = \frac{1}{2\pi}\nabla\psi \times \nabla\phi = \frac{(r - r_{axis})\psi_{rr}\hat{z} - z\psi_{zz}\hat{r}}{2\pi r_{axis}} \qquad (2.39)$$

so

$$B_r = -z\psi_{zz}/2\pi r_{axis}. \qquad (2.40)$$

Thus, in the vicinity of the magnetic axis Eq.(2.37) becomes

$$q_{axis} = -\frac{B_\phi^{axis}}{\psi_{zz}} \oint \frac{dr}{z}. \qquad (2.41)$$

However, on a flux surface Eq.(2.38) can be written as

$$\frac{c^2}{2} = \psi - \psi_0 = \frac{1}{2}(r - r_{axis})^2 \psi_{rr} + \frac{1}{2}z^2 \psi_{zz} \qquad (2.42)$$

where c is a constant. Solving for z gives

$$z = \sqrt{\frac{c^2 - (r - r_{axis})^2 \psi_{rr}}{\psi_{zz}}}. \qquad (2.43)$$

Defining $\sin\eta = (r - r_{axis})\psi_{rr}^{1/2}/c$, so that $\psi_{rr}^{1/2} dr = c\cos\eta\, d\eta$, Eq.(2.41) may be integrated to give

$$q_{axis} = \frac{2\pi B_\phi^{axis}}{\sqrt{\psi_{rr}\psi_{zz}}}. \tag{2.44}$$

The toroidal current density can be written as

$$\mu_0 J_\phi = r\nabla\phi \cdot \nabla \times \left(\frac{1}{2\pi}\nabla\psi \times \nabla\phi\right) = -r\nabla \cdot \left(\frac{1}{2\pi r^2}\nabla_{pol}\psi\right). \tag{2.45}$$

On the magnetic axis $\partial\psi/\partial r$ vanishes and so

$$\mu_0 J_\phi^{axis} = -\frac{1}{2\pi r_{axis}}(\psi_{rr} + \psi_{zz}). \tag{2.46}$$

Dividing Eq.(2.44) by Eq.(2.46) gives

$$q_{axis} = \frac{\left(\sqrt{\psi_{rr}/\psi_{zz}} + \sqrt{\psi_{zz}/\psi_{rr}}\right)}{r_{axis}} \frac{B_\phi^{axis}}{\mu_0 J_\phi^{axis}} \tag{2.47}$$

showing that q_{axis} depends on the flux surface ellipticity $\varepsilon = \psi_{rr}/\psi_{zz}$ in the vicinity of the magnetic axis as well as on the ratio $\mu_0 J_\phi^{axis}/B_\phi^{axis}$. We shall see that the ratio $\mu_0 \mathbf{J} \cdot \mathbf{B}/B^2 \equiv \lambda$ is a very important quantity in the theory of spheromaks. Using this definition and noting that the poloidal field vanishes on the magnetic axis, Eq.(2.47) can be written as

$$q_{axis} = \frac{\varepsilon^{1/2} + \varepsilon^{-1/2}}{\lambda_{axis} r_{axis}}. \tag{2.48}$$

Furthermore, if the flux surfaces in the vicinity of the magnetic axis are circular, then $\varepsilon = 1$ leading to the simple expression

$$q_{axis} = \frac{2}{\lambda_{axis} r_{axis}} \text{ for circular flux surfaces.} \tag{2.49}$$

2.10 The plasma as a magnetic flux conserver

Hot plasmas are very collisionless. This suggests that the plasma acts as a near-perfect electrical conductor so that in the plasma frame the electric field should be close to zero. Consider the magnetic flux through some arbitrary surface S in the frame of the plasma. If this flux were to change in time, then Faraday's law states that there would be an electric field along the perimeter of S. This is in contradiction to the properties of a perfect conductor and so it must be concluded that the flux through S cannot be time-dependent. Hence, the plasma can be considered as a flux conserver or equivalently, magnetic field lines can be considered as being frozen into the plasma.

The frozen-in flux concept is very useful, but breaks down when electron inertia and other non-MHD effects become significant. To see this, consider a standard definition of a perfect conductor[73] "The charges inside a perfect conductor are assumed to be so mobile that they move instantly in response to the fields, no matter how rapid, and always produce the correct surface charge density in order to produce zero electric field inside the perfect conductor". Thus, the extent to which a plasma behaves as a perfect conductor depends on the ability of electrons (the lighter species) to move around rapidly and shield out any applied electric field. If the field changes faster than the electrons can respond, the electrons will not be able to shield out the field; this is important at high frequencies. For low frequencies, the electrons respond to an applied transient field, but if they are truly collisionless, they do not stop when they reach the precise positions required for shielding. Because of finite inertia, the electrons overshoot the precise position required for shielding, causing various oscillations and waves to develop. Ideal MHD ignores this overshooting and pretends that electrons exactly shield electric fields in the plasma frame.

2.11 The condition for frozen-in flux

The partial differential equation corresponding to the frozen-in flux condition will now be established for an arbitrary solenoidal vector $\mathbf{Q}$ and then applied to the particular case of the magnetic field. Let $S(t_0)$ be an arbitrary surface in the plasma at some initial time $t = t_0$. Now suppose that the plasma moves with fluid velocity $\mathbf{U}(\mathbf{x}, t)$ so that at time $t_0 + \Delta t$ plasma elements comprising $S(t_0)$ move to new positions and form a new, deformed surface $S(t_0 + \Delta t)$. For small Δt the two surfaces are related by

$$S(t_0 + \Delta t) = S(t_0) + \Delta t \oint_C \mathbf{U} \times d\mathbf{l} \qquad (2.50)$$

where $d\mathbf{l}$ is the differential element of length along C, the contour following the perimeter of S. At $t = t_0$ the flux through the initial surface is defined as

$$\Phi(t_0) = \int_{S(t_0)} \mathbf{Q}(\mathbf{x}, t_0) \cdot d\mathbf{s} \qquad (2.51)$$

while at the later time $t = t_0 + \Delta t$, the flux through the displaced and possibly deformed surface is

$$\Phi(t_0 + \Delta t) = \int_{S(t_0+\Delta t)} \mathbf{Q}(\mathbf{x},t_0+\Delta t)\cdot d\mathbf{s}. \tag{2.52}$$

Using $\mathbf{Q}(\mathbf{x},t_0+\Delta t) = \mathbf{Q}(\mathbf{x},t_0) + \Delta t \partial \mathbf{Q}/\partial t$ and Eq.(2.50) in a Taylor expansion of Eq. (2.52) to first order in Δt, it is seen that

$$\begin{aligned}\Phi(t_0+\Delta t) &= \int_{S(t_0+\Delta t)} \mathbf{Q}(\mathbf{x},t_0)\cdot d\mathbf{s} + \Delta t \int_{S(t_0)} \frac{\partial \mathbf{Q}}{\partial t}\cdot d\mathbf{s} \\ &= \int_{S(t_0)} \mathbf{Q}(\mathbf{x},t_0)\cdot d\mathbf{s} + \Delta t \oint_C \mathbf{Q}(\mathbf{x},t_0)\cdot \mathbf{U}\times d\mathbf{l} \\ &\quad + \Delta t \int_{S(t_0)} \frac{\partial \mathbf{Q}}{\partial t}\cdot d\mathbf{s}. \end{aligned} \tag{2.53}$$

In order for the flux to be frozen into the fluid, i.e., $\Phi(t_0+\Delta t) = \Phi(t_0)$, the following condition must therefore hold:

$$\oint_C \mathbf{Q}(\mathbf{x},t_0)\cdot \mathbf{U}\times d\mathbf{l} + \int_{S(t_0)} \frac{\partial \mathbf{Q}}{\partial t}\cdot d\mathbf{s} = 0. \tag{2.54}$$

Using Stokes theorem, the first term can be expressed as

$$\oint_C \mathbf{Q}\cdot \mathbf{U}\times d\mathbf{l} = \oint_C \mathbf{Q}\times \mathbf{U}\cdot d\mathbf{l} = \int_{S(t_0)} \nabla\times(\mathbf{Q}\times\mathbf{U})\cdot d\mathbf{s} \tag{2.55}$$

so that

$$\int_{S(t_0)} \left[\nabla\times(\mathbf{Q}\times\mathbf{U}) + \frac{\partial \mathbf{Q}}{\partial t} \right]\cdot d\mathbf{s} = 0. \tag{2.56}$$

Since $S(t_0)$ was arbitrary, the integrand must vanish, i.e.,

$$\frac{\partial \mathbf{Q}}{\partial t} = \nabla\times(\mathbf{U}\times\mathbf{Q}); \tag{2.57}$$

this differential equation corresponds to the frozen-in flux condition for an arbitrary solenoidal vector $\mathbf{Q}$. An important application immediately follows by combining Faraday's law

$$\frac{\partial \mathbf{B}}{\partial t} = -\nabla \times \mathbf{E} \qquad (2.58)$$

and the ideal MHD Ohm's law

$$\mathbf{E} + \mathbf{U} \times \mathbf{B} = 0 \qquad (2.59)$$

to obtain

$$\frac{\partial \mathbf{B}}{\partial t} = \nabla \times (\mathbf{U} \times \mathbf{B}). \qquad (2.60)$$

Thus, magnetic flux is frozen into the plasma if the ideal MHD Ohm's law is applicable.

2.12 Tendency of the plasma to maximize its inductance

Low β plasmas are susceptible to a variety of magnetic instabilities. These instabilities are driven by currents, because without a current, the plasma would be in a minimum-energy vacuum state in which case there would be no free energy. Magnetic energy is equivalent to the inductive energy stored in the plasma if the plasma is considered as a single electric circuit. Thus

$$W = \int \frac{B^2}{2\mu_0} d^3 r = \frac{1}{2} L I^2 \qquad (2.61)$$

where L and I are respectively the inductance and current. However, magnetic flux is just $\Phi = LI$ so the magnetic energy can also be expressed as

$$W = \frac{\Phi^2}{2L}. \qquad (2.62)$$

Magnetic instabilities act to reduce the magnetic energy but must do so in a flux - conserving way. Therefore, in order to reduce the magnetic energy while maintaining constant flux, magnetic instabilities must increase the plasma inductance. Examples are the pinch effect (smaller diameter wires have larger inductance), the hoop force (larger diameter current loops have larger inductance) and the tendency of the plasma to kink (coils have larger inductance than straight wires).

2.13 Cowling's theorem

The Earth's magnetic field is both poloidal and dipolar and so must be produced by a toroidal current loop deep inside the Earth's interior. In an investigation into the origin of this field, Cowling[74] considered what sustains the toroidal current in the presence of the inevitable dissipation due to Ohmic resistance. Any such sustaining mechanism is called a dynamo.

Naively, one might expect it would be possible to formulate a completely axisymmetric model for this problem, i.e., the toroidal current, the dynamo sustaining mechanism, and the loss mechanisms would all be axisymmetric. If this were possible, then one could reasonably expect that some fairly straightforward theory of sustainment could be developed. However, Cowling [74] showed it is impossible to have a steady-state axisymmetric MHD dynamo.

The proof is based on the MHD Ohm's law, namely

$$\mathbf{E} + \mathbf{U} \times \mathbf{B} = \eta \mathbf{J}. \tag{2.63}$$

Since the configuration is assumed axisymmetric, the toroidal current generates nested poloidal flux contours enclosing a magnetic axis at the extremum of the poloidal flux. Sustaining a toroidal current at the magnetic axis would require

$$E_\phi + <\mathbf{U} \times \mathbf{B}>_\phi = \eta J_\phi \tag{2.64}$$

on the magnetic axis. Because steady-state and axisymmetry were assumed, E_ϕ must vanish so the only possible dynamo term is the $<\mathbf{U} \times \mathbf{B}>_\phi$ term and clearly only poloidal fields contribute to the cross product. However, at the magnetic axis the poloidal magnetic field vanishes showing it is not possible to have a steady-state axisymmetric dynamo.

Spheromaks involve sustained axisymmetric equilibria having internal toroidal currents. Cowling's theorem shows that the dynamo which sustains toroidal current on the magnetic axis cannot be axisymmetric. As will be discussed in more detail later, it seems that the time-average of recurring non-axisymmetric instabilities provides the dynamo action. The excellent confinement properties of an axisymmetric system will necessarily be diminished when the dynamo fields are non-axisymmetric. The essential question is to what extent this necessary but undesirable departure from axisymmetry can be tolerated.

CHAPTER 3
Magnetic Helicity

Magnetic helicity is one of the central concepts underlying spheromak physics. Like many profound concepts, magnetic helicity can be understood from several different points of view. For example, magnetic helicity is closely related to both force-free currents and the topological theory of knots. Magnetic helicity is important for two reasons:

1. Magnetic helicity is a measure of the topological properties of the configuration and is more fundamental than any specific geometrical property.
2. Magnetic helicity is a robust invariant with respect to microscopic dissipative processes. This means that magnetic helicity is nearly perfectly conserved even when the system contains substantial small-scale turbulence and reconnection. Other invariants, energy in particular, are not robust and decay rapidly in the presence of small-scale turbulence and reconnection.

Our discussion of helicity will begin with a formal definition and then continue with the derivation of a conservation equation for magnetic helicity. Arguments for why magnetic helicity is a robust invariant will be postponed until the next chapter (cf. Sec.4.2). Before embarking on our discussion of helicity, we first consider (cf. next section) certain subtle but important topological issues regarding the application of Gauss's and Stokes's theorems. Consideration of these issues now will prevent confusion later when helicity is discussed for non-trivial topologies.

3.1 The issue of analyticity in Gauss's and Stokes's theorems

Gauss's and Stokes's theorems are valid only for functions that are continuous and differentiable (i.e., analytic) in a compact (i.e., simply connected) volume of interest. This requirement can be violated by functions which depend on a geometric angle; such behavior is closely related to the complex analysis concepts of analyticity, poles, residues, and branch cuts. Care must be exercised to avoid spurious paradoxes whenever a function depends on an angle.

This issue can be elucidated using the specific example of the scalar function

$$f(\mathbf{r}) = \phi \qquad (3.1)$$

where ϕ is the azimuthal angle in a cylindrical coordinate system (r, ϕ, z). The function $f(\mathbf{r})$ is not analytic on the z axis (i.e., at $r = 0$), because ϕ is not defined on the z axis; furthermore, the gradient ∇f is also not defined on the z axis so all vector operations involving $\nabla \phi$ are not defined on the z axis. Thus, it is necessary to avoid any mathematical operation involving f or ∇f on the z axis. This non-analyticity on the z axis is obvious if one expresses ϕ in Cartesian coordinates so

$$f(x,y,z) = \tan^{-1} y/x \qquad (3.2)$$

which clearly is not differentiable at $x = 0, y = 0$. The question of whether f is analytic depends on how the volume of interest is defined. If the volume of interest includes the z-axis, then f is not analytic but if the z axis is excluded, then f is analytic. Both Gauss's theorems and Stokes's theorems do not apply to functions of f when the volume of interest includes the z axis, but do apply if the volume is simply connected and excludes the z axis. We define $\mathbf{F} = \nabla f$ and consider Gauss's and Stokes's theorems individually.

Consider application of Gauss's theorem to $\mathbf{F}$ over (i) the volume of a sphere with center at the origin and (ii) the volume of a torus with center at the origin. The surfaces of both the sphere and the torus are normal to $\hat{\phi}$ and so $\int_S \mathbf{F} \cdot d\mathbf{s} = 0$ for both the sphere and the torus. On the other hand $\nabla \cdot \mathbf{F} = r^{-1} \partial \phi / \partial \phi = r^{-1}$ is non-zero. Thus, $\int_S \mathbf{F} \cdot d\mathbf{s} \neq \int_V d^3 r \nabla \cdot \mathbf{F}$ showing that Gauss's law does not apply when $\mathbf{F}$ is non-analytic. If we make a cut in the torus, as in Fig.3.1, then two new surfaces S_{ϕ_c} and $S_{\phi_c + 2\pi}$ are introduced at the cut and the volume of interest becomes simply connected. Gauss's law may be applied to the cut torus, because $\mathbf{F}$ is analytic in the volume of the cut torus which is simply connected. Since ϕ differs by 2π on the two cut surfaces, there is a net contribution from S_{ϕ_c} and $S_{\phi_c + 2\pi}$ and one obtains for the cut torus $\int_S \mathbf{F} \cdot d\mathbf{s} = \int_V d^3 r \nabla \cdot \mathbf{F}$.

For Stokes's theorem, consider the vector identity $\nabla \times \nabla f = 0$. If we integrate over a circular surface enclosing the origin then $\int_S d\mathbf{s} \cdot \nabla \times \nabla f = 0$. On the other hand, if as shown in Fig.3.2(a), we integrate along the perimeter of a circle enclosing the z axis, then $\oint d\mathbf{l} \cdot \nabla f = \int_0^{2\pi} (r d\phi)(1/r) = 2\pi$. Thus, $\int_S d\mathbf{s} \cdot \nabla \times \mathbf{F} \neq \oint_C \mathbf{F} \cdot d\mathbf{l}$. If we exclude the origin and make a cut, then as shown in Fig.3.2(b), the perimeter integral involves an outer counterclockwise portion and an inner clockwise portion which cancel so that $\int_S d\mathbf{s} \cdot \nabla \times \mathbf{F} = \oint_C \mathbf{F} \cdot d\mathbf{l}$. One interpretation for this situation is to state that the vector identity $\nabla \times \nabla f = 0$ has no meaning on the z axis, because ∇f is not defined on the z axis.

These issues are of importance to gauge invariance. The magnetic field and vector potential are related by $\mathbf{B} = \nabla \times \mathbf{A}$. Using Stokes's theorem, the flux can be expressed as $\Phi = \int_S \mathbf{B} \cdot d\mathbf{s} = \oint \mathbf{A} \cdot d\mathbf{l}$. The vector potential is undefined with respect to a function of integration called the gauge function, i.e., one can make a gauge transformation $\mathbf{A}' = \mathbf{A} + \nabla \Lambda$ without affecting the magnetic field since $\mathbf{B} = \nabla \times (\mathbf{A} + \nabla \Lambda) = \nabla \times$

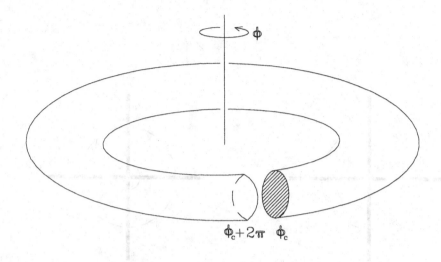

Fig.3.1 The doubly connected torus is cut to become simply connected; the cut creates two new surfaces. These face each other at the cut and are at ϕ_c and $\phi_c + 2\pi$ respectively.

A. However, this is permissible only if Λ is analytic everywhere in the region of interest. In particular, if $\Lambda = \phi$ then $\oint_C (\mathbf{A}+\nabla\phi) \cdot d\mathbf{l} \neq \oint_C \mathbf{A} \cdot d\mathbf{l}$ for contours C encircling the origin as in Fig.3.2(a). Thus, contrary to what is often stated, the gauge potential is not completely arbitrary because it must be analytic within the region of interest. The gauge potential Λ must be a conservative potential because we require that $\oint_C \nabla\Lambda \cdot d\mathbf{l} = 0$ for any contour C within the region of interest in order for flux to be gauge-invariant. The potential $\Lambda = \phi$ is non-conservative because each time the origin is encircled, Λ increases by 2π. From now on, all gauge functions Λ will be assumed to be conservative potentials, i.e., will satisfy the condition $\oint_C \nabla\Lambda \cdot d\mathbf{l} = 0$ for any contour C in the volume of interest.

3.2 Definition of magnetic helicity

Magnetic flux is the surface integral of the magnetic field, and using Stokes's theorem, can also be expressed in terms of the line integral of the vector potential around the perimeter of the surface,

$$\Phi = \int_S \mathbf{B} \cdot d\mathbf{s} = \oint_C \mathbf{A} \cdot d\mathbf{l}. \tag{3.3}$$

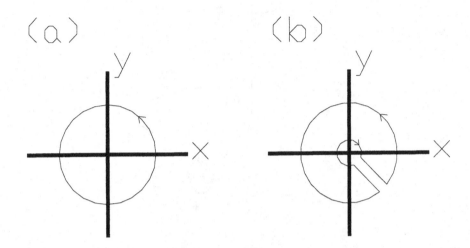

Fig.3.2 (a) Contour integration path that encircles the z axis. (b) Contour integration path that does not encircle the z axis.

Here C refers to the counterclockwise path around the perimeter of the surface S. As discussed in the previous section, an arbitrary gauge transformation may be applied to $\mathbf{A}$ with the proviso that the gauge function is a conservative potential within the volume of interest.

The magnetic helicity K in the volume V is defined to be

$$K = \int_V \mathbf{A} \cdot \mathbf{B} \, d^3 r. \qquad (3.4)$$

The physical meaning of this definition can be understood by considering the simple example[70] of two thin, linked, untwisted flux tubes embedded in a volume V as sketched in Fig.3.3. There are no other flux tubes in volume V so $\mathbf{B}$ vanishes outside the volumes occupied by the two flux tubes. Flux tube #1 occupies volume V_1 and has flux Φ_1 while flux tube #2 occupies volume V_2 and has flux Φ_2.

Evaluation of K is straightforward. Since the integrand in Eq.(3.4) vanishes in the region outside the two flux tubes, the integrand can be expressed as

$$K = K_1 + K_2 \qquad (3.5)$$

3.2 Definition of magnetic helicity

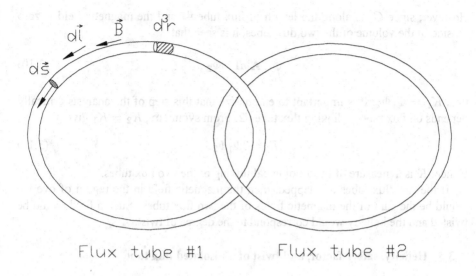

Fig.3.3 Two linked flux tubes. The directions of $d\mathbf{l}$, $d\mathbf{s}$, and $\mathbf{B}$ are all the same and the differential element of volume of a flux tube is $d^3r = d\mathbf{l}\cdot d\mathbf{s}$.

where

$$K_1 = \int_{V_1} \mathbf{A}\cdot\mathbf{B}\, d^3r, \qquad K_2 = \int_{V_2} \mathbf{A}\cdot\mathbf{B}\, d^3r. \tag{3.6}$$

To evaluate K_1 note that $d^3r = d\mathbf{l}\cdot d\mathbf{s}$ where $d\mathbf{l}$ is an element of length along flux tube #1 and $d\mathbf{s}$ is the surface cross section of flux tube #1. Since $d\mathbf{l}$ and $d\mathbf{s}$ are by definition parallel to $\mathbf{B}$, the integrand can be re-arranged to be

$$\mathbf{A}\cdot\mathbf{B}\, d^3r = \mathbf{A}\cdot\mathbf{B}\, d\mathbf{l}\cdot d\mathbf{s} = (\mathbf{A}\cdot d\mathbf{l})\,(\mathbf{B}\cdot d\mathbf{s}) \tag{3.7}$$

so

$$K_1 = \oint_{C_1}\int_{S_1} (\mathbf{A}\cdot d\mathbf{l})\,(\mathbf{B}\cdot d\mathbf{s}) \tag{3.8}$$

where the surface S_1 is the cross-section of flux tube #1 and the contour C_1 is along the length of flux tube #1. By definition of a flux tube, the flux $\Phi_1 = \int_{S_1} \mathbf{B}\cdot d\mathbf{s}$ is invariant along C_1 and so may be factored out of the integral giving

$$K_1 = \Phi_1 \oint_{C_1} \mathbf{A}\cdot d\mathbf{l}. \tag{3.9}$$

However, since C_1 is along the length of flux tube #1 and the magnetic field is zero outside of the volume of the two flux tubes, it is seen that

$$\oint_{C_1} \mathbf{A} \cdot d\mathbf{l} = \Phi_2 ; \qquad (3.10)$$

thus $K_1 = \Phi_1 \Phi_2$. It is important to emphasize that this step of the analysis critically depends on flux tube #1 linking flux tube #2. From symmetry, $K_2 = K_1$ giving

$$K = 2\Phi_1 \Phi_2. \qquad (3.11)$$

Hence, K is a measure of the topological linking of the two flux tubes.

If the two flux tubes overlapped, then the magnetic field in the region of overlap would be the sum of the magnetic fields of the two flux tubes. Such a field would be twisted and the helicity would correspond to the degree of twist.

3.3 Helicity, safety factor, and twist of an isolated flux tube

Consider a set of nested flux surfaces as described in Sec.2.9. The toroidal flux through a flux surface is

$$\Phi(\psi) = \oint_{C_s(\psi)} \mathbf{A}_{pol} \cdot d\mathbf{l} \qquad (3.12)$$

where the contour $C_s(\psi)$ is on the flux surface ψ and goes one time the *short way* around the torus. If θ denotes the poloidal angle, then $d\theta = d\mathbf{l} \cdot \nabla \theta$ where $d\mathbf{l}$ is in the poloidal direction and so Eq.(3.12) implies

$$\mathbf{A}_{pol} = \frac{\Phi}{2\pi} \nabla \theta. \qquad (3.13)$$

Similarly, the poloidal flux is

$$\psi = \oint_{C_l(\psi)} \mathbf{A}_{tor} \cdot d\mathbf{l} \qquad (3.14)$$

where the contour $C_l(\psi)$ is on the flux surface ψ and goes one time the *long way* around the torus. If ϕ denotes the toroidal angle, the toroidal vector potential is

$$\mathbf{A}_{tor} = \frac{\psi}{2\pi} \nabla \phi. \qquad (3.15)$$

An increment of length in the toroidal direction can be expressed as $dl_\phi = d\phi/|\nabla \phi|$, an increment of length in the poloidal direction can be expressed as $dl_\theta = d\theta/|\nabla \theta|$ and

3.3 Helicity, safety factor, and twist of an isolated flux tube

an increment of length in the direction of $\nabla\psi$ can be expressed as $dl_\psi = d\psi/|\nabla\psi|$. Since the poloidal, toroidal and $\nabla\psi$ directions are mutually orthogonal, a volume element[75] is

$$d^3r = dl_\theta dl_\phi dl_\psi = \frac{d\theta d\phi d\psi}{\nabla\theta \times \nabla\phi \cdot \nabla\psi}. \quad (3.16)$$

The sense of the angles is chosen so that θ, ϕ, ψ form a right handed coordinate system and therefore $\nabla\theta \times \nabla\phi \cdot \nabla\psi$ is positive.

The net vector potential (toroidal and poloidal) is

$$\mathbf{A} = \frac{1}{2\pi}(\Phi\nabla\theta + \psi\nabla\phi) \quad (3.17)$$

and the magnetic field is

$$\mathbf{B} = \frac{1}{2\pi}(\nabla\Phi \times \nabla\theta + \nabla\psi \times \nabla\phi). \quad (3.18)$$

Using $\nabla\Phi = \Phi'\nabla\psi$ the helicity can be expressed as

$$\begin{aligned}
K &= \int \mathbf{A} \cdot \mathbf{B} d^3r \\
&= \frac{1}{4\pi^2} \int (\Phi\nabla\theta + \psi\nabla\phi) \cdot (\nabla\Phi \times \nabla\theta + \nabla\psi \times \nabla\phi) d^3r \\
&= \frac{1}{4\pi^2} \int_0^{2\pi}\int_0^{2\pi}\int_0^\psi (\Phi\nabla\theta \cdot \nabla\psi \times \nabla\phi + \psi\nabla\phi \cdot \nabla\Phi \times \nabla\theta)\frac{d\theta d\phi d\psi}{\nabla\theta \times \nabla\phi \cdot \nabla\psi} \\
&= \int_0^\psi (-\Phi + \psi\Phi') d\psi. \quad (3.19)
\end{aligned}$$

The first term can be integrated by parts

$$\int_0^\psi \Phi d\psi = \Phi\psi - \int_0^\psi \psi\Phi' d\psi. \quad (3.20)$$

Using Eq.(2.28) the helicity in the volume occupied by the magnetic surface labeled by ψ can be expressed in terms of the safety factor $q(\psi)$ as

$$K = 2\int_0^\psi q(\psi)\psi d\psi - \Phi\psi. \quad (3.21)$$

There is an arbitrariness in the choice of the ψ and Φ origins. Normally the toroidal flux Φ is the flux through a circle centered on the magnetic axis, so $\Phi = 0$ corresponds to the magnetic axis. Sometimes $\psi = 0$ is taken to be at the magnetic axis in which case the poloidal flux builds up from the magnetic axis and reaches a maximum at the wall, while other times $\psi = 0$ is taken to be zero at the wall in which case the poloidal flux is at its maximum on the magnetic axis. Because of possible confusion, it is important to specify the choice.

We will set $\psi = 0$ at the wall; this is consistent with our previous convention and has the advantage that the term $\Phi\psi$ vanishes on both the wall and on the magnetic axis.

Because $q = d\Phi/d\psi$, a constant q profile implies $\Phi = q(\psi - \psi_{max})$; the constant of integration has been chosen to make Φ vanish on the magnetic axis. Thus, for $q = const.$, integration of Eq.(3.21) from $\psi = 0$ to $\psi = \psi_{max}$ shows that the helicity contained in the volume V is given by

$$K = q\psi_{max}^2. \qquad (3.22)$$

Also

$$\frac{dK}{d\psi} = q\psi - \Phi. \qquad (3.23)$$

Alternatively, the second term in the last line of Eq.(3.19) can be integrated by parts

$$\int_0^\psi \psi d\Phi = \Phi\psi - \int_0^\psi \Phi d\psi \qquad (3.24)$$

so that K can be written as

$$K = \Phi\psi - 2\int_0^\psi \Phi d\psi. \qquad (3.25)$$

If we consider an isolated flux tube and define the ψ origin to be on the exterior surface (i.e., the wall) of the flux tube, then the helicity content of a flux tube will be

$$K_{tube} = 2\int_0^\Phi T\Phi d\Phi \qquad (3.26)$$

where $T = d\psi/d\Phi$ is the number of turns in the poloidal direction per turn in the toroidal direction and so is a measure of the flux tube twist. Here we have changed the limits of integration to be functions of Φ, and noted that Φ is at its maximum when $\psi = 0$ and vice-versa. If T (which is just the inverse of q) is constant, then

$$K_{tube} = T\Phi^2. \qquad (3.27)$$

The twist T is also called $\bar{\iota}$ (pronounced "iota-bar"), a name that originated in stellarator research.

3.4 Gauge invariance

The quantity $\mathbf{A} \cdot \mathbf{B}$ is sometimes called the helicity density, but has no local physical meaning because the vector potential is undetermined with respect to a gauge. The issue of gauge invariance is also critical to the definition of the integral quantity K and, in fact, K is gauge-invariant only if V completely encloses all flux tubes; i.e., only if no portion of any flux tube penetrates the surface S bounding V. To see this, let us make a gauge transformation so that $\mathbf{A}' = \mathbf{A} + \nabla \chi$ and evaluate the helicity in the new gauge, i.e.,

$$\begin{aligned} K' &= \int_V \mathbf{A}' \cdot \mathbf{B}\, d^3r \\ &= K + \int_V \nabla\chi \cdot \mathbf{B}\, d^3r \\ &= K + \int_V \nabla \cdot (\chi \mathbf{B})\, d^3r \\ &= K + \int_S d\mathbf{s} \cdot \chi \mathbf{B}\,. \end{aligned} \quad (3.28)$$

If all flux tubes are fully enclosed by V, then no flux tube penetrates S and so $d\mathbf{s} \cdot \mathbf{B} = 0$ everywhere on S. Therefore $K' = K$ showing that the total magnetic helicity is gauge-invariant if no field lines penetrate S.

Gauge ambiguity corresponds to ambiguity in counting flux linkages. If flux tubes leaving V have linkages external to V, these topological features are not taken into account by Eq.(3.4). Figure 3.4 shows this problem graphically. The two flux tubes shown in Fig. 3.4 are linked once inside V, but are also linked outside V. An integral over V cannot measure how many times the flux tubes are linked.

3.5 Relative helicity

This problem of linkage ambiguity (or equivalently gauge dependence) is resolved by defining a *relative helicity*[39, 76, 77] which depends only on quantities defined inside V. The relative helicity is gauge invariant and physically meaningful because it is independent of properties external to V.

We now derive an expression for the relative helicity using the method of Berger and Field[76] and then relate this expression to the more abstract representation proposed by Finn and Antonsen[77] (the logic of the Berger and Field derivation is more intuitive, but the Finn and Antonsen representation is more convenient to use in actual calculations). The first step is to introduce a second volume V_b external to the volume of

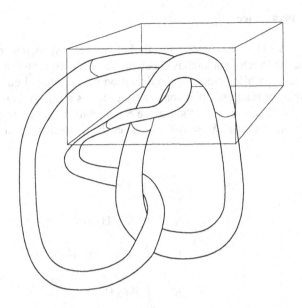

Fig.3.4 If V is the volume of the box shown here, then an integral over V cannot count the number of linkages of the two flux tubes, because this integration does not take into account the linkage external to the box.

interest, now labeled V_a. The volume V_b is defined so that the sum of the two volumes $V = V_a + V_b$ has no open field lines, i.e., $\mathbf{B} \cdot d\mathbf{s} = 0$ on its surface but $\mathbf{B} \cdot d\mathbf{s} \neq 0$ on the interface surface between V_a and V_b. Often V_b is taken to be all space except V_a, but this is not necessary as long as $\mathbf{B} \cdot d\mathbf{s} = 0$ on the surface bounding V. The surfaces bounding V_a, V_b, and V are denoted S_a, S_b, and S respectively, and the unit vector normal to S_a is $\hat{n}_a$, etc.

We define a hypothetical reference magnetic field $\mathbf{B}_{ref}$ which (i) equals $\mathbf{B}$ in V_b, (ii) has the same normal boundary conditions on S_a as does $\mathbf{B}$, and (iii) acts as a convenient benchmark field inside V_a. The simplest magnetic field satisfying prescribed normal boundary conditions on a surface is the vacuum magnetic field $\mathbf{B}_{vac}$ (also called potential magnetic field). The vacuum field satisfies the current-free condition $\nabla \times \mathbf{B}_{vac} = 0$ and can be expressed as the gradient of a potential, $\mathbf{B}_{vac} = \nabla \chi_{vac}$ where $\nabla^2 \chi_{vac} = 0$. A convenient embodiment of the reference magnetic field and its associated vector po-

3.5 Relative helicity

tential is

$$\mathbf{B}_{ref} = \begin{cases} \mathbf{B}_{vac} & \text{in } V_a \\ \mathbf{B} & \text{in } V_b \end{cases}$$

$$\mathbf{A}_{ref} = \begin{cases} \mathbf{A}_{vac} & \text{in } V_a \\ \mathbf{A}+\nabla h & \text{in } V_b \end{cases} \quad (3.29)$$

where $\nabla \times \mathbf{A}_{vac} = \mathbf{B}_{vac}$ in V_a, $\nabla \times \mathbf{A} = \mathbf{B}$ in V_b, and $\hat{n}_a \cdot \mathbf{B} = \hat{n}_a \cdot \mathbf{B}_{vac}$ on S_a. The geometry of the fields can be shown schematically as

	$V=V_a+V_b$		$V=V_a+V_b$	
	V_a	V_b	V_a	V_b
actual fields	$\mathbf{B}$	$\mathbf{B}$	$\mathbf{A}$	$\mathbf{A}$
reference fields	$\mathbf{B}_{vac}$	$\mathbf{B}$	$\mathbf{A}_{vac}$	$\mathbf{A}+\nabla h$
	magnetic fields		vector potentials	

where the boxes denote the volumes and the vertical line between the boxes represents the interface surface between V_a and V_b. An integral over V will not suffer from gauge ambiguity or fail to count any linkages because, by assumption, V has no open field lines. Thus both $K_v = \int_V \mathbf{A} \cdot \mathbf{B} d^3 r$ and $K_v^{ref} = \int_V \mathbf{A}_{ref} \cdot \mathbf{B}_{ref} d^3 r$ are gauge invariant. To demonstrate gauge invariance we make separate gauge transformations for $\mathbf{A}$ and $\mathbf{A}_{ref}$ so that $\mathbf{A}' = \mathbf{A}+\nabla f$ and $\mathbf{A}'_{ref} = \mathbf{A}_{ref}+\nabla g$ and define the relative helicity as

$$K_{rel} = \int_V (\mathbf{A}+\nabla f) \cdot \mathbf{B} d^3 r - \int_V (\mathbf{A}_{ref}+\nabla g) \cdot \mathbf{B}_{ref} d^3 r. \quad (3.30)$$

Because each term is gauge invariant (i.e., independent of f and g respectively), K_{rel} is gauge invariant and so we may discard the terms involving f and g since they vanish upon integration over V.

We now separate the volume integrals into integrals over V_a and V_b, so that

$$\begin{aligned} K_{rel} &= \int_{V_a} (\mathbf{A} \cdot \mathbf{B} - \mathbf{A}_{ref} \cdot \mathbf{B}_{ref}) d^3 r \\ &+ \int_{V_b} (\mathbf{A} - \mathbf{A}_{ref}) \cdot \mathbf{B} d^3 r \end{aligned} \quad (3.31)$$

Using $\mathbf{A} - \mathbf{A}_{ref} = \nabla h$ in V_b, Eq.(3.31) becomes

$$K_{rel} = \int_{V_a} (\mathbf{A} \cdot \mathbf{B} - \mathbf{A}_{vac} \cdot \mathbf{B}_{vac}) d^3 r$$

$$+ \int_{V_b} \nabla h \cdot \mathbf{B} d^3 r. \tag{3.32}$$

By assumption, the magnetic field in V_b can only penetrate the portion of the surface of V_b which interfaces V_a so that

$$\begin{aligned} K_{rel} &= \int_{V_a} (\mathbf{A} \cdot \mathbf{B} - \mathbf{A}_{vac} \cdot \mathbf{B}_{vac}) \, d^3 r \\ &\quad - \int_{S_{int}} h \mathbf{B} \cdot \hat{n}_a ds \end{aligned} \tag{3.33}$$

where S_{int} is the interface between V_a and V_b. A similar definition for relative helicity (but without h) was provided by Jensen and Chu [39].

Finn and Antonsen [77] defined the relative helicity in a slightly different form as

$$K_{rel}^{FA} = \int_{V_a} d^3 r \, (\mathbf{A} + \mathbf{A}_{vac}) \cdot (\mathbf{B} - \mathbf{B}_{vac}) \tag{3.34}$$

which can also be expressed as

$$\begin{aligned} K_{rel}^{FA} &= \int_{V_a} d^3 r \, (\mathbf{A} \cdot \mathbf{B} - \mathbf{A}_{vac} \cdot \mathbf{B}_{vac}) \\ &\quad + \int_{V_a} \nabla \cdot (\mathbf{A} \times \mathbf{A}_{vac}) \, d^3 r. \end{aligned} \tag{3.35}$$

Equation (3.34) is equivalent to (3.30) since

$$\begin{aligned} \int_{V_a} \nabla \cdot (\mathbf{A} \times \mathbf{A}_{vac}) \, d^3 r &= - \int_{V_b} \nabla \cdot (\mathbf{A} \times \mathbf{A}_{vac}) \, d^3 r \\ &= - \int_{V_b} \nabla \cdot (\nabla h \times \mathbf{A}_{vac}) \, d^3 r \\ &= \int_{V_b} \nabla h \cdot \mathbf{B}_{vac} d^3 r \\ &= \int_{S_{int}} h \mathbf{B}_{vac} \cdot \hat{n}_b ds \\ &= - \int_{S_{int}} h \mathbf{B} \cdot \hat{n}_a ds. \end{aligned} \tag{3.36}$$

In the first line of the above expression we have used the condition that the tangential components of both $\mathbf{A}$ and $\mathbf{A}_{vac}$ must be gradients of potentials on the surface of V so that both $\hat{n} \cdot \nabla \times \mathbf{A} = 0$ and $\hat{n} \cdot \nabla \times \mathbf{A}_{vac} = 0$ on this surface. The advantage of (3.34) is that it involves integration over V_a only and does not require knowledge or evaluation of h. Volumes which are doubly connected must be cut to form simply connected volumes before using Gauss's or Stokes's theorems.

3.6 Simply connected volumes v. doubly connected volumes

A simply (also called singly) connected volume is topologically equivalent to a sphere, whereas a doubly connected volume is topologically equivalent to a torus (i.e., has a hole in the middle and has both a long and a short way of going around the volume). With the exception of the Princeton S1 device, spheromaks to date have had simply connected geometry. Tokamaks and reversed field pinches are doubly connected. In doubly connected situations it is important to recall that Gauss's and Stokes's theorems apply *only* to simply connected volumes so that if one wishes to use these theorems on a torus, then as shown in Fig.3.1, the torus must be cut at some arbitrary toroidal angle ϕ_c to form a simply connected volume. The cut causes the toroidal angle $\phi_c + 2\pi$ to become physically distinct from ϕ_c thus preventing double-valued angles; the toroidal cut is analogous to the branch cut used in complex analysis. Cutting the torus creates two new surfaces S_{ϕ_c} and $S_{\phi_c+2\pi}$, each facing the cut, but on opposite sides. Surface integrals must include the new surfaces S_{ϕ_c} and $S_{\phi_c+2\pi}$ and line integrals must go along the face of the cut.

3.6.1 Relative helicity suitable for doubly connected volumes

When the volume is doubly connected, the Finn and Antonsen definition [77] for relative helicity is more convenient to use because it does not contain the scalar h. We now consider relative helicity for a torus which has $\mathbf{B} \cdot d\mathbf{s} = 0$ on its surface. The torus is cut to obtain a simply connected volume with toroidal magnetic field penetrating S_{ϕ_c} and $S_{\phi_c+2\pi}$. The volume of the cut torus is considered as volume V_a and all space outside the cut torus is considered as volume V_b. By definition, the vacuum field in Eq.(3.34) satisfies the same boundary conditions as the actual field on the surface of the cut torus so that there is a vacuum toroidal field penetrating S_{ϕ_c} and $S_{\phi_c+2\pi}$ and on these surfaces $\hat{n} \cdot \mathbf{B}_{vac} = \hat{n} \cdot \mathbf{B}$. Furthermore, the vacuum field does not penetrate the other surfaces of the torus and so the toroidal flux of the vacuum field is the same as the toroidal flux of the actual field, i.e., $\Phi_{vac} = \Phi$. The reference field in volume V_b was defined to be equal to the actual field in volume V_b. The actual field in V_b is not uniquely defined by specification of the actual field in V_a; however, the poloidal flux just outside the toroidal surface is determined by the toroidal current inside the torus and so we require that the reference field in V_b have the same poloidal flux on the toroidal surface as the actual poloidal flux, i.e., $\psi_{vac} = \psi$.

To prove gauge invariance for the relative helicity defined in Eq.(3.34), let gauge transformations be made so $\mathbf{A}' = \mathbf{A} + \nabla f$ and $\mathbf{A}'_{vac} = \mathbf{A}_{vac} + \nabla g$. The relative helicity can now be evaluated as

$$\begin{aligned}
K'^{FA}_{rel} &= \int d^3r \, (\mathbf{A}' + \mathbf{A}'_{vac}) \cdot (\mathbf{B} - \mathbf{B}_{vac}) \\
&= K^{FA}_{rel} + \int d^3r \, \nabla \cdot [(f+g)(\mathbf{B} - \mathbf{B}_{vac})] \\
&= K^{FA}_{rel} + \int_{\phi_c} d\mathbf{s} \cdot [(f+g)(\mathbf{B} - \mathbf{B}_{vac})] \\
&\quad + \int_{\phi_c + 2\pi} d\mathbf{s} \cdot [(f+g)(\mathbf{B} - \mathbf{B}_{vac})]
\end{aligned} \qquad (3.37)$$

since only the contributions from the toroidal cuts conceivably contribute (cf. Fig.3.1). However, because $\mathbf{B}_{vac} = \mathbf{B}$ on the cut surfaces the surface integral on each cut surface vanishes and so the relative helicity is gauge invariant for a doubly connected volume as well. In essence, once the cut has been made the doubly connected volume becomes equivalent to a singly connected volume with open field lines.

3.7 Helicity conservation equation

Magnetic helicity is a most useful concept because its evolution is geometry - independent. Helicity evolution is described by a conservation equation which comes directly from Maxwell's equations and an Ohm's law. To derive the helicity conservation equation, consider

$$\begin{aligned}
-\frac{\partial (\mathbf{A} \cdot \mathbf{B})}{\partial t} &= (\mathbf{E} + \nabla\phi) \cdot \mathbf{B} + \mathbf{A} \cdot \nabla \times \mathbf{E} \\
&= \mathbf{E} \cdot \mathbf{B} + \nabla \cdot (\phi \mathbf{B}) + \nabla \cdot (\mathbf{E} \times \mathbf{A}) + \mathbf{E} \cdot \nabla \times \mathbf{A} \\
&= 2\mathbf{E} \cdot \mathbf{B} + \nabla \cdot (\phi \mathbf{B} + \mathbf{E} \times \mathbf{A}).
\end{aligned} \qquad (3.38)$$

This can be re-arranged to be in the form of a conservation equation

$$\frac{\partial (\mathbf{A} \cdot \mathbf{B})}{\partial t} + \nabla \cdot (\phi \mathbf{B} + \mathbf{E} \times \mathbf{A}) = -2\mathbf{E} \cdot \mathbf{B}. \qquad (3.39)$$

It is useful to split the electric field on the LHS into its electrostatic and inductive components,

$$\begin{aligned}
\nabla \cdot (\mathbf{E} \times \mathbf{A}) &= -\nabla \cdot [(\nabla\phi + \partial \mathbf{A}/\partial t) \times \mathbf{A}] \\
&= -\nabla \cdot (\nabla\phi \times \mathbf{A}) + \nabla \cdot (\mathbf{A} \times \partial \mathbf{A}/\partial t)
\end{aligned}$$

3.7 Helicity conservation equation

$$\begin{aligned}&= \nabla\phi \cdot \nabla \times \mathbf{A} + \nabla \cdot (\mathbf{A} \times \partial \mathbf{A}/\partial t) \\ &= \nabla \cdot (\phi \mathbf{B} + \mathbf{A} \times \partial \mathbf{A}/\partial t)\end{aligned} \quad (3.40)$$

so that the helicity conservation equation becomes

$$\frac{\partial (\mathbf{A} \cdot \mathbf{B})}{\partial t} + \nabla \cdot \left(2\phi \mathbf{B} + \mathbf{A} \times \frac{\partial \mathbf{A}}{\partial t}\right) = -2\mathbf{E} \cdot \mathbf{B}. \quad (3.41)$$

At this point it is important to specify whether the volume of interest V is fixed or time-dependent. Fixed volumes are relevant to laboratory experiments where the volume is bounded by a metal vacuum chamber; time-dependent volumes are relevant to space plasma situations where plasmas are typically unbounded and may undergo expansion. If there are open field lines, then Eq.(3.41) is inadequate and a conservation equation for relative helicity must be developed.

3.7.1 Fixed volume- simply connected region

If V is fixed then there is no distinction between the partial and total time derivatives of volume integrals, i.e.,

$$\frac{d}{dt} \int_V \mathbf{A} \cdot \mathbf{B} d^3 r = \int_V \frac{\partial}{\partial t} (\mathbf{A} \cdot \mathbf{B}) d^3 r. \quad (3.42)$$

It is tempting to invoke the volume integral of Eq.(3.41) as a helicity conservation equation, but we must recall that when there are open field lines the relative helicity must be used. We now develop a conservation equation for relative helicity in a volume bounded by flux-conserving walls.

The electric field can be decomposed into electrostatic and inductive components $\mathbf{E} = -\nabla\phi - \partial\mathbf{A}/\partial t$ so that

$$\partial \mathbf{A}/\partial t = -\mathbf{E} - \nabla\phi. \quad (3.43)$$

In order to clarify the algebra we define

$$\begin{aligned}\mathbf{E}_\pm &= \mathbf{E} \pm \mathbf{E}_{vac}, & \mathbf{B}_\pm &= \mathbf{B} \pm \mathbf{B}_{vac} \\ \mathbf{A}_\pm &= \mathbf{A} \pm \mathbf{A}_{vac}, & \phi_\pm &= \phi \pm \phi_{vac};\end{aligned} \quad (3.44)$$

note that ϕ_{vac} is effectively a free parameter since it does not appear in (3.34). Following Finn and Antonsen [77], we now calculate the time derivative of (3.34) obtaining

$$\frac{dK_{rel}}{dt} = \int_{V_a} d^3 r \left(\frac{\partial \mathbf{A}_+}{\partial t} \cdot \mathbf{B}_- + \mathbf{A}_+ \cdot \frac{\partial \mathbf{B}_-}{\partial t}\right)$$

$$= -\int_{V_a} d^3r \left[\begin{array}{c} (\mathbf{E}_+ + \nabla\phi_+) \cdot \mathbf{B}_- \\ + \mathbf{A}_+ \cdot \nabla \times \mathbf{E}_- \end{array} \right]$$

$$= -\int_{V_a} d^3r \left(\mathbf{E}_+ \cdot \mathbf{B}_- + \mathbf{E}_- \cdot \nabla \times \mathbf{A}_+ \right)$$

$$\quad -\int_{S_a} d\mathbf{s} \cdot \left(\phi_+ \mathbf{B}_- + \mathbf{E}_- \times \mathbf{A}_+ \right)$$

$$= -\int_{V_a} d^3r \left(\mathbf{E}_+ \cdot \mathbf{B}_- + \mathbf{E}_- \cdot \mathbf{B}_+ \right)$$

$$= -2 \int_{V_a} d^3r \left(\mathbf{E} \cdot \mathbf{B} - \mathbf{E}_{vac} \cdot \mathbf{B}_{vac} \right) \qquad (3.45)$$

where we have used $\mathbf{B}_- \cdot d\mathbf{s} = 0$ on S_a and have chosen $\nabla\phi_{vac}$ so that $\mathbf{E}_- \times d\mathbf{s} = 0$ on S_a.

If S_a is a flux conserver, then $\mathbf{B} \cdot \hat{n}_a$ is constant in time on S_a. Since the vacuum field was defined to have the same normal component on S_a as the actual field, we must also have $\mathbf{B}_{vac} \cdot \hat{n}_a$ constant in time on S_a. Because $\mathbf{B}_{vac}$ inside V_a is completely determined by its normal component on S_a, the entire vacuum field inside V_a must therefore also be constant in time. The time derivative of $\mathbf{A} \cdot \mathbf{B}$ is provided by Eq. (3.41); applying this relation to the vacuum field yields

$$\nabla \cdot (\phi_{vac} \mathbf{B}_{vac}) = -\mathbf{E}_{vac} \cdot \mathbf{B}_{vac} \qquad (3.46)$$

because both $\mathbf{A}_{vac}$ and $\mathbf{B}_{vac}$ are constant in time. Thus (3.45) becomes

$$\frac{dK_{rel}}{dt} = -2 \int_{V_a} d^3r \mathbf{E} \cdot \mathbf{B} - 2 \int_{S_a} d\mathbf{s} \cdot (\phi_{vac} \mathbf{B}_{vac}). \qquad (3.47)$$

Since $\mathbf{B} \cdot \hat{n}_a = \mathbf{B}_{vac} \cdot \hat{n}_a = const.$ on S_a, the tangential components of both $\partial \mathbf{A}/\partial t$ and $\partial \mathbf{A}_{vac}/\partial t$ must vanish on S_a.

Furthermore, because ϕ_{vac} was chosen to make the tangential components of $\mathbf{E}$ and $\mathbf{E}_{vac}$ equal on S_a, the tangential components of $\nabla\phi$ and $\nabla\phi_{vac}$ must also be equal on S_a. Thus ϕ and ϕ_{vac} differ at most by a constant on S_a; i.e., $\phi - \phi_{vac} = c$ where c is constant over S_a. The constant c does not affect Eq.(3.47) because $\int_{S_a} d\mathbf{s} \cdot (c\mathbf{B}_{vac}) = c \int_{S_a} d\mathbf{s} \cdot \mathbf{B}_{vac} = 0$.

The rate of change of relative helicity is therefore

$$\frac{dK_{rel}}{dt} = -2 \int_{V_a} d^3r \mathbf{E} \cdot \mathbf{B} - 2 \int_{S_a} d\mathbf{s} \cdot (\phi \mathbf{B}) \qquad (3.48)$$

3.7 Helicity conservation equation

where we have again used $\mathbf{B} \cdot \hat{n}_a = \mathbf{B}_{vac} \cdot \hat{n}_a$ on S_a. If $\mathbf{E} = 0$ inside the plasma (i.e., the plasma is an ideal conductor) and the bounding surface is an equipotential (i.e., $\phi =$ uniform on S_a), then the relative helicity is conserved inside V_a.

Now suppose there is just one open flux tube and let σ_1 and σ_2 be the two surfaces where this flux tube intercepts the boundary. Let ϕ_1, ϕ_2 be the respective electrostatic potentials of σ_1 and σ_2. In this case, (3.48) becomes

$$\frac{dK_{rel}}{dt} + 2\phi_1 \int_{\sigma_1} \mathbf{B} \cdot d\mathbf{s} + 2\phi_2 \int_{\sigma_2} \mathbf{B} \cdot d\mathbf{s} = -2 \int_{V_a} \mathbf{E} \cdot \mathbf{B} d^3 r. \quad (3.49)$$

Since σ_1 and σ_2 are cross-sections of the same flux tube, we must have

$$\int_{\sigma_1} \mathbf{B} \cdot d\mathbf{s} = -\int_{\sigma_2} \mathbf{B} \cdot d\mathbf{s}. \quad (3.50)$$

Defining $\psi(\sigma_1) = -\int_{\sigma_1} \mathbf{B} \cdot d\mathbf{s}$ as the flux into S_a through σ_1, Eq.(3.49) can be expressed as

$$\frac{dK_{rel}}{dt} = 2(\phi_1 - \phi_2)\psi(\sigma_1) - 2\int \mathbf{E} \cdot \mathbf{B} d^3 r. \quad (3.51)$$

Each term in this equation warrants discussion:

Helicity derivative (first term) – Because the volume is fixed, dK_{rel}/dt describes the rate of change of relative helicity in the entire volume.

Helicity flux (surface integral) – The term $2(\phi_1 - \phi_2)\psi(\sigma_1)$ describes the net flux of relative helicity injected into the system through the bounding surface. The potential difference $\phi_1 - \phi_2$ could be provided by any source of EMF (e.g., battery, capacitor, generator).

Helicity dissipation term (RHS) — According to ideal MHD the parallel electric field is zero. This assumption is an oversimplification for real plasmas where there is ample evidence of parallel electric fields produced by phenomena outside the scope of MHD. The magnitude of helicity dissipation can be estimated using the electron fluid equation of motion,

$$m_e n \frac{d\mathbf{u}_e}{dt} = nq_e [\mathbf{E} + \mathbf{u}_e \times \mathbf{B}] - \nabla P_e - m_e \nu_{ei} (\mathbf{u}_e - \mathbf{u}_i). \quad (3.52)$$

We are interested in phenomena slow compared to the electron cyclotron frequency; this means that the perpendicular component of the left hand side may be dropped since it is small compared to the $\mathbf{u}_e \times \mathbf{B}$ term. However, the parallel component of the left hand side must be retained and since the ion parallel motion is negligible, the parallel electron motion constitutes most of the parallel current. The plasma resistivity is defined as

$\eta = m_e \nu_{ei}/ne^2$ so the electron equation of motion becomes approximately

$$\underbrace{\frac{d}{dt}\left(\frac{c^2}{\omega_{pe}^2}\mu_0 \mathbf{J}_\|\right)}_{\text{electron inertia}} = \mathbf{E} + \mathbf{u}_e \times \mathbf{B} + \underbrace{\frac{\nabla P_e}{ne}}_{\substack{\text{electron}\\\text{pressure}}} - \underbrace{\eta \mathbf{J}}_{\text{resistive}} \qquad (3.53)$$

where the terms have been labeled by their conventional names. Dotting Eq.(3.53) with $\mathbf{B}$ gives

$$\mathbf{E} \cdot \mathbf{B} = \eta \mathbf{J} \cdot \mathbf{B} - \frac{\mathbf{B} \cdot \nabla P_e}{ne} + \mathbf{B} \cdot \frac{d}{dt}\left(\frac{c^2}{\omega_{pe}^2}\mu_0 \mathbf{J}_\|\right). \qquad (3.54)$$

Substituting in Eq.(3.51) gives

$$\frac{dK_{rel}}{dt} = 2(\phi_1 - \phi_2)\psi(\sigma_1)$$
$$-2\int_V \left[\eta \mathbf{J} \cdot \mathbf{B} - \frac{\mathbf{B} \cdot \nabla P_e}{ne} + \mathbf{B} \cdot \frac{d}{dt}\left(\frac{c^2}{\omega_{pe}^2}\mu_0 \mathbf{J}_\|\right)\right] d^3r. \qquad (3.55)$$

For typical plasma equilibria, pressure gradients are normal to the magnetic field so $\mathbf{B} \cdot \nabla P_e = 0$. The inertial term has the same scaling as the helicity derivative $\partial(\mathbf{A}_\| \cdot \mathbf{B})/\partial t$ and so must be compared to the helicity derivative. Eliminating the common factors (i.e., time derivative and $\mathbf{B}\cdot$) in making this comparison shows that the inertial term compares in order of magnitude to the relative helicity derivative as

$$\frac{c^2}{\omega_{pe}^2}\mu_0 \mathbf{J}_\| : \mathbf{A}_\|. \qquad (3.56)$$

Using Coulomb gauge (i.e., $\nabla \cdot \mathbf{A} = 0$) the parallel current can be expressed as

$$\mu_0 \mathbf{J}_\| = \hat{B}\hat{B} \cdot [\nabla \times (\nabla \times \mathbf{A})]$$
$$\approx -\nabla_\perp^2 \mathbf{A}_\| \qquad (3.57)$$

so that the comparison in Eq.(3.56) becomes

$$\frac{c^2}{\omega_{pe}^2}\nabla_\perp^2 \mathbf{A}_\| : \mathbf{A}_\|. \qquad (3.58)$$

Thus, the inertial term compares to the helicity derivative by the factor $c^2/L_\perp^2 \omega_{pe}^2$ where $L_\perp$ is the characteristic scale length associated with the $\nabla_\perp^2$ operator. So long as $L_\perp$ is large compared to the electron skin depth c/ω_{pe}, the electron inertial term may be ignored. We may conclude that for a perfectly conducting plasma with pressure gradients normal to the magnetic field and for phenomena having $L_\perp \gg c/\omega_{pe}$ the helicity conservation equation reduces to

$$\frac{dK_{rel}}{dt} = 2\left(\phi_1 - \phi_2\right)\psi(\sigma_1); \tag{3.59}$$

hence if the bounding surface is both a flux conserver and an equipotential then $K_{rel} =$ constant. This is a very powerful result, because it does not depend on geometric details. If there is finite resistivity, no helicity flux, and all phenomena have scale lengths $L_\perp \gg c/\omega_{pe}$, the helicity conservation equation becomes

$$\frac{dK}{dt} = -2\int \eta \mathbf{J} \cdot \mathbf{B}\, d^3r \tag{3.60}$$

showing that helicity decays resistively when there is no helicity flux through the bounding surface.

3.7.2 Fixed volume - doubly connected region

We now consider a toroidal volume V_a with no open field lines and where all currents, fields, and vector potentials are either toroidal or poloidal. At first sight it appears that we do not need to use relative helicity because there are no open field lines. However, because the volume is doubly connected, it must be cut as in Fig.3.1 to allow use of Gauss's and Stokes's theorems. There are now toroidal fields penetrating the cut surfaces and so the system is equivalent to a simply connected volume with open field lines. Relative helicity must be used and in such a case we construct a vacuum magnetic field which matches the actual field on the cut surfaces. Since the cut surfaces are hypothetical, they are not flux-conserving and so the vacuum magnetic field can be time-dependent. Because there are no open field lines penetrating any poloidal surface, there is no vacuum poloidal magnetic field in V_a and so, using Faraday's law, we conclude that the toroidal component of the vacuum electric field is irrotational in V_a, i.e., $\nabla \times \mathbf{E}_{tor,vac} = 0$ inside V_a.

The actual electric field has both toroidal and poloidal components but no component in the direction of the minor radius. In the derivation of Eq.(3.45) we chose $\nabla\phi_{vac}$ so that $\mathbf{E}_- \times d\mathbf{s} = 0$ on S_a; this means that the vacuum electric field also has no component in the direction of the minor radius on the surfaces facing the cut.

From Eq.(3.45) the rate of relative helicity injection is

$$\left(\frac{dK_{rel}}{dt}\right)_{inj} = 2\int_{V_a} d^3r \mathbf{E}_{vac} \cdot \mathbf{B}_{vac}$$

$$= 2\int E_{tor,vac}dl \int ds B_{tor,vac}$$
$$= 2V\Phi = -2\frac{d\psi}{dt}\Phi \qquad (3.61)$$

where $V = -d\psi/dt$ is the one-turn toroidal (loop) voltage, Φ is the toroidal flux of the actual field, and ψ is the poloidal flux of the actual field[39, 77, 78]. In this derivation we have used:

1. $\int E_{tor,vac}dl$ is independent of minor radius because $\nabla \times \mathbf{E}_{tor,vac} = 0$
2. $\mathbf{E}_{vac}$ has no component in the direction of the minor radius.
3. On the cut surface $\mathbf{B}_{vac,tor} = \mathbf{B}_{tor}$ so $\Phi_{vac} = \Phi$.

Equation (3.61) can be used to interpret Ohmic heating current drive in a tokamak or an RFP as a form of helicity injection; here Φ is the toroidal flux and V is the one-turn loop voltage. Equation (3.61) can also be used to explain the helicity injection scheme used on the Princeton S1 device. In this case, the torus is the external vacuum region while V_b, the region external to the torus, is the plasma region. One can reverse the sign in Eq.(3.61) and consider this as an equation for the rate of helicity ejection from V_a into V_b.

3.7.3 Comparison between helicity injection into simply and double connected volumes

Helicity injection into a simply connected region with a flux conserving boundary can only be done electrostatically and requires electrodes that intercept open field lines. Electrostatic helicity injection is also possible into a doubly connected volume [79].

Helicity injection into a doubly connected region can also be done inductively with an injection rate $-2\Phi d\psi/dt$; this method is not possible for a simply connected region [39]. Unlike electrostatic helicity injection, inductive helicity injection does not require open field lines intercepting a flux conserving boundary (i.e., does not require electrodes).

3.7.4 Time-dependent volume with no open field lines

We now consider a somewhat different situation where a volume with no open field lines is evolving in time. Because there are no open field lines, the ordinary helicity may be used. Since the volume is changing, Leibniz's rule must be used to take into

3.7 Helicity conservation equation

account the time-derivative of the volume,

$$\frac{d}{dt}\int_V \mathbf{A} \cdot \mathbf{B} d^3r = \int_V \frac{\partial}{\partial t}(\mathbf{A} \cdot \mathbf{B}) d^3r + \int_S (\mathbf{A} \cdot \mathbf{B})(\mathbf{U} \cdot d\mathbf{s}) \qquad (3.62)$$

where $\mathbf{U}$ is the velocity of the surface of the volume. Integrating Eq.(3.39) over V gives

$$\int_V \frac{\partial(\mathbf{A} \cdot \mathbf{B})}{\partial t} d^3r + \int_S (\phi \mathbf{B} + \mathbf{E} \times \mathbf{A}) \cdot d\mathbf{s} = -2\int_V \mathbf{E} \cdot \mathbf{B} d^3r. \qquad (3.63)$$

and using Eq.(3.62) to substitute for the first term in Eq.(3.63) gives

$$\frac{d}{dt}\int_V \mathbf{A} \cdot \mathbf{B} d^3r - \int_S (\mathbf{A} \cdot \mathbf{B})(\mathbf{U} \cdot d\mathbf{s}) + \int_S (\phi \mathbf{B} + \mathbf{E} \times \mathbf{A}) \cdot d\mathbf{s}$$

$$= -2\int_V \mathbf{E} \cdot \mathbf{B} d^3r. \qquad (3.64)$$

Using $\mathbf{A} \times (\mathbf{U} \times \mathbf{B}) = (\mathbf{A} \cdot \mathbf{B})\mathbf{U} - (\mathbf{A} \cdot \mathbf{U})\mathbf{B}$ this can be rearranged to give

$$\frac{dK}{dt} - \int_S [\mathbf{A} \times (\mathbf{E} + \mathbf{U} \times \mathbf{B}) + (\mathbf{A} \cdot \mathbf{U} - \phi)\mathbf{B}] \cdot d\mathbf{s}$$

$$= -2\int_V \mathbf{E} \cdot \mathbf{B} d^3r. \qquad (3.65)$$

Equation (3.65) gives the rate of change of helicity of an isolated flux tube having a non-constant volume. If ideal MHD is a valid approximation, i.e., $\mathbf{E} + \mathbf{U} \times \mathbf{B} = 0$, then both terms involving $\mathbf{E}$ in Eq.(3.65) vanish and since $\mathbf{B} \cdot d\mathbf{s} = 0$ on the sides of the flux tube the helicity inside this flux-tube with time-dependent volume is conserved.

3.7.5 Comparison of helicity conservation to energy conservation

The helicity conservation equation is reminiscent of the Poynting energy conservation equation obtained by manipulating Faraday's law and the pre-Maxwell Ampere's law. If we dot the former with $\mathbf{B}$ and the latter with $\mathbf{E}$, we obtain

$$\begin{aligned}\mathbf{B} \cdot \nabla \times \mathbf{E} &= -\frac{\partial}{\partial t}\left(\frac{B^2}{2}\right) \\ \mathbf{E} \cdot \nabla \times \mathbf{B} &= \mu_0 \mathbf{E} \cdot \mathbf{J}\end{aligned} \qquad (3.66)$$

Subtracting these equations gives

$$\frac{\partial}{\partial t}\left(\frac{B^2}{2\mu_0}\right) + \nabla \cdot \left(\frac{\mathbf{E} \times \mathbf{B}}{\mu_0}\right) = -\mathbf{E} \cdot \mathbf{J}. \qquad (3.67)$$

If the plasma happens to be stationary and if the MHD Ohm's law $\mathbf{E} = \eta \mathbf{J}$ is valid then the RHS becomes $-\eta J^2$, the conventional Ohmic dissipation. This is a conservation equation for magnetic energy $B^2/2\mu_0$; the energy flux is the Poynting flux $\mathbf{E} \times \mathbf{B}/\mu_0$ and the rate of energy dissipation is $-\mathbf{E} \cdot \mathbf{J}$.

3.8 Single species helicity

There is a more fundamental way[80]-[83] of looking at helicity. We consider the fluid equation of motion for the species σ

$$m_\sigma \left[\frac{\partial \mathbf{u}_\sigma}{\partial t} + \mathbf{u}_\sigma \cdot \nabla \mathbf{u}_\sigma \right] = q_\sigma \left[\mathbf{E} + \mathbf{u}_\sigma \times \mathbf{B} \right] - \frac{\nabla P_\sigma}{n_\sigma}. \quad (3.68)$$

This may be expressed as

$$m_\sigma \left[\frac{\partial \mathbf{u}_\sigma}{\partial t} + \nabla \left(\frac{u_\sigma^2}{2} \right) - \mathbf{u}_\sigma \times (\nabla \times \mathbf{u}_\sigma) \right]$$
$$= q_\sigma \left[-\nabla \phi - \frac{\partial \mathbf{A}}{\partial t} + \mathbf{u}_\sigma \times (\nabla \times \mathbf{A}) \right] - \frac{\nabla P_\sigma}{n_\sigma} \quad (3.69)$$

or after regrouping

$$\frac{\partial}{\partial t} (m_\sigma \mathbf{u}_\sigma + q_\sigma \mathbf{A}) - \mathbf{u}_\sigma \times [\nabla \times (m_\sigma \mathbf{u}_\sigma + q_\sigma \mathbf{A})] = -\nabla \left(\frac{m_\sigma u_\sigma^2}{2} + q_\sigma \phi \right) - \frac{\nabla P_\sigma}{n_\sigma}. \quad (3.70)$$

We will assume that the pressure gradient is parallel to the density gradient (barotropic assumption) so that

$$\frac{\nabla P_\sigma}{n_\sigma} = \nabla \left(\int \frac{dP_\sigma}{n_\sigma} \right) \quad (3.71)$$

and define

$$h = \frac{m_\sigma u_\sigma^2}{2} + q_\sigma \phi + \int \frac{dP_\sigma}{n_\sigma}. \quad (3.72)$$

We define the canonical momentum of a fluid element to be

$$\mathbf{P}_\sigma = m_\sigma \mathbf{u}_\sigma + q_\sigma \mathbf{A} \quad (3.73)$$

so that Eq.(3.70) becomes

$$\frac{\partial \mathbf{P}_\sigma}{\partial t} + (\nabla \times \mathbf{P}_\sigma) \times \mathbf{u}_\sigma = -\nabla h. \quad (3.74)$$

3.8 Single species helicity

The canonical circulation is defined to be

$$\mathbf{Q}_\sigma = \nabla \times \mathbf{P}_\sigma = m_\sigma \mathbf{\Omega}_\sigma + q_\sigma \mathbf{B} \tag{3.75}$$

where $\mathbf{\Omega}_\sigma = \nabla \times \mathbf{u}_\sigma$ is the fluid vorticity. Thus $\mathbf{Q}_\sigma$ is a solenoidal vector and is characterized by flux tubes analogous to magnetic flux tubes.

The curl of Eq.(3.74) gives

$$\frac{\partial \mathbf{Q}_\sigma}{\partial t} + \nabla \times (\mathbf{Q}_\sigma \times \mathbf{u}_\sigma) = 0 \tag{3.76}$$

which is exactly the requirement for frozen-in $\mathbf{Q}_\sigma$ flux (see Eq.(2.57)). Thus flux tubes of canonical circulation $\mathbf{Q}_\sigma$ are frozen into the moving fluid. This is a more precise statement of the concept of frozen-in magnetic flux because particle mass is taken into account.

The canonical helicity in a canonical-circulation flux tube is defined to be

$$K_\sigma = \int_V \mathbf{P}_\sigma \cdot \nabla \times \mathbf{P}_\sigma d^3 r = \int_V \mathbf{P}_\sigma \cdot \mathbf{Q}_\sigma d^3 r \tag{3.77}$$

where the integral is over the volume of the $\mathbf{Q}_\sigma$ flux tube. We consider the general situation in which the $\mathbf{Q}_\sigma$ flux tube convects with the fluid and calculate the rate of change of K_σ in the volume V of the moving flux tube,

$$\frac{dK_\sigma}{dt} = \int_V \left[\frac{\partial \mathbf{P}_\sigma}{\partial t} \cdot \mathbf{Q}_\sigma + \mathbf{P}_\sigma \cdot \frac{\partial \mathbf{Q}_\sigma}{\partial t} \right] d^3 r + \int_S (\mathbf{P}_\sigma \cdot \mathbf{Q}_\sigma)(\mathbf{u}_\sigma \cdot d\mathbf{s}) \tag{3.78}$$

where the last term describes the effect of changes in V due to motion of flux tube walls. Using Eq.(3.76) this becomes

$$\frac{dK_\sigma}{dt} = \int_V \left\{ \frac{\partial \mathbf{P}_\sigma}{\partial t} \cdot \mathbf{Q}_\sigma + \mathbf{P}_\sigma \cdot \nabla \times (\mathbf{u}_\sigma \times \mathbf{Q}_\sigma) + \nabla \cdot [\mathbf{P}_\sigma \cdot \mathbf{Q}_\sigma \mathbf{u}_\sigma] \right\} d^3 r \tag{3.79}$$

or using Eq.(3.74)

$$\frac{dK_\sigma}{dt} = \int_V \nabla \cdot \left\{ -h\mathbf{Q}_\sigma + (\mathbf{u}_\sigma \times \mathbf{Q}_\sigma) \times \mathbf{P}_\sigma + \mathbf{P}_\sigma \cdot \mathbf{Q}_\sigma \mathbf{u}_\sigma \right\} d^3 r$$

$$= \int_S d\mathbf{s} \cdot \mathbf{Q}_\sigma (-h + \mathbf{u}_\sigma \cdot \mathbf{P}_\sigma) . \tag{3.80}$$

However $d\mathbf{s} \cdot \mathbf{Q}_\sigma = 0$ by assumption (the integral is over the volume of a canonical-circulation flux tube) and so we obtain

$$\frac{dK_\sigma}{dt} = 0. \tag{3.81}$$

This is a stronger and more accurate statement than Eq.(3.60) because the system does not have to be bounded and because finite mass is taken into account.

The electron helicity is

$$K_e = \int (m_e \mathbf{u}_\sigma + q_e \mathbf{A}) \cdot \nabla \times (m_e \mathbf{u}_\sigma + q_e \mathbf{A}) \, d^3r. \tag{3.82}$$

Using Eq.(3.57) and again letting $\nabla_\perp \sim L_\perp^{-1}$ a comparison of the magnetic term $q_e \mathbf{A}_\parallel$ to the mechanical term $m_e \mathbf{u}_\sigma$ shows

$$\frac{|q_e \mathbf{A}_\parallel|}{|m_e \mathbf{u}_{\parallel\sigma}|} \sim \frac{\mu_0 n q_e^2 L_\perp^2}{m_e} = \frac{L_\perp^2 \omega_{pe}^2}{c^2}. \tag{3.83}$$

Thus, if the flux tube perpendicular dimension $L_\perp$ is much larger than the electron collisionless skin depth c/ω_{pe}, the magnetic term dominates the mechanical term and the electron helicity is essentially identical to the magnetic helicity.

To summarize: When the plasma is collisionless and the dimensions under consideration are much larger than c/ω_{pe}, magnetic helicity is conserved within a closed three dimensional flux surface no matter what complicated phenomena might be occurring within the volume bounded by that flux surface. This is true when the volume enclosed by the flux surface is fixed (e.g., by bounding the plasma with a rigid flux conserving surface such as a metal wall) and also when the flux surface moves around with the fluid.

3.9 Magnetic reconnection

We noted earlier that highly conducting plasmas are, in general, good flux conservers so that magnetic field lines are frozen into the plasma. While generally true, this statement breaks down on certain cleavage surfaces where the logic of the frozen-in argument fails and stresses can cause motion of field lines across plasma. Magnetic reconnection is a complicated subject and still not fully understood. However, the net result of magnetic reconnection is that flux surfaces and field lines undergo topological transformation. One flux tube might break into two, or two flux tubes might combine into one. Early models of magnetic reconnection showed how this phenomenon can result from the resistive diffusion of magnetic fields across a thin boundary layer and predicted that the time scale of magnetic reconnection would be of the order of the geometric mean of the Alfvén time and the resistive diffusion time. However, experimental measurements often show that magnetic reconnection occurs considerably faster than this rate. An important question to be asked is what happens to the magnetic helicity, the magnetic energy, and the magnetic flux when there is magnetic reconnection. The answer to this question has a profound effect on the behavior of spheromaks.

3.10 Geometric interpretation of magnetic helicity

The magnetic helicity of a configuration is a topological property and therefore remains unchanged during any *continuous* deformation of the configuration even though the deformation may substantially change the geometry. Appropriately chosen continuous deformations show that a given amount of helicity can be manifested geometrically in three equivalent forms [76, 84]. These are shown in Fig.3.5 and are:

1. *writhe*, the amount of twist of a single flux tube as in Eq.(3.27),
2. *linking* of two flux tubes (as in links on a chain),
3. *crossing* of a segment of one flux tube with another segment (as in a highway overpass).

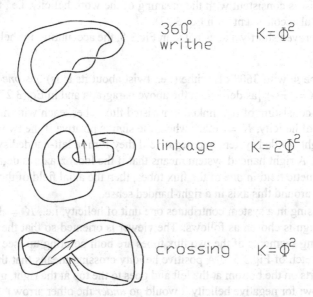

Fig.3.5 Writhe, linking, and crossing with associated helicity for flux tubes with flux Φ.

Continuous deformation of a given flux tube makes it possible for the helicity to appear in any one of the above three forms. The helicity is not a local quantity; the entire configuration must be taken into account in order to measure the helicity. Nevertheless, by categorizing contributions to the helicity as writhes, linkages, or crossings, it is possible to localize contributions to the total helicity to a certain extent. The relationship to the mathematical definition $K = \int d^3r\, \mathbf{A} \cdot \mathbf{B}$ can be seen because as was shown in the derivation of Eq.(3.11) the helicity of two linked, untwisted flux tubes is $K = 2\Phi_a \Phi_b$ where a, b label the two flux tubes. Because the writhe of a single field line is meaningless (like the rotation of a point), descriptions based on the structure of individual field lines are inadequate for quantifying helicity.

The connection between writhe, linking, and crossing can be understood by proceeding through a sequence of continuous deformations. We start by defining one 'unit' of helicity as the helicity of a flux tube Φ with 360^0 of twist (i.e., writhe). Thus, $K = \pm \Phi^2$ is one unit of helicity (one 360^0 twist); the plus sign is chosen for right handed twist and the minus sign for left handed twist. If the flux tube has 720^0 of twist, it would have two units of helicity, i.e., $K = \pm 2\Phi^2$. If it is a Mobius strip with only 180^0 of twist (i.e., a half twist) then it will have half a unit of helicity, $K = \pm \frac{1}{2}\Phi^2$. This counting of twist is consistent with the meaning of the word helicity, i.e., the measure of twist, and is also consistent with Eq.(3.27).

As will be proved shortly and as noted in Fig.3.5, the accounting for helicity is such that:

1. A flux tube Φ with 360^0 of writhe (i.e., twist about its axis) has *one* unit of helicity, $K = \pm \Phi^2$ [as defined in the above paragraph and in Eq.(3.27)].
2. A system consisting of two linked, untwisted flux tubes, each with flux Φ, has *two* units of helicity, $K = \pm 2\Phi^2$ where the sign is positive if the two flux tubes form a right-handed system and negative if they form a left-handed system [see Eq.(3.11]. A right handed system means that if the local z axis is defined by the axial magnetic field in one of the flux tubes, then the axial field of the other flux tube goes around this axis in a right-handed sense.
3. Each crossing in a system contributes one unit of helicity, i.e., $K = \pm \Phi^2$ where the plus sign is chosen as follows: The viewer is oriented so that the fields at the crossing segments of the two flux tubes are both pointing up (see arrows in bottom sketch of Fig.3.5). A positive helicity crossing means that the arrow which starts on the bottom at the left and goes to the top at the right, goes *over* the other arrow; for negative helicity it would go *under* the other arrow.

The validity of this accounting will now be demonstrated. First, imagine that a twisted flux tube is squashed so that instead of having a round cross section, it has a very thin elliptical cross section. Imagine that the ellipse is so thin that it can be represented by a piece of paper or by a ribbon (the thickness of the paper or ribbon corresponds to the minor diameter of the ellipse). These arguments are best followed by taking an actual length of ribbon, twisting it in a right handed sense to have $+360°$ of writhe, and then connecting the ends together to form a closed loop as in the top of Fig.3.5. It is helpful to mark the ribbon beforehand with arrows along its axis to indicate the sense of the magnetic field. The twisted ribbon will now be imagined to represent a thin twisted magnetic flux tube.

Manipulation of the flux tube: Because it has $360°$ of right-handed twist, the flux tube initially has $K = +\Phi^2$.

1. Deform the flux tube into a figure eight shape so that there is no writhe, but instead one crossing as in the bottom of Fig.3.5. Verify that the crossing helicity is $K = +\Phi^2$; i.e., if the configuration is oriented so that the arrows at the crossing both go up, then the arrow which starts from the bottom left goes over the other arrow. This demonstrates that helicity is non-local and that deformation can make helicity appear either as writhe or as crossing.

2. Mentally visualize that the flux tube consists of two parallel smaller flux tubes, each with flux $\Phi/2$ as shown in the top of Fig.3.6. These two smaller flux tubes will now be separated from each other. Take scissors and cut along the length of the flux tube so that the two smaller flux tubes are separated from each other (before cutting it is helpful to draw parallel arrows on either side of the cut, so that the sense of the magnetic field after cutting can be determined). After separation, it will be seen that the two smaller flux tubes are linked with each other and that each has $360°$ of right hand twist (writhe). Since separation did not change the topology, helicity was conserved throughout the separation process (the two smaller flux tubes could have been merely adjacent to each other showing that the deformation is continuous). We now use the helicity accounting rules: Each of the small flux tubes has flux $\Phi/2$ and each has a $360°$ turn so the writhe helicity of each is $K_{writhe} = (\Phi/2)^2 = \Phi^2/4$. The two small flux tubes are linked in a right handed sense, so that the helicity due to linkage is $K_{linkage} = 2(\Phi/2)^2 = \Phi^2/2$. The total helicity for the system is the sum of the individual writhe helicities plus the linkage helicity, i.e., $K_{final} = 2K_{writhe} + K_{linkage} = 2\Phi^2/4 + \Phi^2/2 = \Phi^2$. This verifies the accounting rules since the separation was continuous and conserved total helicity. This also makes clear that helicity is not a local quantity but is globally conserved when the system undergoes continuous transformations.

3.11 Magnetic reconnection and helicity conservation

Magnetic reconnection is often described in terms of individual magnetic field lines breaking and reconnecting. If we consider the difference between an individual field line with zero cross-section and an individual field line as the limit of a flux tube with

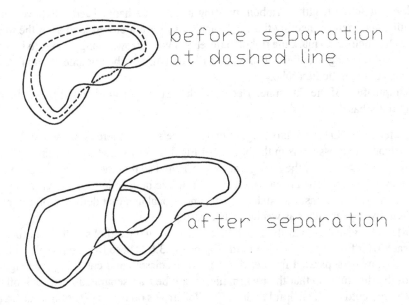

Fig.3.6 Top: Flux tube before cutting along dashed line. Bottom: Linked flux tubes, each with half of original flux after cutting along dashed line and separating.

infinitesimal diameter, it is clear that the zero-cross section model is deficient because it fails to take writhe into account. Zero cross-section corresponds to a point and it is meaningless to talk about the rotation angle of a point. It is therefore essential to consider magnetic reconnection in terms of thin flux tubes (and not field lines) so that helicity can be properly quantified.

We now consider what happens when, as in Fig.3.7, two straight (untwisted) flux tube segments initially cross and then reconnect. The projection of the field in the plane of the dashed square shown in Figs.3.7(a) and (b) is shown in Fig.3.7(c) and it is seen that the reconnection corresponds to formation of an x-point in the plane of projection. The crossing before reconnection corresponds to an initial helicity $K_{initial} = \Phi^2$. After reconnection, each of the newly formed flux tube segments has a half twist with the same handedness[84, 85]. Thus, each of the newly formed flux tube segments has helicity $K = \Phi^2/2$ since each has a half twist. The sum of the final helicities is $K_{final} = \Phi^2$ and so helicity is conserved.

Now consider the geometry sketched in Fig.3.8(a). A highly twisted open flux tube sits above a much less twisted closed flux tube. The ends of the open flux tube intercept

3.11 *Magnetic reconnection and helicity conservation* 65

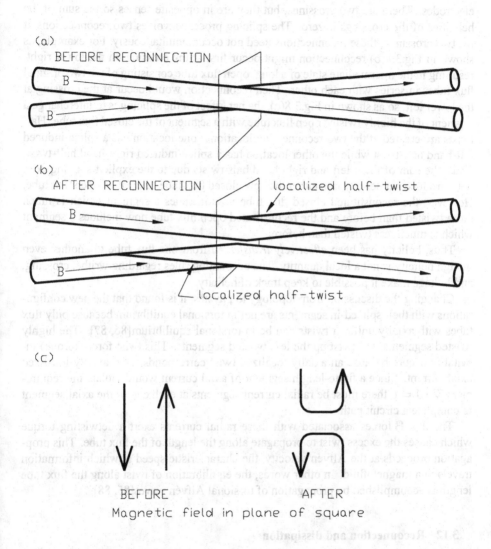

Fig.3.7 (a) Before reconnection: two straight flux tubes cross each other. (b) After reconnection flux tubes have a half-twist with same sense. (c) View of projection of magnetic field onto plane of square in (a) and (b), x-point is developed.

electrodes. There are two crossings, but they are in opposite senses so the sum of the helicities of the crossings is zero. The splicing process involves two reconnections at the two crossings; these reconnections need not occur simultaneously. For example, as shown in Fig.3.8(b) reconnection might occur first at the crossing on the lower right, resulting in the intermediate state of a long, open flux tube consisting of the two original flux tubes in series with each other. Then reconnection would occur at the crossing at the lower left, so as shown in Fig.3.8(c), the net effect of the splicing is to interchange a segment of the highly twisted open flux tube with a segment of the closed flux tube. Half twists are created at the two reconnection locations: one location has a splice-induced left-hand half-twist while the other location has a splice-induced right-hand half-twist. Thus, the sum of these left and right hand half-twists due to the explicit splicing does not introduce net helicity into either the new closed flux tube or the new open flux tube. However, the reconstituted closed flux tube now includes a segment which is much more twisted than before and the reconstituted open flux tube now includes a segment which is much less twisted than before.

Thus, helicity has been *effectively transferred* from one flux tube to another even though helicity is not a local quantity. The accounting rules regarding writhe, crossing, and linkage make it possible to keep track of helicity.

Changing the discussion from topology to physics, it is found that the new configurations with their spliced-in segments are not in torsional equilibrium because only flux tubes with axially uniform twists can be in torsional equilibrium[86, 87]. The highly twisted segments will twist up the less twisted segments. This twist force (torque) inevitably occurs because an axially localized twist corresponds to an axially localized axial current. Since a finite-length segment of axial current would violate the requirement $\nabla \cdot \mathbf{J} = 0$ there must be radial current segments at each end of the axial segment to complete a circuit path.

The $\mathbf{J} \times \mathbf{B}$ forces associated with these radial currents exert a detwisting torque which causes the excess twist to propagate along the length of the flux tube. This propagation proceeds at the Alfvén velocity, the characteristic speed at which information travels in a magnetofluid. In other words, the equilibration of twist along the flux tube length is accomplished by propagation of torsional Alfvén waves[85, 88].

3.12 Reconnection and dissipation

Magnetic reconnection necessarily involves some dissipation of magnetic energy at the location of reconnection. There are various ways to establish this assertion. We provide a simple argument here.

Suppose one starts with two identical twisted flux tubes with 360^0 writhe, each with flux Φ_0, and thus each with helicity $K_0 = \Phi_0^2$. The magnetic energy will be some

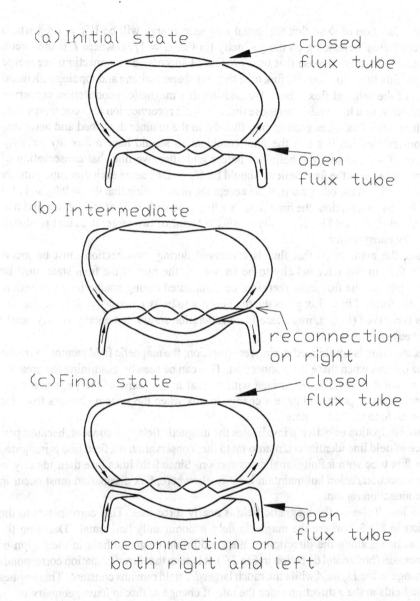

Fig.3.8 Helicity transfer from an open flux tube to a closed flux tube: (a) Initial state, (b) reconnection at bottom right crossing, (c) reconnection at both crossings, yielding final state. After the last step, torsional Alfven waves will equilibrate the twist along the length of each flux tube.

quadratic function of Φ so that the initial magnetic energy will be $W_0 = \alpha \Phi_0^2$ where α is a coefficient depending on the geometry (in fact $\alpha = 1/2L$ where L is the circuit inductance). Further suppose that by a process of successive reconnections we merge these two flux tubes to form one flux tube that has shape, volume and topology identical to each of the original flux tubes. We showed that magnetic reconnection conserves helicity. Now, as a hypothesis, suppose that magnetic reconnection also conserves flux. Combining two flux tubes each having flux Φ_0 in the manner described and assuming flux conservation implies that the final combination should have a flux $\Phi_f = 2\Phi_0$. However, if we consider the helicities before and after, we find that conservation of helicity implies that the final helicity should be $H_f = 2\Phi_0^2$ since each flux tube initially had $H_0 = \Phi_0^2$. On the other hand, if we accept the presumption that $\Phi_f = 2\Phi_0$ and also recall that, by assumption, the final form is a flux tube with 360^0 of writhe, we find that the final helicity should be $H_f = \Phi_f^2 = 4\Phi_0^2$, a factor of two larger than that predicted by helicity conservation.

Thus, the assumption that flux is conserved during reconnection must be incorrect. In fact, in order for helicity to be conserved, the flux of the final state must be $\Phi_f = \sqrt{2}\Phi_0$. Some flux must therefore be annihilated during magnetic reconnection. Using this form of final flux gives the final energy to be $W_f = \alpha \Phi_f^2 = 2W_0$ so that energy is conserved (in fact, more careful consideration shows that energy is only nearly conserved).

Because flux is annihilated during reconnection, the magnetic field cannot be frozen into the plasma when there is reconnection. This can be seen by examining the creation of an x-point in Fig.3.7. A surface with normal in the plane of the page and passing through the x-point will have a change in flux when the system changes from the 'before' state to the 'after' state.

Flux dissipation effectively invalidates the magnetic field line concept, because persistence of field line identity is tantamount to flux conservation if a field line is imagined to be a flux tube with infinitesimal cross section. Since field lines lose their identity in the reconnection region but maintain identity elsewhere, flux dissipation must occur in the reconnection region.

In a low β plasma the magnetic field is nearly force-free. This corresponds to the situation in Fig.3.7 where the magnetic field is dominantly horizontal. Denoting the z axis as being along the direction of the equilibrium magnetic field in the region of reconnection (horizontal direction in Fig.3.7), it is seen that flux dissipation corresponds to a change in the B_y field while the much larger B_z field remains constant. This implies electric fields in the z direction since the rate of change of flux in fixed geometry is

$$\frac{d\Phi}{dt} = \frac{d}{dt}\int_S \mathbf{B} \cdot d\mathbf{s} = \int_S \frac{\partial \mathbf{B}}{\partial t} \cdot d\mathbf{s} = -\oint_C \mathbf{E} \cdot d\mathbf{l}. \quad (3.84)$$

Here S is the surface linked by the B_y magnetic fields undergoing reconnection and so the contour C includes segments along z. This local flux dissipation implies a local dissipation of magnetic energy and also an electric field parallel to the magnetic field.

CHAPTER 4
Relaxation of an Isolated Configuration to the Taylor State

4.1 Introduction

From the discussion in the previous chapter we see that magnetic helicity is conserved during reconnection, whereas magnetic flux is not. Furthermore, the "unfreezing" of field lines from plasma requires some kind of dissipation of magnetic energy. In the 1950's Woltjer[4] noted that both magnetic helicity and magnetic energy are exactly conserved in a non-dissipative plasma, and conjectured that in a slightly dissipative plasma, magnetic helicity would decay at a substantially slower rate than magnetic energy. As a result of this difference in decay rates, the magnetic helicity can be considered as being conserved on the time scale of the magnetic energy decay.

It is clear that magnetic energy cannot decay to zero if the helicity is to be conserved, because if $\mathbf{B}$ became zero, then the helicity would also become zero. Woltjer characterized this constrained decay of energy as a variational problem. By minimizing the magnetic energy subject to the constraint that magnetic helicity is conserved, he showed that the resulting magnetic field satisfies the simple force-free equation $\nabla \times \mathbf{B} = \lambda \mathbf{B}$ where λ is a constant. Taylor[18] extended Woltjer's hypothesis by arguing that *magnetic reconnection conserves global helicity* and so causes an isolated configuration to relax to the lowest energy state consistent with (i) conservation of magnetic helicity and (ii) satisfying relevant boundary conditions.

Taylor applied his model to the reversed field pinch (RFP), a toroidal laboratory device which had the mysterious property[89] of spontaneously developing a reversed toroidal field near the wall. Before Taylor's work it was realized that the RFP magnetic fields satisfied the force-free equilibrium $\nabla \times \mathbf{B} = \lambda \mathbf{B}$ and in particular when modeled in straight cylindrical geometry, satisfied Lundquist's[2] axisymmetric Bessel function solutions $B_\| = \bar{B} J_0(\lambda r)$ and $B_\phi = \pm \bar{B} J_1(\lambda r)$. However, it was not known why the RFP attained this force-free state.

Taylor's generalization of Woltjer's argument was to postulate that turbulent reconnection causes the RFP to relax to the lowest energy force-free state consistent with the initial helicity inventory. This generalization provides an immediate explanation for the mysterious field reversal of the RFP: the lowest energy state is just the Lundquist solution, the axial component of which becomes negative when λr exceeds 2.4, i.e., the first root of J_0. Taylor's model was very successful in describing RFP behavior and has since been applied with similar success to many other magnetically dominated (i.e., low β) turbulent plasma configurations both in the laboratory and in space. Wells[14, 15] made arguments rather similar to Taylor's, but in the context of the conical theta pinch. Although Taylor proposed that the reconnection processes would be turbulent (i.e., random) experimental observations indicate that the reconnection processes are often quite regular and periodic, and behave somewhat like a repetitive pumping process.

The spheromak is a simply connected Taylor state whereas the RFP is a doubly connected Taylor state which can link external conductors. This topological difference means that RFP's can have a toroidal field produced by linked external coils, whereas the toroidal field in the spheromak is produced entirely by internal currents and must vanish on the bounding surface.

Before embarking on the derivation of the Taylor state, it is important to distinguish two quite different situations, namely isolated configurations and driven configurations. In both cases the plasma occupies a volume V bounded by a surface S. An isolated configuration has $\mathbf{B} \cdot d\mathbf{s} = 0$ everywhere on S; in other words no magnetic field lines penetrate S. In contrast a driven configuration has finite $\mathbf{B} \cdot d\mathbf{s}$ on at least some portion of S. This chapter will discuss relaxation in isolated configurations and the next chapter will describe driven configurations. The mathematical analysis of these two situations, while similar, is not identical; the distinction is between the solutions to a homogeneous partial differential equation and the solutions to an inhomogeneous partial differential equation. Driven situations can be further classified into situations with strong coupling from source to load and situations with weak coupling (see Chapters 11 and 12).

4.2 Helicity decay rate v. magnetic energy decay rate

The central concept underlying relaxation theory is that if dissipative dynamics are confined to a microscopic scale, the helicity decay rate is negligible compared to the magnetic energy decay rate. The reason for this assertion can be understood by comparing the dimensions of helicity and magnetic energy. Helicity scales as $B^2 L^4$ whereas magnetic energy scales as $B^2 L^3$ where L is the linear dimension. If a given phenomenon involving attenuation of B is scaled down, the magnetic energy decreases as L^3 whereas the helicity decreases as L^4 so that the ratio of energy decay to helicity decay scales as L^{-1}. Thus, the smaller the scale, the bigger is the ratio of energy loss to helicity loss. A collection of microscopic dissipative events will therefore destroy magnetic energy in much greater proportion than helicity compared to the situation where the same energy is dissipated on a macroscopic scale. Spatially complex phenomena such as turbulence will therefore dissipate magnetic energy preferentially compared to helicity. Magnetic reconnection is an intrinsically small-scale phenomenon and in the limit that the reconnection scale becomes infinitesimal, helicity dissipation during reconnection becomes negligible compared to magnetic energy dissipation. If there is no microscopic activity, then magnetic helicity and magnetic energy decay at the same rate (see Sec.13.3.3).

4.3 Derivation of the isolated Taylor state

Consider a low β plasma bounded by a closed, simply-connected, flux-conserving, perfectly conducting surface S (e.g., a box with perfectly conducting walls) with all field lines initially tangent to the walls. The surface S is shown schematically in Fig.4.1; by assumption $\mathbf{B} \cdot ds = 0$ everywhere on S. Suppose that this plasma is initially not in equilibrium and undergoes various complicated, perhaps turbulent, microscopic reconnection instabilities. Because of the low β assumption, the only free energy source is the energy stored in the non-vacuum part of the magnetic field. However, an isolated configuration has no vacuum magnetic field (i.e., $\mathbf{B} \cdot ds = 0$ everywhere on S) and so the entire magnetic field energy of this isolated configuration is free energy. As discussed in Sec.4.2, microscopic instabilities conserve helicity but not magnetic energy; hence the magnetic reconnection events alter the magnetic topology but not the helicity. Because $\mathbf{B} \cdot ds = 0$ everywhere on S, the helicity defined by Eq.(3.4) is automatically gauge invariant and there is no need here to introduce the concept of relative helicity.

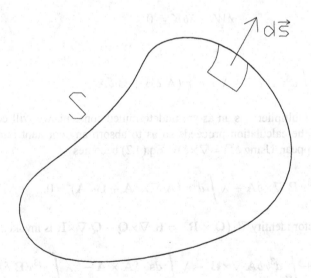

Fig.4.1 Simply connected surface S bounding volume V. It is assumed that $\mathbf{B} \cdot ds = 0$ everywhere on S, i.e., no field lines poke out of S.

After the instabilities have dissipated all available free energy, the plasma will be in a steady state. This state is called the relaxed state and is the minimum energy state consistent with (i) the prescribed boundary condition that $\mathbf{B}\cdot d\mathbf{s} = 0$ everywhere on S and (ii) the initial helicity inventory.

The boundary condition $\mathbf{B}\cdot d\mathbf{s} = 0$ means that there is no magnetic flux through every infinitesimal patch on the conducting surface S. The flux through each patch can be expressed as $d\psi = \oint_C \mathbf{A}\cdot d\mathbf{l}$ where the contour C is around the perimeter of ds. Since $d\psi = 0$ we conclude that the component of $\mathbf{A}$ tangential to S vanishes, or at most is the gradient of a scalar function. However, since S is a perfect conductor, the tangential component of the electric field on S must be zero, and so there can be no change in the tangential component of the vector potential, i.e., $\delta \mathbf{A}_\parallel = 0$.

The problem reduces to minimizing the total magnetic energy $W = \int d^3r\, B^2/2\mu_0$ subject to (i) the constraint that the helicity $K = \int d^3r\, \mathbf{A}\cdot \mathbf{B}$ is conserved and (ii) the boundary condition that $\delta \mathbf{A}_\parallel = 0$ on S. This is a standard variational problem: Minimizing W corresponds to setting the variation $\delta W = 0$ while helicity conservation imposes a constraint which can be taken into account by adding a Lagrange multiplier[90]. The variational problem with constraint included is therefore

$$\delta W - \lambda \delta K = 0 \tag{4.1}$$

or

$$\int d^3r\, \mathbf{B}\cdot\delta\mathbf{B} - \lambda \int d^3r\, (\mathbf{A}\cdot\delta\mathbf{B} + \mathbf{B}\cdot\delta\mathbf{A}) = 0. \tag{4.2}$$

Since the Lagrange multiplier λ is an as-yet undetermined constant, we will continuously redefine it as the calculation proceeds so as to absorb any constant numerical factors which may appear. Using $\delta\mathbf{B} = \nabla\times\delta\mathbf{A}$, Eq.(4.2) becomes

$$\int d^3r\, \mathbf{B}\cdot\nabla\times\delta\mathbf{A} - \lambda \int d^3r\, (\mathbf{A}\cdot\nabla\times\delta\mathbf{A} + \mathbf{B}\cdot\delta\mathbf{A}) = 0. \tag{4.3}$$

At this point, the vector identity $\nabla\cdot(\mathbf{Q}\times\mathbf{R}) = \mathbf{R}\cdot\nabla\times\mathbf{Q} - \mathbf{Q}\cdot\nabla\times\mathbf{R}$ is invoked giving

$$\int_S d\mathbf{s}\cdot(\delta\mathbf{A}\times\mathbf{B}) + \int_V d^3r\, \delta\mathbf{A}\cdot\nabla\times\mathbf{B} - \lambda \int_S d\mathbf{s}\cdot\delta\mathbf{A}\times\mathbf{A} - 2\lambda\int_V d^3r\, \mathbf{B}\cdot\delta\mathbf{A} = 0. \tag{4.4}$$

The surface integrals vanish because $\delta\mathbf{A}_\parallel = 0$. Regrouping the remaining terms gives

$$\int d^3r\, \delta\mathbf{A}\cdot(\nabla\times\mathbf{B} - \lambda\mathbf{B}) = 0. \tag{4.5}$$

Since $\delta \mathbf{A}$ was arbitrary inside V, the integrand in Eq.(4.5) must vanish; thus

$$\nabla \times \mathbf{B} = \lambda \mathbf{B} \qquad (4.6)$$

is the differential equation which must be satisfied inside V. Magnetic fields satisfying Eq.(4.6) are called Taylor states, or sometimes Woltjer-Taylor states.

Using Ampere's law, Eq.(4.6) can be expressed as

$$\mu_0 \mathbf{J} = \lambda \mathbf{B} \qquad (4.7)$$

showing that the configuration is force-free, i.e., $\mathbf{J} \times \mathbf{B} = 0$. Because $\beta \to 0$ has been assumed, the force-free property is to be expected. However, what is new here is that the proportionality coefficient λ is spatially uniform. As will be shown in Sec. 6.2, this spatial uniformity has physical significance relating to kink stability.

Taking the curl of Eq.(4.6) gives the vector Helmholtz equation

$$\nabla^2 \mathbf{B} + \lambda^2 \mathbf{B} = 0. \qquad (4.8)$$

It is now seen that λ plays the role of an eigenvalue and λ^{-1} is related to the characteristic linear dimensions of the system. In an isolated system only certain discrete values of λ are allowed, because only for these values will the solution to Eq.(4.8) satisfy the prescribed boundary conditions.

4.4 Relationship between helicity, energy, eigenvalue

The magnetic energy of the isolated Taylor state described by Eq.(4.6) can be expressed as

$$\begin{aligned} W &= \frac{1}{2\mu_0} \int \mathbf{B} \cdot \nabla \times \mathbf{A} d^3 r \\ &= \frac{1}{2\mu_0} \int [\nabla \cdot (\mathbf{A} \times \mathbf{B}) + \mathbf{A} \cdot \nabla \times \mathbf{B}] d^3 r \\ &= \frac{1}{2\mu_0} \int [\nabla \cdot (\mathbf{A} \times \mathbf{B}) + \lambda \mathbf{A} \cdot \mathbf{B}] d^3 r. \end{aligned} \qquad (4.9)$$

However, integrating Eq.(4.6) gives

$$\mathbf{B} = \lambda \mathbf{A} + \nabla f \qquad (4.10)$$

where f is an arbitrary gauge potential. The first term in the last line of Eq.(4.9) can be evaluated as

$$\int d^3r \nabla \cdot (\mathbf{A} \times \mathbf{B}) = \int d^3r \nabla \cdot (\mathbf{A} \times \nabla f) = \int d^3r \, \nabla f \cdot \mathbf{B}$$
$$= \int d^3r \, \nabla \cdot (f\mathbf{B}) = \int_S d\mathbf{s} \cdot f\mathbf{B} = 0 \quad (4.11)$$

since, by assumption $\mathbf{B} \cdot d\mathbf{s} = 0$ over all of S for an isolated configuration. Thus Eq.(4.9) can be recast as

$$\lambda = \frac{\int B^2 d^3r}{\int \mathbf{A} \cdot \mathbf{B} d^3r} = 2\mu_0 \frac{W}{K}. \quad (4.12)$$

This provides an important alternate interpretation for λ, namely $\lambda \sim$ energy/helicity of an isolated configuration. For a given helicity, the energy of a system is proportional to λ and so, for a given helicity, the lowest energy state of a system will be the state with the smallest allowed λ.

These seemingly arcane properties of λ have an important relevance to real-world dynamics because these properties prescribe the ultimate state towards which instabilities inevitably drive an isolated system. In effect, λ describes the end but not the means. If the system is not in the eigenstate having the smallest allowed λ for the given helicity, then the system has free energy that could be expended in relaxing to this ultimate state. Because energy is proportional to λ, the system is always unstable with respect to helicity-conserving transitions from large allowed λ to smaller allowed λ; it can only be stable when it is in the state with the lowest allowed λ for the given helicity. A relation between energy, twist and λ can be obtained using Eqs.(4.12) and (3.26), namely

$$W = \frac{\lambda}{\mu_0} \int_0^\Phi T(\Phi) \Phi d\Phi. \quad (4.13)$$

4.4.1 Equality of poloidal and toroidal field energies in an isolated axisymmetric spheromak

An isolated spheromak is simply connected and thus has no toroidal field at the wall. This is because if there were a toroidal field at the wall, there would have to be a net z-directed current in the spheromak, but this would be impossible in an isolated simply connected configuration. Thus, an isolated spheromak is an axisymmetric Taylor state with zero toroidal field on S. We will now prove that the toroidal and poloidal field energies are equal for this situation. This is demonstrated by invoking the spheromak boundary condition $\mathbf{B}_{tor} = 0$ on S and also using Eq.(4.6) to give

$$\begin{aligned} W_{tor} - W_{pol} &= \frac{1}{2\mu_0} \int_V \left(B_{tor}^2 - B_{pol}^2 \right) d^3r \\ &= \frac{1}{2\mu_0 \lambda} \int_V \left(\mathbf{B}_{tor} \cdot \nabla \times \mathbf{B}_{pol} - \mathbf{B}_{pol} \cdot \nabla \times \mathbf{B}_{tor} \right) d^3r \\ &= \frac{1}{2\mu_0 \lambda} \int_V \nabla \cdot \left(\mathbf{B}_{pol} \times \mathbf{B}_{tor} \right) d^3r \\ &= \frac{1}{2\mu_0 \lambda} \int_S d\mathbf{s} \cdot \left(\mathbf{B}_{pol} \times \mathbf{B}_{tor} \right) = 0. \end{aligned} \quad (4.14)$$

The fact that the curl of a toroidal field is poloidal (and vice versa) has been used in the second line of Eq.(4.14).

4.5 Cylindrical force-free states

We now assume that S is a cylinder and use cylindrical coordinates r, ϕ, z to describe the system. A set of orthonormal eigenfunction solutions to Eqs.(4.6) and (4.8) will now be constructed by assuming $\mathbf{B} \sim \exp(im\phi + ikz)$. The z component of Eq.(4.8) can be written as

$$\frac{\partial^2 B_z}{\partial r^2} + \frac{1}{r}\frac{\partial B_z}{\partial r} + \left(\lambda^2 - k^2 - \frac{m^2}{r^2} \right) B_z = 0 \quad (4.15)$$

which has the solution

$$B_z = \bar{B} J_m(\gamma r) e^{im\phi + ikz} \quad (4.16)$$

where

$$\gamma = \sqrt{\lambda^2 - k^2}. \quad (4.17)$$

The other components of the magnetic field are not independent of B_z and are determined from Eq.(4.6) as follows. Both the magnetic field and the ∇ operator are split into components parallel and perpendicular to z so that Eq. (4.6) becomes

$$(\nabla_\perp + ik\hat{z}) \times (\mathbf{B}_\perp + B_z \hat{z}) = \lambda (\mathbf{B}_\perp + B_z \hat{z}). \quad (4.18)$$

The perpendicular component can be written as

$$\lambda \mathbf{B}_\perp - ik\hat{z} \times \mathbf{B}_\perp = \nabla_\perp B_z \times \hat{z}. \quad (4.19)$$

Crossing the above equation with $\hat{z}$ gives

$$ik\mathbf{B}_\perp + \lambda \hat{z} \times \mathbf{B}_\perp = \nabla_\perp B_z. \tag{4.20}$$

Equations (4.19) and (4.20) comprise a system of two linear equations in the unknowns $\mathbf{B}_\perp$ and $\hat{z} \times \mathbf{B}_\perp$. Solving this system for $\mathbf{B}_\perp$ gives

$$\mathbf{B}_\perp = \frac{\lambda \nabla_\perp B_z \times \hat{z} + ik\nabla_\perp B_z}{\lambda^2 - k^2}. \tag{4.21}$$

Using Eq.(4.16) the r and ϕ components are

$$\begin{aligned} B_r &= \frac{i\bar{B}}{\gamma^2}\left(\frac{m\lambda}{r}J_m(\gamma r) + k\gamma J'_m(\gamma r)\right)e^{im\phi+ikz} \\ B_\phi &= -\frac{\bar{B}}{\gamma^2}\left(\frac{mk}{r}J_m(\gamma r) + \lambda\gamma J'_m(\gamma r)\right)e^{im\phi+ikz}. \end{aligned} \tag{4.22}$$

Taking the real parts of the three components gives

$$\begin{aligned} B_r(r,\phi,z) &= -\frac{\bar{B}}{\gamma}\left(\frac{m\lambda}{\gamma r}J_m(\gamma r) + kJ'_m(\gamma r)\right)\sin(m\phi+kz) \\ B_\phi(r,\phi,z) &= -\frac{\bar{B}}{\gamma}\left(\frac{mk}{\gamma r}J_m(\gamma r) + \lambda J'_m(\gamma r)\right)\cos(m\phi+kz) \\ B_z(r,\phi,z) &= \bar{B}J_m(\gamma r)\cos(m\phi+kz); \end{aligned} \tag{4.23}$$

these are called the Chandrasekhar-Kendall functions[6].

The lowest order solution is obtained by setting $m = 0, k = 0$ giving the Lundquist [2] solution (also called the Bessel Function Model or BFM)

$$\begin{aligned} B_r(r,\phi,z) &= 0 \\ B_\phi(r,\phi,z) &= \bar{B}J_1(\lambda r) \\ B_z(r,\phi,z) &= \bar{B}J_0(\lambda r) \end{aligned} \tag{4.24}$$

which is independent of both ϕ and z. Despite its simplicity, Eq.(4.24) has been found to describe a large number of experimental observations with impressive accuracy. It is most appropriate for configurations that can be approximated as long axisymmetric cylinders so that $\lambda \gg k$. This turns out to be a reasonable first approximation for

4.6 Comparison of minimum energy states in a long cylinder

the straight-cylinder approximation of a large aspect torus. The next most complicated solution has finite k and $m = 0$; these are appropriate for finite length cylinders.

4.6 Comparison of minimum energy states in a long cylinder

Although force-free equilibria represent configurations with a local minimum of energy for a given helicity, there might be more than one such state for a given helicity. These states could be distinct and separated so that while each is a local energy minimum, one could have lower energy than the other. In such a case a plasma initially in a higher energy force-free state would be unstable with respect to a helicity-conserving transformation into an allowable lower energy force-free state.

Consider a plasma in a very long flux-conserving cylindrical shell with radius a. Initially no magnetic field penetrates the shell, i.e., initially $B_r = 0$ on the shell. Since the shell is flux conserving, the radial boundary condition will be $B_r = 0$ for all times. The plasma is assumed to be initially in a Woltjer-Taylor state with eigenvalue λ_1 and will be unstable with respect to turbulent relaxation into a different Woltjer-Taylor state having allowed eigenvalue λ_2 if $\lambda_2 < \lambda_1$ because in this case the instability would lower the system energy while conserving helicity[18, 71, 91]. The adjective "allowed" means that the fields computed with the specified eigenvalue satisfy all prescribed boundary conditions.

The $m = 0, k = 0$ state given by Eq.(4.24) automatically satisfies the flux-conserving boundary condition that $B_r = 0$ at $r = a$. There is no constraint on λ for these solutions so any λ is possible in principle.

Now consider states having $m \neq 0$. From Eq.(4.23) the $B_r = 0$ wall boundary condition becomes

$$m\lambda J_m(\gamma a) + ka\gamma J'_m(\gamma a) = 0. \tag{4.25}$$

We define $x = \gamma a$ and $\Lambda = \lambda a$ so from Eq.(4.17), $ka = \sqrt{\Lambda^2 - x^2}$. With these definitions Eq.(4.25) can be expressed as

$$\Lambda \frac{m J_m(x)}{x J'_m(x)} = -\sqrt{\Lambda^2 - x^2} \tag{4.26}$$

which may be solved for Λ to give

$$\Lambda = \frac{x}{\sqrt{1 - \left(\frac{m J_m(x)}{x J'_m(x)}\right)^2}}. \tag{4.27}$$

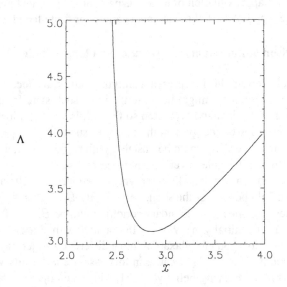

Fig.4.2 Plot of Λ v. x

Figure 4.2 plots Λ v. x for $m = 1$ and shows that Λ has a minimum of 3.1 when $x = 2.8$; this corresponds to $ka = \sqrt{\lambda^2 - x^2} = 1.3$ while for larger m, Λ has a higher minimum.

Thus, if a large aspect torus starts with $m = 0$, $k = 0$ symmetry and λ increases above 3.1, the system could lower its energy by relaxing to an $m = 1$ state with $ka = 1.3$. Therefore the λ of a large aspect axisymmetric torus can only lie in the range $0 < \lambda \leq 3.1$; if a larger λ is attempted, the system would not be in a lowest energy state and will shed energy until it relaxes into the non-axisymmetric, $m = 1$, $ka = 1.3$ state.

4.7 Spheromaks in spherical geometry

Spheromaks are most often modeled using cylindrical geometry because there is usually cylindrical symmetry. However, the force-free equation can also be solved in spherical geometry and in fact the name spheromak originates from Rosenbluth and Bussac's analysis of tilt instability in spherical geometry. We now briefly consider spherical force-free states and so will use spherical coordinates r, θ, ϕ.

In spherical geometry the solution of Eq. (4.6) can be expressed in terms of the

4.7 Spheromaks in spherical geometry

scalar χ [19]

$$\mathbf{B} = \lambda \mathbf{r} \times \nabla \chi + \nabla \times (\mathbf{r} \times \nabla \chi) \qquad (4.28)$$

where χ satisfies the scalar Helmholtz equation

$$\nabla^2 \chi + \lambda^2 \chi = 0. \qquad (4.29)$$

To verify that this is indeed a solution take the curl of Eq.(4.28) to obtain

$$\nabla \times \mathbf{B} = \lambda \nabla \times (\mathbf{r} \times \nabla \chi) + \nabla \times (\nabla \times (\mathbf{r} \times \nabla \chi)) \qquad (4.30)$$

or

$$\nabla \times \mathbf{B} = \lambda \nabla \times (\mathbf{r} \times \nabla \chi) + \nabla \nabla \cdot (\mathbf{r} \times \nabla \chi) - \nabla^2 (\mathbf{r} \times \nabla \chi). \qquad (4.31)$$

However, $\nabla \cdot (\mathbf{r} \times \nabla \chi) = 0$ and temporarily using Cartesian coordinates,

$$\begin{aligned}\nabla^2 (\mathbf{r} \times \nabla \chi) &= \nabla^2 \mathbf{r} \times \nabla \chi + \mathbf{r} \times \nabla^2 \nabla \chi + 2 \left(\sum_i \frac{\partial}{\partial x_i} \mathbf{r} \times \frac{\partial}{\partial x_i} \nabla \chi \right) \\ &= \mathbf{r} \times \nabla^2 \nabla \chi \end{aligned} \qquad (4.32)$$

since $\nabla^2 \mathbf{r} = 0$ and $\sum_i \partial \mathbf{r}/\partial x_i \times \nabla \partial \chi/\partial x_i = \nabla \times \nabla \chi = 0$. Thus, Eq.(4.31) becomes

$$\nabla \times \mathbf{B} = \lambda \nabla \times (\mathbf{r} \times \nabla \chi) - \mathbf{r} \times \nabla \nabla^2 \chi. \qquad (4.33)$$

Using Eq.(4.29) to replace the Laplacian in the last term gives

$$\nabla \times \mathbf{B} = \lambda \nabla \times (\mathbf{r} \times \nabla \chi) + \lambda^2 \mathbf{r} \times \nabla \chi \qquad (4.34)$$

i.e.,

$$\nabla \times \mathbf{B} = \lambda \mathbf{B}. \qquad (4.35)$$

The solutions of Eq.(4.29) in spherical geometry are products of spherical Bessel functions and associated Legendre functions,

$$\chi_m^n = j_m(\lambda r) P_m^n(\cos\theta) e^{in\phi}. \qquad (4.36)$$

The $m = 0, n = 0$ solution $\chi_0^0 = j_0(kr)$ is trivial and of no interest because $\mathbf{r} \times \nabla j_0(kr) = 0$. The lowest order non-trivial solution has $m = 1, n = 0$ and using $j_1(z) =$

$-d(z^{-1}\sin z)/dz = z^{-2}\sin z - z^{-1}\cos z$ and $P_1^0 = \cos\theta$ is

$$\chi_1^0(r,\theta) = -B_0 a \left(\frac{\sin(\lambda r)}{\lambda^2 r^2} - \frac{\cos(\lambda r)}{\lambda r} \right) \cos\theta. \tag{4.37}$$

The constant factor $B_0 a$ has been introduced to have dimensionally correct fields and the minus sign has been chosen to have a positive toroidal field. The radial component of the magnetic field is found by dotting Eq.(4.28) with $\hat{r} = \nabla r$ to obtain

$$\begin{aligned} B_r &= \nabla r \cdot \nabla \times (\mathbf{r} \times \nabla \chi) \\ &= \nabla \cdot ((\mathbf{r} \times \nabla \chi) \times \nabla r) \\ &= \nabla \cdot (r \nabla_\perp \chi) \\ &= r \nabla_\perp^2 \chi \end{aligned} \tag{4.38}$$

where $\perp$ refers to the direction orthogonal to the r direction. For the mode given by Eq.(4.37) this becomes

$$\begin{aligned} B_r &= \frac{1}{r\sin\theta} \frac{\partial}{\partial\theta}\left(\sin\theta \frac{\partial}{\partial\theta}(B_0 a j_1(\lambda r)\cos\theta)\right) \\ &= 2B_0 \frac{a}{r} j_1(\lambda r)\cos\theta. \end{aligned} \tag{4.39}$$

If the spheromak is contained inside a spherical flux conserver of radius a then the radial magnetic field must vanish at $r=a$, giving the boundary condition $j_1(\lambda a)=0$. Thus λa must be a root of j_1. The smallest allowed λ corresponds to the lowest energy state and since the smallest finite root of j_1 is 4.493, the Taylor state has $\lambda = 4.493/a$.

Because the configuration is axisymmetric, the force-free magnetic field can also be expressed in terms of the poloidal flux function as

$$\mathbf{B} = \frac{1}{2\pi}(\nabla\psi \times \nabla\phi + \lambda\psi\nabla\phi) \tag{4.40}$$

where the toroidal component has been chosen so that $\nabla \times \mathbf{B}_{tor} = \lambda \mathbf{B}_{pol}$.

The functional form of the poloidal flux can be determined by explicit evaluation of the flux of B_r through a dome-shaped surface of constant radius r subtending a conical angle θ about the z axis. The poloidal flux intercepted by this surface is

$$\begin{aligned} \psi(r,\theta) &= \int_0^\theta r d\theta\, 2\pi r \sin\theta\, B_r \\ &= \int_0^\theta r d\theta\, 2\pi r \sin\theta\, 2B_0 \frac{a}{r} j_1(\lambda r)\cos\theta \end{aligned}$$

4.7 Spheromaks in spherical geometry

$$= 2\pi r a B_0 j_1(\lambda r) \sin^2 \theta \tag{4.41}$$

The components of the magnetic field are therefore [16, 19]

$$B_r = 2B_0 \frac{a}{r} j_1(\lambda r) \cos \theta$$
$$B_\theta = -B_0 \frac{a}{r} \frac{\partial}{\partial r}(r j_1(\lambda r)) \sin \theta$$
$$B_\phi = \lambda a B_0 j_1(\lambda r) \sin \theta. \tag{4.42}$$

Surfaces of constant ψ are plotted in Fig. 4.3; the circle flux surface $r/a = 1$ corresponds to where $j_1(\lambda r) = 0$.

For purposes of calculating the safety factor on axis, it is convenient to express this flux function in cylindrical coordinates (which will be used in the remainder of this section)

$$\psi = \frac{2\pi r^2 B_0 a}{\lambda (r^2 + z^2)} \left(\frac{\sin(\lambda \sqrt{r^2 + z^2})}{(\lambda \sqrt{r^2 + z^2})} - \cos(\lambda \sqrt{r^2 + z^2}) \right). \tag{4.43}$$

The magnetic axis is where B_θ vanishes, i.e., at the maximum of $r j_1(\lambda r)$. This maximum occurs at $\lambda r = 2.75$ so the magnetic axis is at $r_0 = 2.75/\lambda = 0.61a$.

The ratio of second derivatives at the magnetic axis is

$$\left(\frac{\psi_{rr}}{\psi_{zz}} \right)_{axis} = 2.76. \tag{4.44}$$

Thus the ellipticity ratio of the flux surfaces in the vicinity of the magnetic axis is $\sqrt{\psi_{rr}/\psi_{zz}} = 1.66$ and so, using $\mu_0 J_\phi = \lambda B_\phi$, Eq. (2.47) gives

$$q_{axis} = \frac{\sqrt{\psi_{rr}/\psi_{zz}} + \sqrt{\psi_{zz}/\psi_{rr}}}{\lambda r_0} = 0.82. \tag{4.45}$$

It is significant that q_{axis} is below unity. Also, when the ellipticity ratio is increased, q_{axis} will increase. As will be discussed later, instability occurs when q_{axis} reaches unity, and so there is a maximum allowed ellipticity.

Typically the safety factor of a spheromak is near unity at the magnetic axis and decreases on going towards the wall. This is quite different from tokamaks which have a safety factor near unity at the magnetic axis and increasing monotonically on going outwards to the wall. Since the average toroidal field at the wall of a tokamak is about the same as the toroidal field on the magnetic axis, the q profile of a tokamak is determined mainly by the toroidal current profile. In contrast, for a spheromak the toroidal

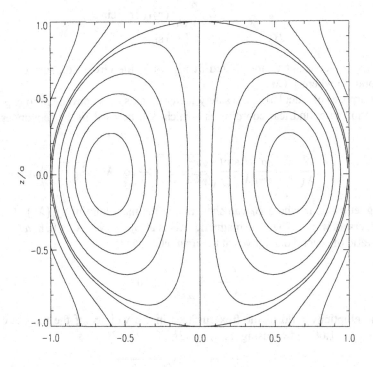

Fig.4.3 Spheromak flux surfaces in spherical geometry: Surfaces of constant ψ as viewed in plane $\phi = 0$. Straight vertical line is geometric axis, magnetic axis is at $r/a = 0.61$, $z = 0$. Wall is on circle of radius unity.

4.7 Spheromaks in spherical geometry

field goes to zero at the wall and so the toroidal field profile has a strong influence on q.

At first sight it seems that q at the wall of a spheromak should be zero because the toroidal field is zero at the wall. However, one must remember that q is the number of times a field line goes around toroidally for each time it goes around poloidally. Suppose one follows the trajectory of a field line, starting from a point just inside the wall. Since the toroidal field is zero at the wall, this field line will be purely poloidal so long as it is adjacent to the wall. However, if one follows this field line as it goes poloidally to the geometric axis, once it reaches the geometric axis it heads up (or down) the geometric axis and so is no longer adjacent to the wall. On the geometric axis, the solution of the force-free equation is essentially the small argument limit of the Lundquist solution Eq.(1.2), so in the vicinity of the geometric axis $B_\phi/B_z \simeq \lambda r/2$. The trajectory of a magnetic field line is given by $r d\phi/B_\phi = dz/B_z$ so the increment in ϕ on going up the geometric axis is $\Delta\phi = 2a d\phi/dz = \lambda a$. Thus, the number of toroidal turns will be $\lambda a/2\pi$ so $q_{wall} \approx \lambda a/2\pi = 0.715$ using $\lambda = 4.493/a$ as determined from the wall boundary condition.

Thus, the spheromak has very little shear since there is very little difference between $q_{wall} = 0.715$ and $q_{axis} = 0.82$; this has led to concern that the spheromak would be vulnerable to pressure-driven interchange modes (see discussion in Sec.6.3 on p.103).

CHAPTER 5
Relaxation in Driven Configurations

An isolated spheromak decays with time. In order to sustain a spheromak there has to be some driving mechanism to replenish the energy and helicity consumed by dissipative processes in a plasma. In a simply connected system this driving mechanism is obtained by imposing a voltage across the ends of open field lines as shown in Fig.5.1. The current driven by this voltage results in a power influx and it will be shown that there is also a helicity influx. Thus, analysis of a driven system involves consideration of the situation shown in Fig.5.1 where open field lines intercept a gapped flux-conserving wall.

In Chapter 3 it was shown that relative helicity must be used to characterize systems having open field lines. The volume of interest was denoted as volume V_a with bounding surface S_a, and it was assumed that volume V_a adjoined a volume V_b with surface S_b such that the total volume $V_a + V_b$ had no open field lines. In this chapter we assume that V_a is the finite volume of interest (i.e., the spheromak plasma) and that V_b is the rest of space so that surface S_b is at infinity. The Finn-Antonsen representation of relative helicity will be used and since this representation does not make explicit reference to V_b or S_b, from now on we drop the subscripts and refer to V_a and S_a as simply V and S.

5.1 Taylor relaxation in systems with open field lines

The open-field-line situation shown schematically in Fig.5.1 is characterized by:

1. A bounding surface S such that on each patch ds, the differential flux $\mathbf{B}\cdot d\mathbf{s}$ is constant in time but may be non-zero.
2. Gaps segmenting S where each segment can be biased to a different electrostatic potential.

The Taylor relaxation analysis of the previous chapter can be generalized under these circumstances provided the relative helicity defined by Eq.(3.34) is used instead of the simple helicity defined by Eq.(3.4). This is because the open field lines cause the simple helicity to become gauge-dependent and hence ambiguous, whereas the relative helicity is gauge-invariant and therefore unambiguous

The system is assume to evolve such that

1. The relative helicity defined by Eq.(3.34) is conserved on the time scale of MHD

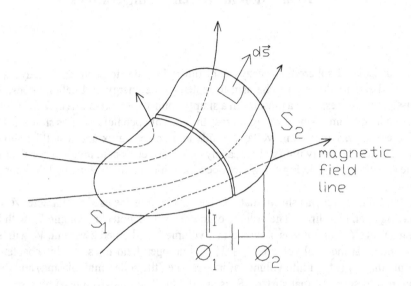

Fig.5.1 Driven configuration: Here $\mathbf{B}\cdot d\mathbf{s}$ is finite because field lines penetrate the surface. Furthermore the surface has a gap so that it is decomposed into two segments, S_1 and S_2, each of which is an equipotential. A potential difference $\phi_1 - \phi_2$ is maintained by a power supply and a current I flows from the power supply to S_1 and then back from S_2. In general, there may also be interior field lines that do not penetrate the surface.

    instabilities.
2. The magnetic energy within V is minimized by MHD instabilities.
3. The magnetic flux $\mathbf{B}\cdot d\mathbf{s}$ through each patch of the surface S does not change with time; i.e., if the time variation of the magnetic field is $\delta \mathbf{B}$, then $\delta \mathbf{B}\cdot d\mathbf{s} = 0$ for every patch $d\mathbf{s}$ of the bounding surface S.

In Sec. 3.5 we showed that relative helicity requires construction of the vacuum magnetic field and vector potential satisfying the same normal boundary conditions on S as does the actual magnetic field; i.e., $\mathbf{B}_{vac}$ is a field which satisfies $\nabla \times \mathbf{B}_{vac} = 0$ within V and $\mathbf{B}_{vac}\cdot d\mathbf{s} = \mathbf{B}\cdot d\mathbf{s}$ on each $d\mathbf{s}$ of S. Condition 3 above means that the variation of the tangential component of the vector potential must vanish on each patch $d\mathbf{s}$ since $0 = d\mathbf{s}\cdot\delta\mathbf{B} = d\mathbf{s}\cdot\nabla\times\delta\mathbf{A}$. Because $\mathbf{B}_{vac}\cdot d\mathbf{s} = \mathbf{B}\cdot d\mathbf{s}$, the variation of the tangential component of the vector potential for the vacuum magnetic field must also vanish,

5.1 Taylor relaxation in systems with open field lines

i.e., $d\mathbf{s}\cdot\nabla\times\delta\mathbf{A}_{vac} = 0$.

Conditions 1-3 can be expressed as a variational problem, i.e., for prescribed boundary conditions (condition 3), we minimize the magnetic energy W (condition 2), subject to the constraint that the relative helicity is conserved (condition 1). The constraint appears as a Lagrange multiplier λ for the variation in K_{rel} and so the variational problem is

$$\delta W - \lambda \delta K_{rel} = 0. \qquad (5.1)$$

For convenience we write the relative helicity, Eq.(3.34) as

$$K_{rel} = \int_V d^3 r\, \mathbf{A}_+\cdot\mathbf{B}_- \qquad (5.2)$$

where

$$\begin{aligned} \mathbf{A}_\pm &= \mathbf{A} \pm \mathbf{A}_{vac} \\ \mathbf{B}_\pm &= \mathbf{B} \pm \mathbf{B}_{vac}. \end{aligned} \qquad (5.3)$$

Using these definitions, Eq.(5.1) can be expanded as

$$\begin{aligned} 0 = &\int_V \mathbf{B}\cdot(\nabla\times\delta\mathbf{A})\, d^3r \\ &- \lambda \int_V (\delta\mathbf{A}_+\cdot\mathbf{B}_- + \mathbf{A}_+\cdot\delta\mathbf{B}_-)\, d^3r \end{aligned} \qquad (5.4)$$

where some constant factors have been absorbed into λ.

We showed that $\delta\mathbf{B}_{vac}\cdot d\mathbf{s} = 0$ on S, i.e., the normal component of the vacuum magnetic field must be invariant on S. However, the vacuum magnetic field has the property that its normal component on the bounding surface completely determines the field within that surface and so we must conclude that the entire vacuum field within V must also be invariant, i.e., $\delta\mathbf{B}_{vac} = 0$ inside V. Except for a possible gauge transformation (which is of no consequence, since K_{rel} is gauge invariant) $\mathbf{A}_{vac}$ is also entirely determined by the normal component of $\mathbf{B}_{vac}$ on the surface and thus, $\delta\mathbf{A}_{vac} = 0$ inside V as well. Setting to zero all terms in Eq.(5.4) involving $\delta\mathbf{A}_{vac}$ and $\delta\mathbf{B}_{vac}$ gives

$$0 = \int_V \mathbf{B}\cdot(\nabla\times\delta\mathbf{A})\, d^3r - \lambda \int_V [\delta\mathbf{A}\cdot\mathbf{B}_- + \mathbf{A}_+\cdot(\nabla\times\delta\mathbf{A})]\, d^3r. \qquad (5.5)$$

After integration by parts on the two curl terms, this becomes

$$0 = \int_V \delta\mathbf{A}\cdot(\nabla\times\mathbf{B})\, d^3r$$

$$-\lambda \int_V [\delta \mathbf{A} \cdot (\mathbf{B} - \mathbf{B}_{vac}) + \delta \mathbf{A} \cdot \nabla \times (\mathbf{A} + \mathbf{A}_{vac})] \, d^3r \qquad (5.6)$$

where all surface integrals vanish because the tangential component of $\delta \mathbf{A}$ vanishes on the surface. The two vacuum terms cancel each other and absorbing a factor of 2 into λ, this becomes

$$\int_V \delta \mathbf{A} \cdot (\nabla \times \mathbf{B} - \lambda \mathbf{B}) \, d^3r = 0. \qquad (5.7)$$

Since $\delta \mathbf{A}$ is arbitrary within V, we must have

$$\nabla \times \mathbf{B} = \lambda \mathbf{B} \qquad (5.8)$$

where $\mathbf{B}$ satisfies the prescribed normal boundary conditions on S. Any initial configuration will relax to a state characterized by Eq.(5.8). Although this formalism appears quite similar to the formalism for the isolated system, it must be remembered that the surface boundary conditions are different. Furthermore, the interpretation of λ turns out to be different: for an isolated system λ is a dependent parameter whereas for the driven system λ is an independent parameter.

5.1.1 Relation between energy and helicity for system with open field lines

The vacuum field satisfies

$$\nabla \times \mathbf{B}_v = 0 \qquad (5.9)$$

Summing Eqs. (5.8) and (5.9) and integrating gives

$$\mathbf{B} + \mathbf{B}_{vac} = \lambda \mathbf{A} + \nabla f \qquad (5.10)$$

where f is some scalar function. If the volume is doubly connected, then f is constrained to be periodic in the direction the long way around the torus.

We define the relative magnetic energy W_{rel} as the change in energy relative to the vacuum field energy,

$$W_{rel} = \frac{1}{2\mu_0} \int_V \left(B^2 - B_{vac}^2 \right) d^3r. \qquad (5.11)$$

The relative magnetic energy is non-negative because the vacuum magnetic field is the lowest energy field for the prescribed boundary conditions.

Using Eq.(5.10) the relative energy can be evaluated as

$$2\mu_0 W_{rel} = \int_V (\mathbf{B} - \mathbf{B}_{vac}) \cdot (\lambda \mathbf{A} + \nabla f) \, d^3r$$

$$= \lambda \int_V \mathbf{A} \cdot (\mathbf{B} - \mathbf{B}_{vac}) \, d^3r + \int d^3r \nabla \cdot [f(\mathbf{B} - \mathbf{B}_{vac})]$$

$$= \lambda \int_V \mathbf{A} \cdot (\mathbf{B} - \mathbf{B}_{vac}) \, d^3r. \tag{5.12}$$

Equation (5.12) shows that λ can be expressed as

$$\lambda = \frac{2\mu_0 W_{rel}}{\int_V \mathbf{A} \cdot (\mathbf{B} - \mathbf{B}_{vac}) \, d^3r}. \tag{5.13}$$

5.2 Helicity injection

Since $\mathbf{B} \cdot d\mathbf{s} = 0$ is constant in time for each $d\mathbf{s}$, the following quantities associated with the vacuum magnetic field are also all constant in time:

1. The entire vacuum field;
2. The vacuum field magnetic energy;
3. The currents external to V which create the vacuum field.
4. The vector potential $\mathbf{A}_{vac}$ associated with the vacuum field.

The conservation equation for relative helicity, Eq.(3.48) showed that

$$\frac{dK_{rel}}{dt} = -2 \int_S \phi \mathbf{B} \cdot d\mathbf{s} - 2 \int_V \mathbf{E} \cdot \mathbf{B} \, d^3r \tag{5.14}$$

where the first term on the right hand side describes helicity injection through the bounding surface and the second term describes helicity dissipation within the volume. This relation is of considerable interest because it can be used to develop a rather detailed description of the evolution of the plasma produced by a coaxial magnetized plasma gun.

5.2.1 Bounding surface is an equipotential

If S is an equipotential (i.e., ϕ is uniform on S) then the surface integral in Eq.(5.14) vanishes; this situation occurs if the source of a coaxial plasma gun is short-circuited (crowbarred). Equivalently, this would be the case of having a wall with no gaps. If the plasma satisfies the MHD Ohm's law $\mathbf{E} + \mathbf{U} \times \mathbf{B} = \eta \mathbf{J}$, Eq.(5.14) becomes in this short-circuited case

$$\frac{dK_{rel}}{dt} = -2 \int_V \eta \mathbf{J} \cdot \mathbf{B} \, d^3r. \tag{5.15}$$

Using Eq.(5.8) it is seen that when there are no gaps in S, the relative helicity decays as

$$\frac{dK_{rel}}{dt} = -2\lambda \int_V \frac{\eta B^2}{\mu_0} d^3r. \tag{5.16}$$

5.2.2 Bounding surface is not an equipotential

This case corresponds to a driven spheromak gun. The surfaces S_1 and S_2 are the locations where magnetic field lines intercept the wall, say S_1 is where field lines enter the volume V and S_2 is where they exit. S_1 and S_2 are separated by a gap and are maintained at different potentials by an external power supply. The net rate of helicity injection is

$$\left(\frac{dK_{rel}}{dt}\right)_{inj} = -2\int_S \phi \mathbf{B} \cdot d\mathbf{s} = -2\left[\int_{S_1} \phi \mathbf{B} \cdot d\mathbf{s} + \int_{S_2} \phi \mathbf{B} \cdot d\mathbf{s}\right]. \tag{5.17}$$

We define ψ_{open} to be the magnetic flux leaving V at surface S_2, i.e.,

$$\psi_{open} = \int_{S_2} \mathbf{B} \cdot d\mathbf{s} \tag{5.18}$$

and since $\nabla \cdot \mathbf{B} = 0$ the surface integral over S_1 is

$$\int_{S_1} \mathbf{B} \cdot d\mathbf{s} = -\psi_{open}. \tag{5.19}$$

Because the surfaces S_1 and S_2 are equipotentials, the rate of helicity injection into the volume V will be

$$\left(\frac{dK_{rel}}{dt}\right)_{inj} = 2\psi_{open}(\phi_1 - \phi_2) \tag{5.20}$$

where ϕ_1 is the potential on S_1 and ϕ_2 is the potential on S_2. If $\phi_1 = \phi_2$ then there is no helicity injection. Thus, a perfectly conducting boundary with no gaps can be considered as an insulator (barrier) for helicity.

5.3 Impedance of the driven force-free configuration

Consider the situation where a Taylor state with open field lines is maintained by a voltage difference applied across S_1 and S_2; i.e., λ is constant in time as are all

5.3 Impedance of the driven force-free configuration

magnetic fields. Thus, in steady state Eq.(5.14) becomes

$$\int_S \phi \mathbf{B} \cdot d\mathbf{s} = -\int_V \mathbf{E} \cdot \mathbf{B} d^3 r. \tag{5.21}$$

Using Eq.(5.18) and Ohm's law $\mathbf{E} + \mathbf{U} \times \mathbf{B} = \eta \mathbf{J}$, Eq.(5.21) becomes

$$(\phi_1 - \phi_2)\psi_{open} = \int \eta \mathbf{J} \cdot \mathbf{B} d^3 r. \tag{5.22}$$

However, using the force-free condition, Eq.(5.8), this becomes

$$(\phi_1 - \phi_2)\psi_{open} = \frac{\lambda}{\mu_0} \int \eta B^2 d^3 r. \tag{5.23}$$

If the force-free equation is integrated over S_1 we obtain

$$\int_{S_1} d\mathbf{s} \cdot \nabla \times \mathbf{B} = \lambda \int_{S_1} d\mathbf{s} \cdot \mathbf{B} \tag{5.24}$$

or

$$\mu_0 I = \lambda \psi_{open} \tag{5.25}$$

where I is the current flowing into S_1 and out S_2. Equation (5.25) reveals an important distinction between a driven system and an isolated system. In a driven system λ is determined by the ratio of current to open flux and therefore can be continuously varied by adjusting the external power supply so as to change the current I going to the electrodes. In contrast, for an isolated system λ is a discrete eigenvalue associated with the system geometry and cannot be adjusted. Thus, the λ's for a driven system form a continuum whereas the λ's for an isolated system form a discrete set. In actual fact and as will be discussed in Chapter 12, λ is typically spatially non-uniform in a driven system, but for now we will consider that λ can be approximated as being spatially uniform (the basis for λ non-uniformity and the rational for assuming λ is approximately uniform is discussed in Chapter 8).

Dividing Eq.(5.23) by Eq.(5.25) gives

$$Z = \frac{(\phi_1 - \phi_2)}{I} = \frac{\int d^3 r \, \eta B^2}{\psi_{open}^2} \; ; \tag{5.26}$$

this is the impedance the configuration presents to the external power supply. Equation (5.26) shows that the impedance becomes infinite as the amount of open flux goes to zero. If η is non-uniform, then Eq.(5.26) also shows that the impedance is dominated by the regions of high η; for example if the outer edge of the plasma is resistive, then most of the impedance results from the outer edge and so most of the power will be deposited in this high impedance region. In effect, the force-free equilibrium behaves as if the various internal currents are connected in series. If the resistivity is uniform then the configuration impedance is

$$Z = \frac{2\mu_0 \eta}{\psi_{open}^2} W \qquad (5.27)$$

where $W = \int_V d^3r\, B^2/2\mu_0$ is the total magnetic energy.

CHAPTER 6
The MHD Energy Principle, Helicity, and Taylor States

6.1 Derivation of the MHD Energy Principle

The energy principle[92] is a method for establishing the stability of a given MHD equilibrium without having to solve the detailed dynamics. This chapter will show how Taylor states can be interpreted using the energy principle. To this end, a brief derivation of the energy principle in a form relevant to spheromaks will be provided. In particular, this derivation will emphasize properties of the vector potential and its relation to helicity conservation.

The energy principle is based on the assumption that the plasma can be described by the ideal MHD equations

$$\rho \frac{d\mathbf{U}}{dt} = \mathbf{J} \times \mathbf{B} - \nabla P \tag{6.1}$$

$$\mathbf{E} + \mathbf{U} \times \mathbf{B} = 0 \tag{6.2}$$

$$P \sim \rho^\gamma \tag{6.3}$$

$$\frac{\partial \rho}{\partial t} + \nabla \cdot (\rho \mathbf{U}) = 0 \tag{6.4}$$

$$\nabla \times \mathbf{E} = -\frac{\partial \mathbf{B}}{\partial t} \tag{6.5}$$

$$\nabla \times \mathbf{B} = \mu_0 \mathbf{J} \tag{6.6}$$

$$\nabla \cdot \mathbf{B} = 0. \tag{6.7}$$

These equations must be used with caution because they incorporate several shortcomings:

1. Equation (6.2) omits electron inertial and kinetic effects which affect Alfvén wave propagation and reconnection dynamics.
2. The adiabatic assumption, Eq.(6.3) may be true for ions but is not true for electrons when $\beta > m_e/m_i$, because in this case the Alfvén velocity is slower than the electron thermal velocity.
3. Ideal MHD is based on the rather self-contradictory assumption that the plasma is at once collisionless and highly collisional [93]. The former limit is assumed so that the plasma acts as a perfect conductor, while the latter limit is assumed so

that the pressure can be considered isotropic and the total pressure $P = P_e + P_i$ can be considered as behaving in a simple adiabatic fashion.

4. The MHD model ignores implications of the unequal electron and ion temperatures likely to occur if the plasma is indeed collisionless. The use of a single equation of state, Eq.(6.3), to describe $P = P_e + P_i$ implies that P_e and P_i have the same dynamical behavior and in particular that electrons and ions are in thermal equilibrium so that $T_e = T_i$.

5. Equation (6.3) implies the plasma supports an acoustic wave quite analogous to acoustic waves in neutral gases. However, Vlasov analysis of the more accurate kinetic equations shows that the plasma acoustic wave differs substantively from the acoustic wave in neutral gases and in particular when $T_e = T_i$ the acoustic wave in a collisionless plasma is so heavily Landau damped that it is effectively non-propagating.

It seems best to consider the ideal MHD equations as describing a hypothetical 'magnetofluid' which bears substantial resemblance to a real plasma but is not necessarily exactly the same. Analysis of magnetofluid behavior gives insights relevant to plasma behavior, but these insights are not guaranteed.

The MHD energy principle[92] for a low β plasma bounded by a rigid flux conserver will now be established. In equilibrium, Eq.(6.1) is

$$0 = \mathbf{J}_0 \times \mathbf{B}_0 - \nabla P_0 \qquad (6.8)$$

which shows that

$$\mathbf{B}_0 \cdot \nabla P_0 = 0, \qquad \mathbf{J}_0 \cdot \nabla P_0 = 0. \qquad (6.9)$$

We consider a perturbation where each fluid element makes a displacement $\boldsymbol{\xi}(\mathbf{x}, t)$ from its equilibrium position. The perturbation may be either stable or unstable. As shown schematically in Fig.6.1, a stable perturbation causes an increase in system potential energy whereas an unstable perturbation causes a decrease in system potential energy.

In order to determine the corresponding change in the magnetofluid potential energy when subject to the perturbation, the force associated with the perturbation must be calculated and then integrated. To do this requires several steps. First, Eq.(6.1) is linearized giving

$$\rho_0 \ddot{\boldsymbol{\xi}} = \mathbf{F}(\boldsymbol{\xi}). \qquad (6.10)$$

where

$$\mathbf{F}(\boldsymbol{\xi}) = \mathbf{J}_0 \times \mathbf{B}_1 + \frac{1}{\mu_0} (\nabla \times \mathbf{B}_1) \times \mathbf{B}_0 - \nabla P_1 \qquad (6.11)$$

6.1 Derivation of the MHD Energy Principle

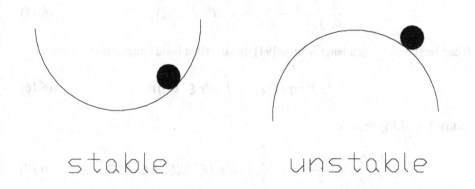

Fig. 6.1 For a stable perturbation (left), the perturbation results in an increase of system potential energy, while for an unstable perturbation (right), the system potential energy is reduced.

and all first order quantities are functions of $\boldsymbol{\xi}$. Since the magnetofluid boundary is assumed rigid, the displacement normal to the boundary must vanish, i.e., $\boldsymbol{\xi}\cdot d\mathbf{s} = 0$ for all surface elements of the bounding surface. This property will be useful when integrating by parts.

Combining Eqs.(6.2) and (6.5) gives the first order magnetic field

$$\mathbf{B}_1 = \nabla \times (\boldsymbol{\xi} \times \mathbf{B}_0) \tag{6.12}$$

which implies the perturbed vector potential is

$$\mathbf{A}_1 = \boldsymbol{\xi} \times \mathbf{B}_0. \tag{6.13}$$

The combination of Eq.(6.4) and (6.3) gives the first order pressure

$$P_1 = -\boldsymbol{\xi}\cdot\nabla P_0 - \gamma P_0 \nabla \cdot \boldsymbol{\xi} \tag{6.14}$$

Both $\mathbf{B}_1$ and P_1 are linear functions of $\boldsymbol{\xi}(\mathbf{x},t)$ thereby demonstrating that the force $\mathbf{F}$ is indeed a linear function of $\boldsymbol{\xi}(\mathbf{x},t)$.

In order to obtain an energy equation, Eq.(6.10) is multiplied by $\dot{\boldsymbol{\xi}}$ and integrated over the entire volume up to the boundary giving

$$\frac{d}{dt}\int_V d^3r \frac{\rho_0 \dot{\xi}^2}{2} = \int_V d^3r\, \dot{\boldsymbol{\xi}} \cdot \mathbf{F}(\boldsymbol{\xi}). \tag{6.15}$$

It can be shown using a lengthy proof[94] that the right hand integral is Hermitian, i.e.,

$$\int_V d^3r\, \boldsymbol{\eta} \cdot \mathbf{F}(\boldsymbol{\xi}) = \int_V d^3r\, \boldsymbol{\xi} \cdot \mathbf{F}(\boldsymbol{\eta}) \tag{6.16}$$

so that Eq. (6.15) becomes

$$\begin{aligned}\frac{d}{dt}\int_V d^3r \frac{\rho_0 \dot{\xi}^2}{2} &= \frac{1}{2}\int_V d^3r \left(\dot{\boldsymbol{\xi}} \cdot \mathbf{F}(\boldsymbol{\xi}) + \boldsymbol{\xi} \cdot \mathbf{F}(\dot{\boldsymbol{\xi}})\right) \\ &= \frac{1}{2}\frac{d}{dt}\int_V d^3r\, \boldsymbol{\xi} \cdot \mathbf{F}(\boldsymbol{\xi}).\end{aligned} \tag{6.17}$$

This can be written as

$$\frac{d}{dt}(\delta T + \delta W) = 0 \tag{6.18}$$

where

$$\delta T = \frac{1}{2}\int_V d^3r\, \rho_0 \dot{\xi}^2 \tag{6.19}$$

is the magnetofluid kinetic energy associated with the perturbation and

$$\delta W = -\frac{1}{2}\int_V d^3r\, \boldsymbol{\xi} \cdot \mathbf{F}(\boldsymbol{\xi}) \tag{6.20}$$

is the corresponding change in the magnetofluid potential energy. Equation (6.18) can be integrated to give

$$\delta T + \delta W = 0; \tag{6.21}$$

the constant of integration has been chosen to be zero because at $t = 0$, the system is assumed to start from rest. The system will be unstable if there exists some possible perturbation which has a negative δW; this is a multi-dimensional analog to the reduction in potential energy of a ball rolling down a hill as in Fig.6.1 (right). Stability

6.1 Derivation of the MHD Energy Principle

analysis becomes a hunt for possible ways that δW could become negative. If δW can be shown to be positive for all possible perturbations, then the system is stable.

The analysis consists essentially of re-arranging δW to be in a more intuitive form; the required manipulation involves a sequence of integration by parts which make repeated use of the vector identity $\mathbf{b}\cdot\nabla\times\mathbf{a} = \nabla\cdot(\mathbf{a}\times\mathbf{b}) + \mathbf{a}\cdot\nabla\times\mathbf{b}$.

The potential energy can be decomposed into three terms

$$\delta W = \delta W_{J_0} + \delta W_{B_1} + \delta W_{P_1} \tag{6.22}$$

where

$$\delta W_{J_0} = -\frac{1}{2}\int_V d^3r\, \boldsymbol{\xi}\cdot\mathbf{J}_0\times\mathbf{B}_1$$

$$\delta W_{B_1} = -\frac{1}{2\mu_0}\int_V d^3r\, \boldsymbol{\xi}\cdot(\nabla\times\mathbf{B}_1)\times\mathbf{B}_0$$

$$\delta W_{P_1} = \frac{1}{2}\int_V d^3r\, \boldsymbol{\xi}\cdot\nabla P_1 \tag{6.23}$$

are the contributions associated with equilibrium current, perturbed magnetic field, and pressure.

The magnetic term can be easily rearranged, giving

$$\delta W_{B_1} = \frac{1}{2\mu_0}\int_V d^3r\, \boldsymbol{\xi}\times\mathbf{B}_0\cdot(\nabla\times\mathbf{B}_1) = \frac{1}{2\mu_0}\int_V d^3r\, B_1^2. \tag{6.24}$$

The pressure term is rearranged as

$$\delta W_{P_1} = \frac{1}{2}\int_V d^3r\, [\boldsymbol{\xi}\cdot\nabla(-\boldsymbol{\xi}\cdot\nabla P_0 - \gamma P_0 \nabla\cdot\boldsymbol{\xi})]$$

$$= \frac{1}{2}\int_V d^3r\, \left[-\boldsymbol{\xi}\cdot\nabla(\boldsymbol{\xi}\cdot\nabla P_0) + \gamma P_0(\nabla\cdot\boldsymbol{\xi})^2\right] \tag{6.25}$$

so that

$$\delta W = \frac{1}{2}\int_V d^3r\, \left[\frac{B_1^2}{\mu_0} + \gamma P_0(\nabla\cdot\boldsymbol{\xi})^2 - \boldsymbol{\xi}\cdot(\mathbf{J}_0\times\mathbf{B}_1 + \nabla(\boldsymbol{\xi}\cdot\nabla P_0))\right]. \tag{6.26}$$

It turns out that only the perpendicular component of $\boldsymbol{\xi}$ matters in the last term. This can be seen by writing $\boldsymbol{\xi} = \boldsymbol{\xi}_\perp + \xi_\parallel B_0^{-1}\mathbf{B}_0$ and considering

$$\begin{aligned}\mathbf{B}_0\cdot(\mathbf{J}_0\times\mathbf{B}_1 + \nabla(\boldsymbol{\xi}\cdot\nabla P_0)) &= -\mathbf{B}_1\cdot\mathbf{J}_0\times\mathbf{B}_0 + \nabla\cdot[\mathbf{B}_0\boldsymbol{\xi}\cdot\nabla P_0] \\ &= -[\nabla\times(\boldsymbol{\xi}\times\mathbf{B}_0)]\cdot\nabla P_0 + \nabla\cdot[\mathbf{B}_0\boldsymbol{\xi}\cdot\nabla P_0] \\ &= \nabla\cdot[\nabla P_0\times(\boldsymbol{\xi}\times\mathbf{B}_0) + \mathbf{B}_0\boldsymbol{\xi}\cdot\nabla P_0]\end{aligned}$$

$$= \nabla \cdot [\boldsymbol{\xi} \, \mathbf{B}_0 \cdot \nabla P_0] = 0 \tag{6.27}$$

where Eq.(6.9) has been used for the last step. Thus, the change in potential energy resulting from the perturbation is

$$\delta W = \frac{1}{2} \int_V d^3 r \left[\frac{B_1^2}{\mu_0} + \gamma P_0 \left(\nabla \cdot \boldsymbol{\xi} \right)^2 - \boldsymbol{\xi}_\perp \cdot \left(\mathbf{J}_0 \times \mathbf{B}_1 + \nabla \left(\boldsymbol{\xi}_\perp \cdot \nabla P_0 \right) \right) \right]. \tag{6.28}$$

Using Eq.(2.19) the equilibrium equation can be written as

$$\mu_0 \nabla P_0 = B_0^2 \, \boldsymbol{\kappa} - \nabla_\perp \left(\frac{B_0^2}{2} \right) \tag{6.29}$$

where $\boldsymbol{\kappa} = \hat{B}_0 \cdot \nabla \hat{B}_0 = -\hat{B}_0 \times \nabla \times \hat{B}_0$. We also recall that $\boldsymbol{\kappa} = -\hat{R}/R$ where R is the local radius of curvature of the magnetic field.

Next the equilibrium current is separated into perpendicular and parallel components $\mathbf{J}_0 = \mathbf{J}_{0\perp} + J_{0\|} B_0^{-1} \mathbf{B}_0$ and $\mathbf{B}_1$ is similarly decomposed. These are used to evaluate the term in Eq.(6.28) involving equilibrium current

$$\begin{aligned}
\boldsymbol{\xi}_\perp \cdot \mathbf{J}_0 \times \mathbf{B}_1 &= \boldsymbol{\xi}_\perp \cdot \left(\mathbf{J}_{0\perp} + J_{0\|} B_0^{-1} \mathbf{B}_0 \right) \times \left(\mathbf{B}_{1\perp} + B_{1\|} B_0^{-1} \mathbf{B}_0 \right) \\
&= \boldsymbol{\xi}_\perp \cdot \left(\mathbf{J}_{0\perp} \times B_{1\|} \mathbf{B}_0 + J_{0\|} \mathbf{B}_0 \times \mathbf{B}_{1\perp} \right) / B_0 \\
&= \boldsymbol{\xi}_\perp \cdot \left(B_{1\|} \nabla P_0 + J_{0\|} \mathbf{B}_0 \times \mathbf{B}_{1\perp} \right) / B_0.
\end{aligned} \tag{6.30}$$

The parallel component of $\mathbf{B}_1$ is

$$\begin{aligned}
B_{1\|} &= \hat{B}_0 \cdot \nabla \times (\boldsymbol{\xi}_\perp \times \mathbf{B}_0) \\
&= \nabla \cdot \left[(\boldsymbol{\xi}_\perp \times \mathbf{B}_0) \times \hat{B}_0 \right] + (\boldsymbol{\xi}_\perp \times \mathbf{B}_0) \cdot \nabla \times \hat{B}_0 \\
&= -\nabla \cdot [B_0 \boldsymbol{\xi}_\perp] - B_0 \boldsymbol{\kappa} \cdot \boldsymbol{\xi}_\perp \\
&= -B_0 \nabla \cdot \boldsymbol{\xi}_\perp - B_0^{-1} \boldsymbol{\xi}_\perp \cdot \nabla \frac{B_0^2}{2} - B_0 \boldsymbol{\kappa} \cdot \boldsymbol{\xi}_\perp \\
&= -B_0 \left(2\boldsymbol{\kappa} \cdot \boldsymbol{\xi}_\perp + \nabla \cdot \boldsymbol{\xi}_\perp \right) + \mu_0 B_0^{-1} \boldsymbol{\xi}_\perp \cdot \nabla P_0
\end{aligned} \tag{6.31}$$

where Eq.(6.29) has been used to eliminate $\nabla_\perp \left(B_0^2/2 \right)$. Substituting for $B_{1\|}$ in Eq.(6.30) gives

$$\begin{aligned}
\boldsymbol{\xi}_\perp \cdot \mathbf{J}_0 \times \mathbf{B}_1 &= (\boldsymbol{\xi}_\perp \cdot \nabla P_0) B_{1\|}/B_0 + \boldsymbol{\xi}_\perp \times \mathbf{B}_0 \cdot \mathbf{B}_{1\perp} J_{0\|}/B_0 \\
&= -(\boldsymbol{\xi}_\perp \cdot \nabla P_0)(2\boldsymbol{\kappa} \cdot \boldsymbol{\xi}_\perp + \nabla \cdot \boldsymbol{\xi}_\perp) + \\
&\quad \mu_0 B_0^{-2} (\boldsymbol{\xi}_\perp \cdot \nabla P_0)^2 - \boldsymbol{\xi}_\perp \times \mathbf{B}_{1\perp} \cdot \mathbf{B}_0 J_{0\|}/B_0
\end{aligned} \tag{6.32}$$

so that Eq.(6.28) becomes

$$\delta W = \frac{1}{2} \int_V d^3x \left[\begin{array}{c} B_1^2/\mu_0 + \gamma P_0 \left(\nabla \cdot \boldsymbol{\xi}\right)^2 + \left(\boldsymbol{\xi}_\perp \cdot \nabla P_0\right) \left(2\boldsymbol{\kappa} \cdot \boldsymbol{\xi}_\perp + \nabla \cdot \boldsymbol{\xi}_\perp\right) \\ -\mu_0 B_0^{-2} \left(\boldsymbol{\xi}_\perp \cdot \nabla P_0\right)^2 \\ + \boldsymbol{\xi}_\perp \times \mathbf{B}_{1\perp} \cdot \mathbf{B}_0 J_{0\parallel}/B_0 - \boldsymbol{\xi}_\perp \cdot \left(\nabla \left(\boldsymbol{\xi}_\perp \cdot \nabla P_0\right)\right) \end{array} \right]. \tag{6.33}$$

This expression can be simplified somewhat by decomposing B_1^2 as

$$\begin{aligned} B_1^2 &= B_{1\perp}^2 + B_{1\parallel}^2 \\ &= B_{1\perp}^2 + \left[-B_0 \left(2\boldsymbol{\kappa} \cdot \boldsymbol{\xi}_\perp + \nabla \cdot \boldsymbol{\xi}_\perp\right) + \mu_0 B_0^{-1} \boldsymbol{\xi}_\perp \cdot \nabla P_0 \right]^2 \end{aligned} \tag{6.34}$$

and also noting that

$$\int_V d^3r \, \boldsymbol{\xi}_\perp \cdot \nabla \left(\boldsymbol{\xi}_\perp \cdot \nabla P_0\right) = -\int_V d^3r \, \left(\nabla \cdot \boldsymbol{\xi}_\perp\right) \left(\boldsymbol{\xi}_\perp \cdot \nabla P_0\right). \tag{6.35}$$

Using these last two relations, the perturbed potential energy can be written in the intuitive form[95, 96, 97]

$$\delta W = \frac{1}{2} \int_V d^3r \left[\begin{array}{c} B_{1\perp}^2/\mu_0 + B_0^2 \left(2\boldsymbol{\kappa} \cdot \boldsymbol{\xi}_\perp + \nabla \cdot \boldsymbol{\xi}_\perp\right)^2/\mu_0 + \gamma P_0 \left(\nabla \cdot \boldsymbol{\xi}\right)^2 \\ - \left(2\boldsymbol{\kappa} \cdot \boldsymbol{\xi}_\perp\right) \left(\boldsymbol{\xi}_\perp \cdot \nabla P_0\right) + \boldsymbol{\xi}_\perp \times \mathbf{B}_{1\perp} \cdot \mathbf{B}_0 J_{0\parallel}/B_0 \end{array} \right]. \tag{6.36}$$

The first three terms in Eq.(6.36) are positive definite and thus stabilizing. The term $-2\left(\boldsymbol{\kappa} \cdot \boldsymbol{\xi}_\perp\right)\left(\boldsymbol{\xi}_\perp \cdot \nabla P_0\right)$ is destabilizing when the pressure gradient and magnetic curvature term $\boldsymbol{\kappa}$ are in the same direction, i.e., when $\boldsymbol{\kappa} \cdot \nabla P_0 > 0$; this is called bad curvature and corresponds to convex magnetic fields at the plasma periphery. Instabilities involving this term are pressure-driven and become important as β increases. If $\beta \to 0$, pressure driven instabilities are of no consequence because there is no free energy stored in the pressure. The term $\boldsymbol{\xi}_\perp \times \mathbf{B}_{1\perp} \cdot \mathbf{B}_0 J_{0\parallel}/B_0$ depends on parallel current $J_{0\parallel}$ and if this term is sufficiently negative to overcome the positive definite terms an instability results. This kind of instability is called a current-driven instability and typically corresponds to kink instabilities.

6.2 Relationship of the energy principle to Taylor states

The Taylor argument applies to plasmas having $\beta \to 0$, i.e., $P_0 \to 0$. In this case, Eq.(6.33) becomes

$$\delta W = \frac{1}{2} \int_V d^3r \left(\frac{B_{1\perp}^2}{\mu_0} + \frac{B_{1\parallel}^2}{\mu_0} - \mathbf{A}_1 \cdot \mathbf{B}_{1\perp} \frac{J_{0\parallel}}{B_0} \right) \tag{6.37}$$

where Eq.(6.13) has been used. Because $\mathbf{B}_{1\|}$ appears in only one term which is positive definite, it is seen that the most unstable modes would have $\mathbf{B}_{1\|} = 0$ and so we may set $\mathbf{B}_{1\|} = 0$ in order to search for instability, giving

$$\delta W = \frac{1}{2} \int_V d^3r \left(\frac{B_{1\perp}^2}{\mu_0} - \mathbf{A}_1 \cdot \mathbf{B}_{1\perp} \frac{J_{0\|}}{B_0} \right). \tag{6.38}$$

Using Eq.(6.2) it is seen that $\mathbf{E} \cdot \mathbf{B} = 0$ for an ideal magnetofluid. If the plasma is isolated and bounded by a flux-conserving conducting wall, the relative helicity is the same as the helicity and as shown by Eq.(3.59), $K = const.$ In the presence of a perturbation K must remain invariant and so

$$\delta K = \int_V d^3r \, (\mathbf{A}_0 + \mathbf{A}_1) \cdot (\mathbf{B}_0 + \mathbf{B}_1) - \int_V d^3r \, \mathbf{A}_0 \cdot \mathbf{B}_0 = 0. \tag{6.39}$$

Since $\boldsymbol{\xi}$ is arbitrary, δK must vanish to each order of $\boldsymbol{\xi}$ and in particular, the term which is second order with respect to $\boldsymbol{\xi}$ must vanish, i.e.,

$$\delta K^{(2)} = \int_V d^3r \, \mathbf{A}_1 \cdot \mathbf{B}_1 = 0 \tag{6.40}$$

However, Eq.(6.13) shows that $\mathbf{A}_1$ is perpendicular to $\mathbf{B}_0$ so this becomes

$$\int_V d^3r \, \mathbf{A}_{1\perp} \cdot \mathbf{B}_{1\perp} = 0. \tag{6.41}$$

Thus, if $J_{0\|}/B_0$ is *spatially uniform*, then $J_{0\|}/B_0$ can be factored out of the integral in Eq.(6.37) leaving

$$\delta W = \frac{1}{2} \int_V d^3r B_{1\perp}^2/\mu_0 \tag{6.42}$$

which is positive definite and therefore corresponds to absolute stability for a $\beta = 0$ plasma. The key requirement for stability is thus $J_{0\|}/B_0 =$ uniform. Since the plasma has zero β, there is no perpendicular equilibrium current and so spatial uniformity of $J_{0\|}/B_0$ corresponds precisely to

$$\mathbf{J} = \lambda \mathbf{B} \tag{6.43}$$

where $\lambda =$uniform. This is just the Taylor state, and so we can conclude (not surprisingly) that these states are absolutely MHD stable. Non-uniformity of $J_{0\|}/B_0$ provides the free energy for current-driven instability. When $J_{0\|}/B_0$ is uniform, there is no more free energy for current-driven instabilities. Thus, the Taylor state can be considered as the final configuration towards which a kink instability evolves.

6.3 Beta limit

While the Taylor argument applies to a zero beta plasma, a real spheromak will have finite β. In fact, the usual reason for making a spheromak is to confine plasma and so the more plasma confined, the better. The question arises as to how large a β can be achieved. When β becomes significant, the term $-\,(2\boldsymbol{\kappa}\cdot\boldsymbol{\xi}_\perp)\,(\boldsymbol{\xi}_\perp\cdot\nabla P_0)$ in δW will be destabilizing when $\boldsymbol{\kappa}\cdot\nabla P_0 > 0$, i.e., when field lines have bad curvature. This instability involves interchanging a flux tube containing high pressure plasma with another flux tube containing low pressure plasma. The two flux tubes contain the same flux so that the magnetic energy is not altered by the interchange. The interchange reduces the system potential energy if it results in volumetric expansion of the flux tube originally containing the higher pressure; this is what happens when there is bad curvature. If the magnetic field is sheared, then adjacent flux tubes have slightly different topology so that the interchange of flux tubes requires some work to contort the magnetic field. Since shear is quantified by the gradient of the safety factor, a large $(q')^2$ will counteract the undesirable effect of bad curvature. We present here a highly simplified non-rigorous analysis showing how shear stabilizes pressure-driven instabilities; the reader interested in a rigorous analysis (outside of the scope of this book) should consult the texts by Bateman[98] or Freidberg[99] or one of the original papers on the subject.

In complicated magnetic geometries, the curvature of field lines varies from good to bad along the length of the field line. Assuming that β is small in spheromaks, the curvature can be estimated using $\mathbf{J}\times\mathbf{B}\approx 0$ together with Eq.(2.19) to show that in a spheromak

$$\boldsymbol{\kappa} \simeq \frac{1}{2}\nabla_\perp \ln B^2. \qquad (6.44)$$

Since B^2 is largest near the geometric axis and generally decreases with radius r, the term $-\,(2\boldsymbol{\kappa}\cdot\boldsymbol{\xi}_\perp)\,(\boldsymbol{\xi}_\perp\cdot\nabla P_0)$ will be destabilizing on the outboard side of the magnetic axis and stabilizing on the inboard side. A typical flux tube has a helical trajectory about the magnetic axis, so that it alternates between being in the good curvature region (inboard of magnetic axis) and in the bad curvature region (outboard region). Figure 6.2 plots contours of B^2 for a cylindrical spheromak superimposed on flux contours. Here B^2 is evaluated using Eqs.(9.12a)-(9.12c) while the flux contours were evaluated from Eq.(9.18); the z axis origin has been shifted in Fig. 6.2 compared to Eqs.(9.12a)-(9.12c) and Eq.(9.18) so that $z = 0$ defines the spheromak midplane. According to the Grad-

Shafranov equation, pressure is constant on flux surfaces so that the flux contours can be considered as pressure contours (maximum pressure on magnetic axis). The small ticks denote the downhill direction of the respective contour plots. From these ticks it is seen that $\kappa \cdot \nabla P_0$ is generally positive except for a small inboard region. Thus, most of the extent of a field line has bad curvature.

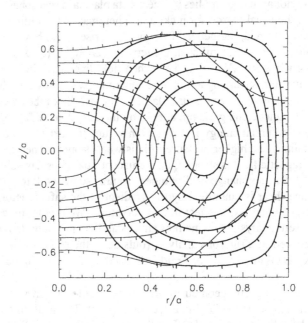

Fig.6.2 Pressure contours (heavier and closed contours) and B^2 contours (lighter and open contours) for a cylindrical spheromak with $h/a = 1.5$; ticks show downhill direction. Midplane is at $z = 0$.

The system can be described by more than one coordinate system and we will use two different coordinate systems depending on which is more convenient. The first coordinate system is geometry-based and involves the unit vectors $\hat{e}_\psi, \hat{\theta}, \hat{\phi}$ where $\hat{e}_\psi = \nabla\psi/|\nabla\psi|$ is the generalization of the minor radius direction and $\hat{\theta} = \hat{\phi} \times \hat{e}_\psi$ is the local poloidal direction[2]. The second coordinate system is field-based and again involves $\hat{e}_\psi$ with the other two unit vectors lying in a magnetic surface. However, in this case the latter two unit vectors are respectively parallel to **B** and perpendicular to **B**. Thus, at

---

[2] The minor radius is the distance from the magnetic axis and the θ is the angle about the magnetic axis.

6.3 Beta limit

each point in space the field-based coordinate system involves (i) the direction of the equilibrium magnetic field (unit vector $\hat{B}$), (ii) the direction of the poloidal flux gradient (unit vector $\hat{e}_\psi = \nabla\psi/|\nabla\psi|$), and a third direction which is orthogonal to the previous two and which lies in the poloidal flux surface, (unit vector $\hat{e}_\eta = \hat{B} \times \hat{e}_\psi$). The fluid displacement vector $\boldsymbol{\xi}$ can therefore be written as

$$\boldsymbol{\xi} = \xi_\parallel \hat{B} + \xi_\psi \hat{e}_\psi + \xi_\eta \hat{e}_\eta \tag{6.45}$$

and we identify $\boldsymbol{\xi}_\perp = \xi_\psi \hat{e}_\psi + \xi_\eta \hat{e}_\eta$ as the component of fluid displacement which is perpendicular to the equilibrium magnetic field. The fluid displacement is a single-valued function of position and so must be periodic in both toroidal and poloidal angles, i.e., on each flux surface, the displacement can be written as

$$\boldsymbol{\xi} = \operatorname{Re}\left[\bar{\boldsymbol{\xi}}(\psi)e^{im\theta+in\phi}\right] \tag{6.46}$$

where m and n are integers and $\bar{\boldsymbol{\xi}}(\psi)$ is a complex vector. Now let us turn attention to Eq.(6.36) and note that this equation has three positive definite terms. The parallel fluid displacement $\xi_\parallel$ appears only in the term $\gamma P_0 (\nabla \cdot \boldsymbol{\xi})^2$ and thus $\xi_\parallel$ is a free parameter that can be set to any desired value. The most pessimistic situation (i.e., situation with smallest δW) would have the least possible value of each of the positive definite terms in δW. Because $\xi_\parallel$ is a free parameter, we choose it to minimize $(\nabla \cdot \boldsymbol{\xi})^2$, i.e., we choose $\xi_\parallel$ to make $\nabla \cdot \boldsymbol{\xi} = 0$. This shows that incompressible modes are the most unstable.

We ignore the current-driven instability term (i.e., the term involving $J_{0\parallel}$ since we are not considering kink stability). We next arrange for $\boldsymbol{\xi}_\perp$ to minimize the term in Eq.(6.36) involving $(2\boldsymbol{\kappa} \cdot \boldsymbol{\xi}_\perp + \nabla \cdot \boldsymbol{\xi}_\perp)^2$ and in particular choose $\boldsymbol{\xi}_\perp$ to satisfy

$$2\boldsymbol{\kappa} \cdot \boldsymbol{\xi}_\perp + \nabla \cdot \boldsymbol{\xi}_\perp = 0. \tag{6.47}$$

which indicates that $\boldsymbol{\xi}_\perp$ must have a spatial gradient of the same magnitude as $\boldsymbol{\kappa}$. What remains is

$$\delta W = \frac{1}{2} \int_V d^3r \left[\frac{B_{1\perp}^2}{\mu_0} - (2\boldsymbol{\kappa} \cdot \boldsymbol{\xi}_\perp)(\boldsymbol{\xi}_\perp \cdot \nabla P_0)\right] \tag{6.48}$$

where

$$\mathbf{B}_{1\perp} = [\nabla \times (\boldsymbol{\xi} \times \mathbf{B})]_\perp = [\mathbf{B}\cdot\nabla\boldsymbol{\xi} - \boldsymbol{\xi}\cdot\nabla\mathbf{B}]_\perp \simeq \mathbf{B}\cdot\nabla\boldsymbol{\xi}_\perp. \tag{6.49}$$

Using Eq.(3.18) to express $\mathbf{B}$, the perpendicular magnetic perturbation becomes

$$\mathbf{B}_{1\perp} = \frac{1}{2\pi} \operatorname{Re}\left\{(\nabla\Phi \times \nabla\theta + \nabla\psi \times \nabla\phi) \cdot (im\nabla\theta + in\nabla\phi)\,\bar{\boldsymbol{\xi}}_\perp e^{im\theta+in\phi}\right\}$$

$$= \frac{1}{2\pi} \text{Re} \left\{ i \left[m \nabla \psi \times \nabla \phi \cdot \nabla \theta + nq \nabla \psi \times \nabla \theta \cdot \nabla \phi \right] \bar{\xi}_\perp e^{im\theta + in\phi} \right\}$$

$$= \frac{1}{2\pi} \text{Re} \left\{ in \left[\frac{m}{n} - q(\psi) \right] \nabla \psi \times \nabla \phi \cdot \nabla \theta \, \bar{\xi}_\perp e^{im\theta + in\phi} \right\}. \quad (6.50)$$

Thus, $\mathbf{B}_{1\perp}$ vanishes on the flux surfaces where $q(\psi) = m/n$; these flux surfaces are called rational surfaces because q is the ratio of two integers on these surfaces. Let us label the rational surfaces as $\psi^{(m,n)}$ and for a particular m, n Taylor expand $q(\psi)$ in the vicinity of the rational surface so that

$$\begin{aligned} q(\psi) &= q(\psi^{(m,n)}) + (\psi - \psi^{(m,n)}) \frac{dq}{d\psi} \\ &= \frac{m}{n} + (\psi - \psi^{(m,n)}) \frac{dq}{d\psi}. \end{aligned} \quad (6.51)$$

Using this expansion it is seen that

$$\mathbf{B}_{1\perp} = -n \left(\psi - \psi^{(m,n)} \right) \frac{dq}{d\psi} \left(\mathbf{B}_{pol} \cdot \nabla \theta \right) \text{Re} \left[i \bar{\xi}_\perp e^{im\theta + in\phi} \right]. \quad (6.52)$$

The element of length in the poloidal direction is $dl_{pol} = d\theta / |\nabla \theta|$ so $\mathbf{B}_{pol} \cdot \nabla \theta = B_{pol} d\theta / dl_{pol}$. If the flux surface is approximately circular, then from Eq.(2.34) it is seen that

$$B_{pol} \simeq \frac{B_\phi}{2\pi qr} \oint dl_{pol} \quad (6.53)$$

and so

$$\mathbf{B}_{pol} \cdot \nabla \theta \simeq \frac{B_\phi}{qr}. \quad (6.54)$$

Equation (6.48) becomes

$$\delta W \approx \frac{1}{4\mu_0} \int_V d^3 r B_\phi^2 |\bar{\xi}_\perp|^2 \left[n^2 \left(\psi - \psi^{(m,n)} \right)^2 \left(\frac{1}{rq} \frac{dq}{d\psi} \right)^2 - 2\kappa \frac{\mu_0 |\nabla P|}{B_\phi^2} \right]. \quad (6.55)$$

Using Eq.(3.16) to express the volume element, Eq.(6.55) becomes

$$\delta W = \frac{1}{4\mu_0} \int_V \frac{d\theta d\phi d\psi}{2\pi |\mathbf{B}_{pol} \cdot \nabla \theta|} |\bar{\xi}_\perp|^2 B_\phi^2 \left[\begin{array}{c} n^2 \left(\psi - \psi^{(m,n)} \right)^2 \left(\frac{1}{rq} \frac{dq}{d\psi} \right)^2 \\ - 2\kappa \frac{\mu_0 |\nabla P|}{B_\phi^2} \end{array} \right]$$

6.3 Beta limit

$$= \frac{1}{8\pi\mu_0} \int_V d\theta d\phi d\psi \, |\bar{\xi}_\perp|^2 \, qr B_\phi \left[\begin{array}{c} n^2 \left(\psi - \psi^{(m,n)}\right)^2 \left(\frac{1}{rq}\frac{dq}{d\psi}\right)^2 \\ - 2\kappa \frac{\mu_0 |\nabla P|}{B_\phi^2} \end{array} \right]. \tag{6.56}$$

We make only a minimal attempt to take into account the volumetric weighting of the various terms and, in particular, assume that all quantities are approximately constant within the volume except for $\left(\psi - \psi^{(m,n)}\right)^2$. The integral of $\left(\psi - \psi^{(m,n)}\right)^2$ over ψ has the minimum value

$$\int_0^{\psi_{axis}} d\psi \left(\psi - \psi^{(m,n)}\right)^2 = \frac{1}{12} \psi_{axis}^3 \tag{6.57}$$

when $\psi^{(m,n)} = \psi_{axis}/2$; this represents the most pessimistic choice (i.e., minimizes the stabilizing term). With these approximations, the stability criterion $\delta W > 0$ becomes

$$\frac{1}{12} \psi_{axis}^3 \left(\frac{1}{rq}\frac{dq}{d\psi}\right)^2 - 2\kappa \frac{\mu_0 |\nabla P|}{B_\phi^2} \psi_{axis} > 0 \tag{6.58}$$

where we have set $n = 1$ to have the most pessimistic (least stable) situation.

Assuming $dq/d\psi$ is uniform, we may express

$$\frac{\psi_{axis}}{q} \frac{dq}{d\psi} \approx \frac{q_{axis} - q_{wall}}{(q_{axis} + q_{wall})/2} \tag{6.59}$$

so the $\delta W > 0$ criterion becomes

$$\left(\frac{q_{axis} - q_{wall}}{(q_{axis} + q_{wall})/2}\right)^2 - 24\kappa r_{axis} \frac{\mu_0 r_{axis} |\nabla P|}{B_\phi^2} > 0. \tag{6.60}$$

Finally, using Eq.(6.44) to approximate $2\kappa \sim 1/r_{axis}$, the stability criterion becomes

$$\left(\frac{q_{axis} - q_{wall}}{(q_{axis} + q_{wall})/2}\right)^2 - 12\beta > 0 \tag{6.61}$$

where $\beta = 2\mu_0 P/B_\phi^2$. Equation (6.61) is essentially the Suydam criterion if the numerical coefficient is 4[100].

On p. 85 it was shown that a spherical spheromak has $q_{axis} = 0.82$ and $q_{wall} = 0.715$; thus for an $n = 1$ mode (i.e., the most unstable mode), the β limit for a spherical spheromak would be only $\sim 0.15\%$. In Sec.9.3.3 it is shown that q_{wall} is proportional to the length of the geometric axis whereas q_{axis} is essentially independent of this length. This means that shortening the length of the geometric axis while keeping other dimensions constant (i.e., "dimpling" the spheromak) increases the shear and indicates that a cylindrical spheromak (more dimpled than a spherical spheromak) has a higher β limit than a spherical spheromak.

The Mercier stability analysis [101] is a rigorous calculation of what we have approximated above and prescribes the amount of shear $(q'/q)^2$ required to stabilize against pressure driven modes for a given β. Analytic evaluation of the Mercier stability limit for the spheromak appears to be unfeasible because of the spheromak's low aspect ratio. Numerical calculations of Mercier limits for cylindrical spheromaks by Mayo and Marklin [102] for various plausible q profiles predict β limits of $\sim 0.5\% - 1.5\%$. If this low β limit is indeed correct, spheromaks would be unattractive as confinement systems. However, actual measurements by Wysocki et al. [65] of the onset of pressure driven instability showed that such instabilities occurred only when the peak electron β is of the order of 20%. It is not clear why the calculated β limit is so much more pessimistic than Wysocki et al.'s experimental observations, but this fortuitous behavior is a major motivation for pursuing the development of spheromaks as confinement devices. One possibility that has been suggested [103] is that present-day spheromaks are in fact continually undergoing low-level pressure-driven interchange instabilities, but because these devices are relatively cold, they are quite resistive so that the Ohmic heating easily replenishes the thermal energy leaking out via interchange instabilities. If this hypothesis is correct, the experimentally observed β limit should decrease as the plasma becomes hotter and less resistive so that the Ohmic heating power for a given current decreases.

CHAPTER 7
Survey of Spheromak Formation Schemes

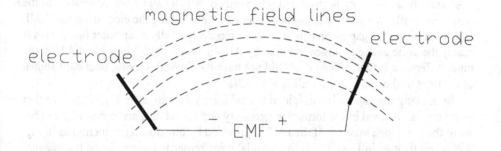

Fig.7.1 Generic spheromak formation scheme: An EMF applied along field lines drives field-aligned current, or equivalently, injects magnetic helicity. When the ratio of field-aligned current to intercepted flux exceeds a threshold, closed field lines are formed, i.e., a spheromak-like configuration is created.

Spheromak configurations have been produced using quite diverse methods; this diversity validates the assertion that the spheromak is a minimum-energy state towards which any initial configuration will relax. As shown in Fig. (7.1) the common feature of all formation schemes is application of an electromotive force (EMF) along initially vacuum magnetic field lines so as to drive field-aligned plasma current. If the current rise-time is very slow compared to the characteristic Alfvén dynamical time scale then, except for occasional brief transients, the system evolves through a sequence of relaxed states. In other words, as the current is ramped up by the applied EMF, the magnetic field and current always remain in a force-free equilibria described by $\nabla \times \mathbf{B} = \lambda \mathbf{B}$.

The net magnetic field is the vector sum of the original vacuum field and the field due to the field-aligned current; this net field is twisted and its twist is an increasing function of λ. When the field-aligned current, or equivalently λ, exceeds a certain threshold, the topology of the force-free equilibrium changes in such a way as to form closed field lines. This topological transformation necessitates magnetic reconnection.

Spheromaks devices are usually axisymmetric, but this is not essential. All that is essential is sufficient helicity injection, i.e., sufficient build-up of λ. Figure 7.2 shows a variety of axisymmetric formation schemes; only the initial vacuum field is shown and not the ultimate spheromak configuration.

The most commonly used and least complex method is the coaxial magnetized plasma gun shown schematically in Fig. 7.2(a). Here an annular cut in the end wall of a cylinder divides the end wall into two concentric electrodes separated by a gap (the terms "wall" and "electrode" will be used interchangeably in this chapter; an electrode is essentially an electrically insulated wall segment which can be biased relative to the remaining wall). A coil creates poloidal magnetic field linking the electrodes, and EMF is applied across the gap by an external power supply, typically a capacitor bank. This is nearly the same as the arrangement used by Alfvén et al.[8] (cf. Figs. 1.2 and 1.3); the main difference is that Alfvén et al. did not have the flux-conserving boundary shown to the right of the gapped end wall in Fig. 7.2(a).

By moving the gap to the middle of the cylinder side wall, as in Fig. 7.2(b), further symmetry is obtained in the formation geometry and in fact, this arrangement is the basis of the z–θ pinch method. If the EMF is provided by transformer induction as in Fig. 7.2(c) and then as in Fig. 7.2(d) the toroidal transformer is moved inside the vacuum chamber so that the plasma surrounds the transformer, one obtains the essential ingredients for the inductive method. These three methods (coaxial, $z-\theta$ pinch, inductive) will now be discussed in more detail. While all these schemes attempt to impose axisymmetry on the plasma, it is important to bear in mind that Cowling's anti-dynamo theorem demonstrates that a spheromak cannot be formed via purely axisymmetric processes. Although the device might be axisymmetric, non-axisymmetric plasma behavior will inevitably occur.

7.1 Magnetized coaxial gun

These devices have two coaxial cylindrical electrodes, called the inner and outer source electrode, which are linked by a vacuum poloidal magnetic field produced by an external coil system. Figure 7.3 shows the arrangement of an early magnetized coaxial gun used at Los Alamos [60] and Fig.7.4 shows the design of the Livermore SSPX coaxial spheromak[55] scheduled to begin operation in 1999. Operation involves the following distinct steps:

1. An electromagnet creates vacuum poloidal magnetic field spanning the inner and outer electrodes; in Fig.7.3 this field is produced by currents in the coils labeled "poloidal field coil" and "bias field coil". While this example used two concentric

7.1 Magnetized coaxial gun

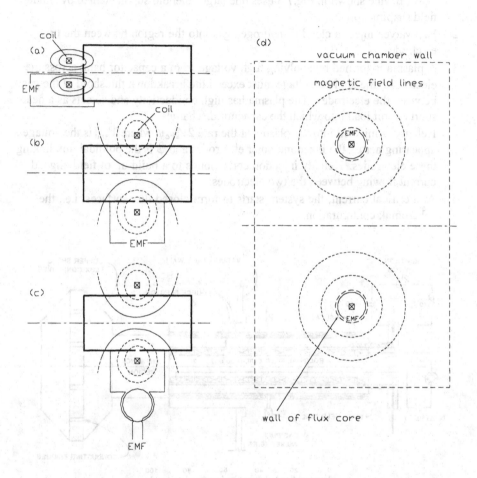

Fig. 7.2 (a) Generic coaxial gun method, (b) moving gap to side wall, (c) using induction to create EMF, (d) locating inductive source in toroidal flux-core inside plasma.

coils, one coil would suffice since all that matters is that there be vacuum poloidal magnetic field linking the inner and outer coaxial electrodes. The more modern SSPX device shown in Fig.7.4 uses one large solenoid supplemented by smaller field shaping coils.

2. Fast valves inject a cloud of hydrogen gas into the region between the two electrodes.
3. A plasma is formed by applying high voltage from a capacitor bank across the electrodes; the applied voltage must exceed the breakdown threshold of the gas between the electrodes. The plasma has high conductivity and so acts as a near short circuit load into which the capacitor discharges.
4. Helicity is injected into the plasma at the rate $2V_{src}\psi$ where V_{src} is the voltage appearing across the inner and outer electrodes and ψ is the poloidal flux linking these electrodes. Helicity injection corresponds to a build-up of field-aligned current flowing between the two electrodes.
5. At a critical current, the system starts to form closed flux surfaces, i.e., the spheromak configuration.

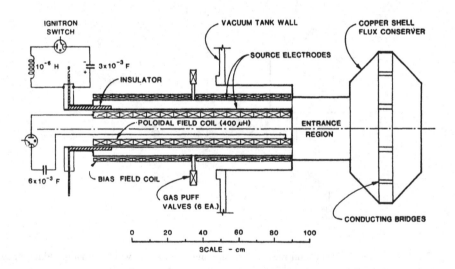

Fig. 7.3 Typical layout of a coaxial plasma gun spheromak (from Jarboe et al.[60]).

We now discuss these steps in more detail.

Vacuum field: The axisymmetric vacuum field is poloidal and is typically estab-

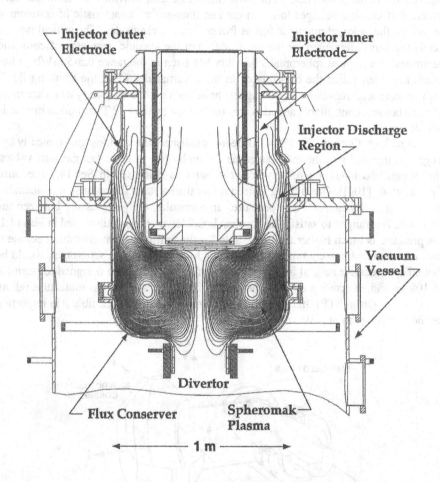

Fig. 7.4 Layout of the Livermore SSPX spheromak (to begin operation in 1999) from Hooper et al.[55].

lished on a much slower time-scale than the rest of the experiment. This slow timescale allows the vacuum field to soak into the electrodes and so provide the finite $\mathbf{B} \cdot \hat{n}$ required for helicity injection. The slow time-scale also provides the technical convenience of keeping voltages low. On the fast time-scale characteristic of spheromak formation, the poloidal magnetic flux is frozen into the electrodes. The critical parameter is the flux intercepted by the electrodes, not the specific geometry of coils and electrodes. For typical spheromaks this flux has been in the range 0.5-5 mWb; it has sometimes been called the bias flux or in more vernacular form, the "stuffing flux". Both air-core and iron-core electromagnets have been used; iron core systems are more efficient but less controllable and also are constrained by the ~ 2 T saturation magnetic field of iron.

Gas puffing- Once the bias flux has been established, the working gas, typically hydrogen, is injected into the inter-electrode region by one or more fast gas puff valves. The design and operation of a typical gas valve is discussed in Sec.14.2 (see also Thomas et al. [104]). Gas puffing is used rather than a static fill because it is desirable to have a spatially non-uniform pressure. In particular, the gas pressure between the electrodes is arranged to satisfy the Paschen breakdown criterion discussed in Sec.14.1; this pressure is much higher than desirable for the entire vacuum chamber because if the vacuum chamber were uniformly at such a high pressure, the spheromak would be choked by excessive neutral background. The fast gas valves are designed to open for 25-100 μs and, depending on the experiment, inject $10^{19} - 10^{20}$ gas molecules (about 0.3-3 cm^3 of gas at STP). In order for the spheromak to be reproducible it is important for the gas valves to operate very consistently.

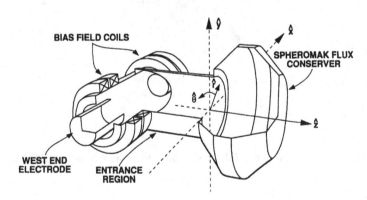

Fig.7.5 Non-axisymmetric spheromak gun used at Los Alamos (from Fernandez et al.[105]).

7.2 Non-axisymmetric gun method

Gas breakdown and plasma formation - During the $\sim 50\,\mu s$ interval that the gas cloud is located between the coaxial electrodes, a high energy capacitor bank is connected across the electrodes by a fast, high-power electronic switch. This switch, capable of holding off 5-25 kV, is typically an ignitron or a triggered spark gap. When currents exceeding the ~ 100 kA capability of a single switch are required, several switches are connected in parallel. The Paschen criterion for the gas breakdown voltage V_{br} has the functional form $V_{br} = f(Pd)$ where P is the gas pressure between electrodes separated by distance d and f is a function which has a minimum at a certain value of Pd. In general f rises steeply to the left of this minimum, but more gently to the right. For hydrogen, the minimum of f is approximately 300 V and occurs when $Pd \sim 1 - 2$ Torr-cm. In order to minimize the fill pressure, spheromak experiments typically operate slightly to the left of the minimum and so involve breakdown voltages much larger than 300 V.

Current ramp–up - Breakdown of the neutral gas results in plasma formation. After several microseconds of electron avalanching, a highly ionized plasma is formed. During this time the load impedance changes from infinite (open circuit) to milliohms or less. When the plasma impedance drops below the characteristic impedance of the capacitor bank, the plasma appears as an effective short circuit to the capacitor bank and the voltage across the electrodes drops from the several kV applied voltage to somewhere between a few hundred volts and a kilovolt. The rate of current rise is limited by the series inductance of the circuit comprised by the capacitor bank, the transmission line, and the load. The current increases sinusoidally with a characteristic rise time of the order $\pi\sqrt{LC}/2$ where L is the total circuit inductance and C is the capacitance.

Helicity injection- The voltage across the electrodes is $V = IZ$ where I is the current and Z is the load impedance. This voltage varies erratically as Z goes up and down as a result of dynamic changes in plasma geometry and topology. The capacitor bank thus acts as a current source, not a voltage source. Helicity is injected into the system because there is a voltage drop across the two electrodes, and the rate of helicity injection is $dK/dt = 2V\psi$ where ψ is the magnetic flux intercepting the electrodes. Since the voltage V appearing across the electrodes is determined mainly by the highly dynamic plasma, it is important to measure V at the electrode location; voltages at other points along the transmission line differ because of inductive drops along the transmission line.

Spheromak formation - As the current rises, the magnetic field becomes twisted and eventually the helicity content of the configuration exceeds the maximum allowed for a force-free equilibrium to exist in an attached state between the two electrodes. The twisted field lines break off, reconnect and form a detached spheromak. The details of this process will be discussed in Chapter 11.

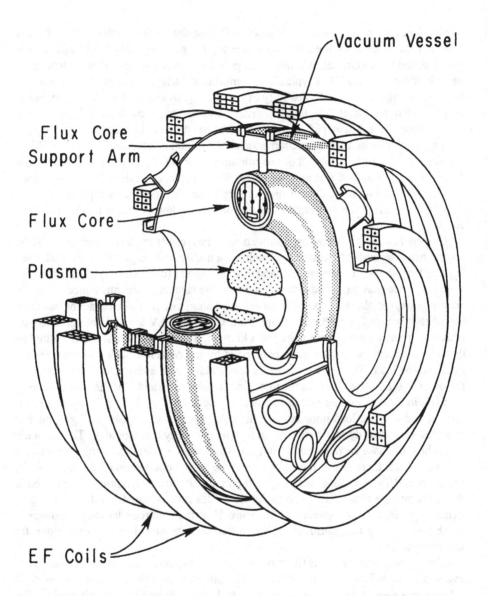

Fig. 7.6 Mechanical layout of S-1 device from Yamada[106]. Note that equilibrium magnetic field is provided by external coils rather than by a close-fitting wall. Also note that the spheromak plasma is quite small compared to the overall device size.

7.2 Non-axisymmetric gun method

To demonstrate that axisymmetry is not essential, the Los Alamos group formed a spheromak using an unstable z-pinch gun[105] as shown in Fig.7.5. The sequence of operation was essentially the same as for the coaxial gun and similar plasmas were produced.

7.3 The inductive method

The inductive method[25], developed at the Princeton Plasma Physics Lab, was first used on the small Proto-S1 spheromak and then later on the large S-1 spheromak[106]. Figure 7.6 shows the mechanical layout while Fig. 7.7 shows the magnetic geometry and formation sequence. The inductive method was motivated by a desire to avoid plasma-electrode contact and thereby eliminate impurity sputtering from electrodes.

Instead of surrounding the plasma with a flux-conserving wall, a separatrix is established using external coils to provide a so-called "vertical" magnetic field which balances the internal hoop forces of the spheromak toroidal current; these coils are marked as "EF coils" in Fig.7.6 (the vertical field is actually horizontal for the orientation of Fig.7.6, but by tradition the field which balances the hoop force is always called the "vertical" field). The vertical field is designed to be identical to the field that would be produced by image currents in a flux-conserving wall if such a wall were located at the separatrix. The expected advantage of this approach is that a spheromak magnetic field geometry would be obtained without having a metal wall in close proximity to the plasma. Furthermore, open field lines outside the separatrix would act as a natural divertor.

Unfortunately, it turns out that a flux-conserving wall not only provides the axisymmetric vertical field required for equilibrium, but also provides transient nonaxisymmetric fields which prevent tilt instabilities. To avoid these instabilities in the S1 device, it became necessary to impose partial walls in critical regions of the separatrix. In effect, a portion of the flux-conserving wall was re-introduced and any expected advantage of minimizing contact with close-fitting walls was lost. The performance of inductively-driven spheromaks was found to be no better than coaxial gun spheromaks but the cost and complexity were significantly higher.

The inductive method employs a "flux-core" which is a vacuum-tight toroid inserted inside a large vacuum tank (see Fig.7.6). The flux-core contains internal toroidal and poloidal field coils and has support leads connecting the coils to external capacitor bank power supplies. From a helicity injection point of view, the flux-core can be considered as an initial store of helicity because it has linked toroidal and poloidal fields. In the spheromak formation sequence the flux-core fields are sequentially changed, and because helicity is conserved, the plasma surrounding the flux core takes up this store of helicity.

The coupling of toroidal and poloidal fields to the region external to the flux-core differ significantly. The toroidal field coil produces toroidal magnetic field *inside* the flux-core but not outside. In contrast, the poloidal field coil produces poloidal magnetic field *outside*, and if the flux-core surface is a poloidal flux surface, no poloidal field inside.

The sequence of operation is as follows:

1. A quasi-static vertical field is established on a slow time scale using external coils (EF coils in Fig.7.6). This vertical field is equivalent to the field that would have been provided by image currents in a flux-conserving wall. The vacuum chamber contains a static gas fill (this may be undesirable, see discussion on p.232).
2. The poloidal magnetic field is ramped up to maximum value so that there is a poloidal field linking the flux core as shown in Fig.7.7(a). This poloidal field is the sum of the external vertical field (from the EF coils) and the field produced by the toroidal current in the B_p coil shown in Fig.7.7(a). The sense of the external vertical field is such as to oppose the flux-core field on the small major radius side; i.e., the current in the EF coil flows in the opposite sense of the current flowing in the flux-core coil.
3. The toroidal field is now ramped up inducing a poloidal electric field which first breaks down the neutral gas to form plasma and, once the plasma is formed, drives poloidal current in the plasma. This plasma poloidal current produces toroidal field in the plasma as shown by the ×'s in Fig.7.7(b).
4. The toroidal current flowing in the flux-core is brought to zero and then reversed. The negative time derivative of poloidal field creates a toroidal electric field which drives toroidal current in the plasma. The current in the B_p coil is crowbarred (i.e., the coil is short-circuited so that the current in it continues to flow subject only to internal resistive losses).
5. Because the plasma is a good conductor and therefore approximates a flux conserver, the sum of the poloidal flux due to the toroidal currents in the plasma and the flux-core should be approximately constant. Since the plasma was initiated when the toroidal current in the flux-core was nearly zero, the plasma toroidal current should flow in the direction opposite that of the toroidal current in the B_p coil. Since anti-parallel currents repel, there will be a repulsive force between the toroidal current in the flux core and the toroidal current in the plasma. The net effect of this force and also the force due to the vertical field is to squeeze the plasma inwards to a smaller major radius than the flux-core.
6. The pushed-in plasma undergoes magnetic reconnection and de-links from the

7.3 The inductive method

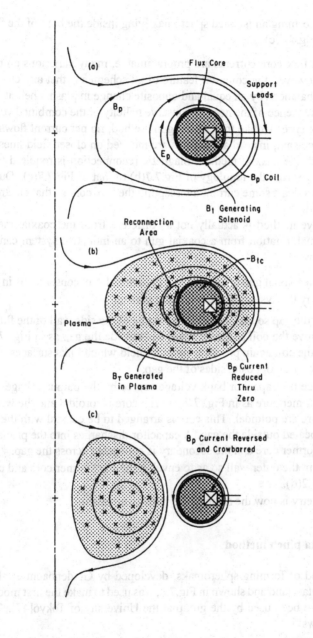

Fig. 7.7 Formation sequence for inductive method (from Yamada et al.[25])

flux-core, forming an isolated spheromak lying inside the bore of the flux-core as shown in Fig.7.7(c).

Because the flux-core currents are programmable, many variations on this scheme can be used. However, the common feature of all schemes is that any change in flux-core helicity is balanced by an equal and opposite change in plasma helicity. This is an immediate consequence of the need to conserve helicity in the combined volume of the plasma and flux-core. Unlike the coaxial gun method, no net current flows from electrodes into the plasma; instead all currents are induced on closed field lines. However, in common with the coaxial method, magnetic reconnection is required to attain the final state, i.e., go from the topology of Fig.7.7(b) to that of Fig.7.7(c). During reconnection there is at least some current intercepting the flux core so that electrode contact is not eliminated.

The inductive method is actually not so different from the coaxial method and a "gedanken" transformation from a coaxial gun to an inductive system can be accomplished as follows

1. Start with a coaxial gun connected to a cylindrical flux conserver as in Fig. 7.2(a).
2. Now modify the system in the following manner:
   (a) Move the gap so that it lies in the middle of the side wall of the flux conserver and move the coil so that it again lies outside the gap as in Fig. 7.2(b). Thus the coil again produces poloidal field which links surfaces of the flux conserver on opposite sides of the gap,
   (b) Replace the capacitor bank voltage source by the output voltage of a transformer core as in Fig. 7.2(c). The core is toroidal and the windings on the core are poloidal. This core is arranged to be coaxial with the cylinder and located outside the gap. A capacitor discharges into the primary of the transformer core and the secondary is connected across the gap.
   (c) Deform the outer wall so as to envelop the transformer core and coil as in Fig. 7.2(d).
3. The geometry is now the same as the inductive system.

7.4 Z-Theta pinch method

This method of forming spheromaks, developed by Goldenbaum et al.[21] at the University of Maryland and shown in Fig.7.8, was used to make the first modern spheromak. It has also been used by the group at the University of Tokyo[47]. The method works as follows:

7.4 Z-Theta pinch method

1. A straight bias B_z field is established using a slow external coil ("pre-ionization").
2. A capacitor bank is applied across two sets of annular electrodes so as to draw an annular current I_z along the initial B_z field. The toroidal magnetic field associated with this current links the annular current channel ("pre-ionization").
3. A fast capacitor bank generates a reversed B_z field (implosion field) on the outside of the bias B_z field. This fast rising field causes a radial compression of both the initial bias field and the field-aligned current I_z ("implosion").
4. The plasma spontaneously reconnects at the end regions near the electrodes so that the inner bias field becomes the inner poloidal field, the outer (reversed) implosion field becomes the outer poloidal field, and the initial axial current now becomes a poloidal current which generates toroidal magnetic field ("reconnection").
5. The plasma relaxes to a minimum energy, Taylor state ("equilibrium").

In summary, the field-aligned current I_z flowing along the initial bias field constitutes an initial helicity inventory while the subsequent stages guide the system to relax to a lowest energy state consistent with this helicity inventory.

A later and much larger version, the Maryland Spheromak (MS) had the I_ϕ coil separated into two portions, a slow bias coil located outside a metal vacuum chamber and a fast reversal coil located inside the vacuum chamber [107]. The MS device was only able to operate at very high densities with the undesirable consequence that the electron temperature was clamped to a low level by impurity radiation (see discussion on p.238).

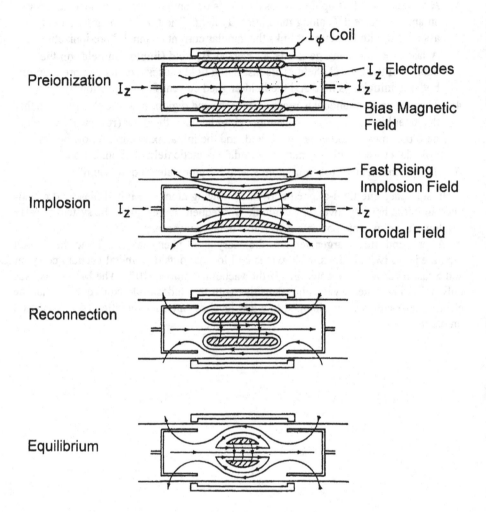

Fig. 7.8 Formation sequence for $z - \theta$ method (from Goldenbaum et al.[21]).

CHAPTER 8
Classification of Regimes: an Imperfect Analogy to Thermodynamics

8.1 Analogy to thermodynamics

Thermodynamics is similar in many ways to Taylor relaxation theory: both thermodynamics and Taylor relaxation theory are based on the postulate that the outcome of complicated dynamics results in the maximization or minimization of some particular physical quantity while some other physical quantity is conserved. In thermodynamics, the entropy of a group of particles is maximized subject to the constraint that the total energy of the group is conserved. In Taylor relaxation theory, the magnetic energy is minimized subject to the constraint that the helicity is conserved. The choice of quantity to be maximized/minimized and the quantity to be conserved is determined by physical arguments outside the scope of the respective theory. Thermodynamics is justified by the expectation that microscopic collisional processes will evolve the system towards a state of maximum disorder, i.e. towards a state of maximum entropy. Taylor relaxation theory is justified by the expectation that microscopic processes dissipate energy much faster than they dissipate helicity.

Because energy is the conserved quantity in thermodynamics and the non-conserved quantity in Taylor relaxation theory, a one-to-one correspondence cannot be made between the two theories. There are nevertheless conceptual analogies relating thermodynamics and Taylor relaxation of an MHD system (and also certain other physical problems, cf. Brown [108]). These analogies can be quite useful for classifying the regimes of the MHD system in terms of the regimes of the more familiar thermodynamic system. Furthermore, simply identifying the regime characterizing a given configuration often provides important insights into qualitative behavior.

8.2 Classification of thermodynamic problems

The most elementary thermodynamic problem is the isolated system shown in Fig. 8.1(a). This system could be in any initial state, but after a long time it settles down to a state of thermodynamic equilibrium, i.e., the state of maximum entropy for the given energy. The equilibrium is characterized by a single parameter, the temperature T, which for a constant volume system is the ratio of the energy to the entropy. Both entropy and energy are extensive quantities, i.e., $1/N^{th}$ of the system contains $1/N^{th}$ of the energy and $1/N^{th}$ of the entropy of the entire system. The temperature is an intensive quantity, i.e., $1/N^{th}$ of the system has the same temperature as the entire system. There is no net heat flux in thermodynamic equilibrium.

A slightly more complicated thermodynamic problem is shown in Fig.8.1(b). Here the system is not isolated, but is in intimate contact with a constant temperature bath. In thermodynamic equilibrium, the system attains the same temperature as the bath. One could also consider the system and bath as one larger, isolated system and obtain the same result. Again there is no net heat flux in equilibrium. There is, however, a strong coupling between the bath and the system; the lack of a net heat flux can be construed as being due to equal heat fluxes flowing to the left and to the right at the interface between the bath and the system.

A still more complicated problem is shown in Fig.8.1(c). Here the system is again not isolated but is now in contact with two geometrically separated constant temperature baths having respective temperatures T_{high} and T_{low}. In this situation, the system cannot attain a uniform temperature and there is a temperature gradient in steady state. There is also a net heat flux from the hot bath through the system to the cold bath. This heat flux is locally determined by the temperature gradient and by a diffusion coefficient, the thermal conductivity. The heat flux is uniform within the system in Fig.8.1(c) because all the heat that enters on the left must exit on the right. A diffusion equation must be solved to give the temperature profile. This system should probably be called a stationary state, rather than a true thermodynamic equilibrium. Characterization of this system requires specifying the thermal conductivity, a quantity determined by microscopic dynamics outside the scope of thermodynamics. If the temperature gradients are gentle, then the heat flux is small and one can approximate small subregions of the system as being in a local thermodynamic equilibrium. As the orifice to the low temperature heat bath is made smaller, the configuration begins to approximate Fig.8.1(b) and so will tend to have a uniform temperature with value T_{high}. On the other hand, if the orifice to the high temperature heat bath is made smaller, then the system will tend to have a uniform temperature with value T_{low}.

Finally, Fig.8.1(d) shows an even more complicated problem. Here the system is not isolated, there is a constant temperature heat source on the left, and the system has numerous leaks through which heat can be lost. Again a diffusion equation must be solved, but now this equation will be multi-dimensional and so the heat flux is non-uniform. If the leaks are large, very little heat reaches the far right of the system resulting in weak coupling of the bath to the far right of the system. If the leaks are small then this system will be similar to Fig.8.1(b). Unlike Fig.8.1(b) where the temperature gradient would be constant (assuming a constant heat conductivity), here the temperature gradient will be steeper on the left and shallower on the right of the figure.

8.3 Analogy between lambda and temperature

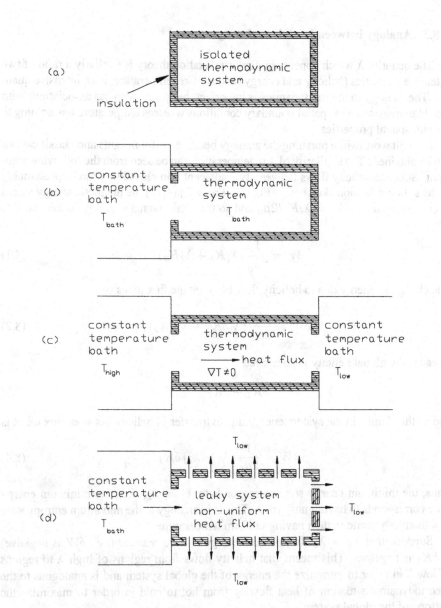

Fig. 8.1 Classificiation of thermodynamic systems: (a) Isolated (surrounded by insulation on all sides), (b) connected to constant temperature heat bath (one location only), (c) located between two heat baths having different temperatures, (d) leaky, connected to constant temperature heat source (left).

8.3 Analogy between lambda and temperature

The quantity λ which appears in Taylor relaxation theory is similarly a ratio of two extensive quantities (helicity and energy) and so, like temperature, is an intensive quantity. The analogy to thermodynamics is imperfect, because λ also has associations with spatial dimensions and spatial boundary conditions whereas temperature has nothing to do with spatial properties.

It is still worthwhile pursuing the analogy because useful insights and classifications can be obtained. The similarity of λ to temperature can be seen from the following argument. Suppose helicity flows between two adjacent thin closed flux tubes, presumably in the splicing fashion sketched in Fig.3.8. From Eq.(4.12) the magnetic energy stored in each flux tube is $W_i = \lambda_i K_i / 2\mu_0$ and so the total magnetic energy in the two flux tubes is

$$W = \frac{1}{2\mu_0} \left(\lambda_1 K_1 + \lambda_2 K_2 \right). \tag{8.1}$$

The change in energy due to helicity flow between the flux tubes is

$$\delta W = \frac{1}{2\mu_0} \left(\lambda_1 \delta K_1 + \lambda_2 \delta K_2 \right). \tag{8.2}$$

Because the global helicity is conserved

$$\delta K_1 + \delta K_2 = 0 \tag{8.3}$$

and so the change in the system energy due to transfer of helicity between flux tubes is

$$\delta W = \frac{1}{2\mu_0} \left(\lambda_1 - \lambda_2 \right) \delta K_1. \tag{8.4}$$

Thus, the minimum energy state corresponds to $\lambda_1 = \lambda_2$; i.e., the minimum energy state corresponds to having uniform λ. This is in analogy to the maximum entropy state of a thermodynamic system having uniform temperature.

Suppose that $\lambda_1 > \lambda_2$. In this case energy will be reduced (i.e., δW is negative) if δK_1 is negative. This means that helicity flows from regions of high λ to regions of low λ in order to minimize the energy of the global system and is analogous to the thermodynamic situation of heat flowing from hot to cold in order to maximize the entropy of the global system.

Thus, just as temperature gradients cause heat to flow from hot to cold, λ gradients cause helicity to flow from regions of high λ to regions of low λ. In both cases gradients of the intensive parameter (T or λ) cause a flux of the conserved extensive parameter.

8.4 Strong and weak coupling

These fluxes are related to the gradients by a diffusion coefficient so

$$\text{thermal flux} = -\kappa \nabla T \tag{8.5}$$

and

$$\text{helicity flux} = -D_K \nabla \lambda. \tag{8.6}$$

Here κ is the thermal conductivity and D_K is a diffusion coefficient for helicity. Chapter 12 shows that appropriately phased MHD fluctuations transport helicity and so we expect D_K to be proportional to the amplitude of magnetic fluctuations. The dynamical behavior underlying helicity diffusion differs from the dynamical behavior underlying classical heat diffusion; helicity diffusion results from the nonlinearity of quasi-coherent macroscopic fluctuations whereas heat diffusion comes from the statistical averaging of a large number of microscopic particle collisions.

Just as the thermodynamic equilibrium of an isolated system [cf. Fig. 8.1(a)] has uniform temperature, we expect that the relaxed state of an isolated magnetohydrodynamic system has uniform λ. Pushing the analogy further, we might expect that just as a thermodynamic system in intimate contact with a constant temperature bath will assume the temperature of the bath [cf. Fig. 8.1(b)], we might expect that a magnetohydrodynamic system in intimate contact with a uniform λ bath will assume the same λ as the bath.

Similarly, just as a thermodynamic system located between two baths having different temperatures, will develop a temperature gradient and steady state heat flux, we expect that a magnetohydrodynamic system connected between two constant λ reservoirs will develop a λ gradient. Furthermore, there will be a steady-state helicity flux determined by the local gradient of λ and the helicity diffusion coefficient.

Finally, just as the leaky thermodynamic system in Fig. 8.1(d) has temperature gradients and non-uniform flux, we expect that a leaky magnetohydrodynamic system will have both λ gradients and a non-uniform helicity flux.

Since helicity can only cross an interface if (i) there are field lines penetrating the interface and (ii) there are differences of electrostatic potential on the interface, we conclude that an interface which is perfectly conducting acts as a helicity insulator if it has no gaps. In other words, helicity can only cross an interface if there is an electric field tangential to the interface. Perfect conductors act as helicity insulators provided there are no gaps. If there are two electrically isolated perfect conductors, then helicity can flow through the gap between the conductors because a potential drop can develop across the gap.

8.4 Strong and weak coupling

Figure 8.1(d) describes the situation of trying to heat a large, long house with a furnace on the left. The temperature will be high near the furnace but if the heat leaks are large, the temperature quickly falls off as one moves down the house away from the furnace. The furnace is weakly coupled to the rest of the house, because the heat must travel a long path and there is plenty of opportunity for the heat to be lost along the way. The loading of the furnace (i.e., the amount of fuel consumed) is hardly affected by the temperature in the last room [far right leaks in Fig.8.1(d)].

On the other hand, if the heat leaks in Fig.8.1(d) are small, the house will heat up quickly and the loading of the furnace will be affected by the temperatures attained in the house. Once the house approaches the furnace temperature, very little heat flows from the furnace to the house and so the furnace loading will be reduced.

Let us compare these two situations to an electric circuit. We define a load impedance as the ratio of the furnace temperature to the heat flux. In the first case (large, long house) the load impedance remains essentially constant whereas in the second case the load impedance increases as less heat is drawn from the furnace when the house temperature approaches the furnace temperature. In the limit that the house temperature equals the furnace temperature, the load impedance is infinite.

In the MHD context, if a source of given λ is connected to a spatially extended lossy system [like Fig.8.1(d)], then we expect a λ gradient to develop and in steady-state there will be a helicity flux from the source. This system will be weakly coupled because there is a negligible flux of helicity back into the source from the load. On the other hand, if a source of given λ is connected to a system which has minimal losses [like Fig.8.1(d) with very small leaks], only a slight λ gradient will be developed so that the stationary state will have nearly uniform λ.

8.5 Overview of next five chapters

The next five chapters will discuss spheromak regimes related to these various situations. Chapters 9 and 10 discuss isolated systems [i.e., systems analogous to Fig.8.1(a)]. Chapter 11 discusses strongly coupled systems; these are systems with minimal losses in intimate contact with a source having a prescribed λ [in analogy to Fig.8.1(b) if a slight leak is added]. Chapter 12 describes weakly coupled configurations where there is substantial distance between the source and distributed sinks so there is both a λ gradient and a non-uniform helicity flux [as in Fig.8.1(d)]. Chapter 12 also discusses how helicity is transported by MHD fluctuations. Chapter 13 discusses spheromak confinement properties; these properties determine the size of the leaks and hence the operating regime.

CHAPTER 9
Analysis of Isolated Cylindrical Spheromaks

9.1 Flux, current, magnetic field, helicity and energy

We now analyze an isolated axisymmetric cylindrical spheromak bounded by a closed perfectly conducting cylindrical flux conserver having radius a and length h. This configuration is shown schematically in Fig.9.1. Because the flux conserver is closed and perfectly conducting, the flux conserver acts as an insulator (barrier) for magnetic helicity (see p.92). We also assume that magnetic field lines do not penetrate the flux conserver so that $\mathbf{B} \cdot d\mathbf{s} = 0$ for each surface element $d\mathbf{s}$ of the flux conserver; this means that $B_z = 0$ at $z = 0$ and $z = h$ and also that $B_r = 0$ at $r = a$. The thermodynamic analog of this system would be an isolated system which has come to equilibrium [cf. Fig. 8.1(a)] and so has a uniform temperature. Here the isolated equilibrium relaxes to a Taylor state and has uniform λ.

The force-free equation with uniform λ is

$$\nabla \times \mathbf{B} = \lambda \mathbf{B} \tag{9.1}$$

and has solutions of the form[71]

$$\mathbf{B} = \lambda \nabla \chi \times \nabla z + \nabla \times (\nabla \chi \times \nabla z) \tag{9.2}$$

where

$$\nabla^2 \chi + \lambda^2 \chi = 0. \tag{9.3}$$

The validity of Eq.(9.2) can be checked by direct substitution into Eq.(9.1) and then using Eq.(9.3). Expanding Eq.(9.3) and invoking axisymmetry gives

$$\frac{1}{r}\frac{\partial}{\partial r}r\frac{\partial \chi}{\partial r} + \frac{\partial^2 \chi}{\partial z^2} + \lambda^2 \chi = 0. \tag{9.4}$$

Using Eq.(9.2) the components of the magnetic field are

$$B_r = \frac{\partial^2 \chi}{\partial r \partial z} \tag{9.5a}$$

$$B_\phi = -\lambda \frac{\partial \chi}{\partial r} \tag{9.5b}$$

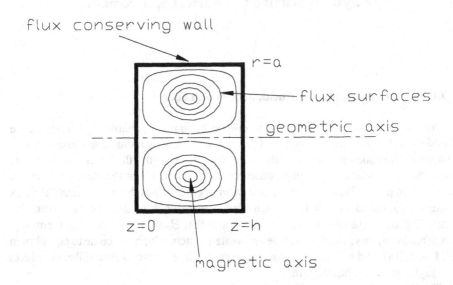

Fig.9.1 Isolated cylindrical spheromak with radius a, length h. Toroidal field lines link the geometric axis while poloidal field lines link the magnetic axis.

$$B_z = \frac{\partial^2 \chi}{\partial z^2} + \lambda^2 \chi. \qquad (9.5c)$$

By assumption B_z vanishes at both $z = 0$ and $z = h$ indicating that solutions which are periodic in z must be used. We therefore assume solutions of the form $\exp(\pm ikz)$ where k is to be determined. With this assumption Eq.(9.4) becomes a Bessel's equation of order zero and has solutions $\chi \sim J_0(\gamma r)$ where $\gamma = \sqrt{\lambda^2 - k^2}$. The solution can be expressed entirely in terms of λ and γ as

$$\chi \sim J_0(\gamma r) \exp(\pm i \sqrt{\lambda^2 - \gamma^2} z). \qquad (9.6)$$

Using Eq.(9.5a) and the radial boundary condition $B_r = 0$ at $r = a$ requires $J_0'(\gamma a) = -J_1(\gamma a) = 0$; this condition forces γ to be quantized at the values

$$\gamma_n = x_{1n}/a \qquad (9.7)$$

where x_{1n} are roots of J_1.

9.1 Flux, current, magnetic field, helicity and energy

The z dependence can now be expressed as

$$B_z = \gamma^2 \chi \tag{9.8}$$

using Eq.(9.6) and Eq.(9.5c). Because B_z is required to vanish at both $z = 0$ and $z = h$, the axial wavenumber k must also be quantized to satisfy the end wall boundary conditions. Thus, we must have

$$\chi \sim J_0(\gamma_n r) \sin(k_m (z - h)) \tag{9.9}$$

where

$$k_m = m\pi/h. \tag{9.10}$$

Substitution of Eq.(9.9) into Eq.(9.3) gives

$$\lambda_{nm} = \pm\sqrt{\gamma_n^2 + k_m^2}. \tag{9.11}$$

Since λ is proportional to the ratio of energy to helicity [see Eq.(4.12)], the lowest energy state for a given helicity will be the $m = n = 1$ state. It is convenient[23] to express all fields in terms of the poloidal field at $r = 0$, $z = h/2$, i.e., in terms of $B_z(0, h/2)$. Denoting this reference field as B_0, Eqs.(9.5a)-(9.5c) show the minimum energy, force-free fields (i.e., Taylor state) of an isolated configuration are

$$B_r = B_0 \frac{\pi}{\gamma_1 h} J_1(\gamma_1 r) \cos(k_1 (z - h)) \tag{9.12a}$$

$$B_\phi = -B_0 \frac{\lambda}{\gamma_1} J_1(\gamma_1 r) \sin(k_1 (z - h)) \tag{9.12b}$$

$$B_z = -B_0 J_0(\gamma_1 r) \sin(k_1 (z - h)) \tag{9.12c}$$

with corresponding eigenvalue

$$\lambda = \pm\sqrt{\frac{x_{11}^2}{a^2} + \frac{\pi^2}{h^2}}. \tag{9.13}$$

Also from Eq.(9.8)

$$\chi = -\frac{B_0}{\gamma_1^2} J_0(\gamma_1 r) \sin(k_1 (z - h)). \tag{9.14}$$

We now define a poloidal flux function ψ which is the magnetic flux intercepted by a circle of radius r at axial location z. The poloidal magnetic field can be expressed in terms of the poloidal flux as

$$\mathbf{B}_{pol} = \frac{1}{2\pi} \nabla \psi \times \nabla \phi. \tag{9.15}$$

Since $\nabla \times \mathbf{B}_{tor} = \lambda \mathbf{B}_{pol}$, the toroidal magnetic field must be

$$\mathbf{B}_{tor} = \frac{1}{2\pi} \lambda \psi \nabla \phi \tag{9.16}$$

so that the total magnetic field is

$$\mathbf{B} = \frac{1}{2\pi} \left(\nabla \psi \times \nabla \phi + \lambda \psi \nabla \phi \right). \tag{9.17}$$

Equating Eqs.(9.12b) and (9.16) shows that the poloidal flux function is

$$\psi = -\frac{2\pi}{\gamma_1} B_0 r J_1(\gamma_1 r) \sin\left(k_1 \left(z - h\right)\right); \tag{9.18}$$

this flux function may also obtained by direct integration of $B_z(r,z)$ over a circle of radius r.

The local maximum of ψ defines the magnetic axis; this is at the point where $d(r J_1(\gamma_1 r))/dr = 0$ and $z = h/2$. Using the Bessel relation [109],

$$\frac{1}{s} \frac{d}{ds} (s J_1(s)) = J_0(s) \tag{9.19}$$

it is seen that the radial location of the magnetic axis occurs when $\gamma_1 r = x_{01}$, the first root of J_0. Thus, the magnetic axis radius is at

$$r_{axis} = \frac{x_{01}}{\gamma_1} = \frac{x_{01}}{x_{11}} a = 0.63 a \tag{9.20}$$

and so the maximum poloidal flux (i.e., the value at the magnetic axis) is

$$\psi_{axis} = 2\pi B_0 \frac{x_{01}}{\gamma_1^2} J_1(x_{01}). \tag{9.21}$$

The toroidal flux at the wall is found by integrating B_ϕ over the minor cross section,

$$\begin{aligned}\Phi_{wall} &= \int_0^h dz \int_0^a dr\, B_\phi(r,z) \\ &= -\frac{\lambda}{\gamma_1} B_0 \int_0^h dz \int_0^a dr\, J_1(\gamma_1 r) \sin\left(k_1 \left(z - h\right)\right)\end{aligned}$$

9.1 Flux, current, magnetic field, helicity and energy

$$= 2B_0 \frac{\lambda}{k_1 \gamma_1^2} [1 - J_0(x_{11})]. \tag{9.22}$$

Since $\mu_0 J_\phi = \lambda B_\phi$, the total toroidal current is

$$I_\phi = 2B_0 \frac{\lambda^2}{\mu_0 k_1 \gamma_1^2} [1 - J_0(x_{11})]. \tag{9.23}$$

[Recall that on p.44 the origin of the toroidal flux was set to be at the magnetic axis and the origin of the poloidal flux was set to be at the wall; these respective origins were selected to eliminate integrated terms in Eq.(3.21) and so obtain a simple relationship between twist, helicity, and safety factor.]

The ratio of toroidal flux at the wall to poloidal flux at the magnetic axis is

$$\frac{\Phi_{wall}}{\psi_{axis}} = \frac{\lambda}{\pi k_1} \frac{[1 - J_0(x_{11})]}{x_{01} J_1(x_{01})}. \tag{9.24}$$

Using Ampere's law to relate the toroidal magnetic field to the poloidal current I linked by a toroidal field line gives $B_\phi = \mu_0 I / 2\pi r$. This poloidal current flows outside the poloidal flux surface ψ associated with the toroidal field line and will be denoted $I_{pol}^{out}(\psi)$. Equation (9.16) gives $B_\phi = \lambda \psi / 2\pi r$ and so, comparing these two expressions for toroidal magnetic field shows that

$$I_{pol}^{out}(\psi) = \lambda \psi / \mu_0. \tag{9.25}$$

The poloidal current flowing inside a magnetic flux surface (i.e., between the flux surface and the magnetic axis) will be

$$I_{pol}^{in} = \lambda(\psi_{max} - \psi)/\mu_0. \tag{9.26}$$

Equations (9.2) and (9.17) give the magnetic field in two complementary ways, i.e., in terms of the Helmholtz function χ or in terms of the flux function ψ. This duality of representation can be used to find the vector potential. Separating the magnetic field into toroidal and poloidal components for both these representations gives

$$\begin{aligned} \mathbf{B}_{pol} &= \nabla \times (\nabla \chi \times \nabla z) = \nabla \times \left(\frac{1}{2\pi} \psi \nabla \phi \right) \\ \mathbf{B}_{tor} &= \nabla \times (\lambda \chi \nabla z) = \frac{1}{2\pi} \lambda \psi \nabla \phi \end{aligned} \tag{9.27}$$

from which toroidal and poloidal vector potentials can be identified, namely,

$$\mathbf{A}_{pol} = \lambda \chi \nabla z$$

$$\mathbf{A}_{tor} = \frac{1}{2\pi}\psi\nabla\phi. \tag{9.28}$$

An arbitrary gauge function ∇f can also be added so that the complete vector potential is

$$\mathbf{A} = \lambda\chi\nabla z + \frac{1}{2\pi}\psi\nabla\phi + \nabla f. \tag{9.29}$$

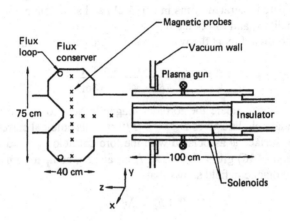

Fig.9.2 Coordinate system of Beta-II spheromak from Turner et al. [23].

The helicity can now be evaluated and is found to be

$$\begin{aligned} K &= \lambda\int_0^a dr \int_0^h dz \left(\chi\frac{\partial\psi}{\partial r} - \psi\frac{\partial\chi}{\partial r}\right) \\ &= -2\lambda\int_0^a dr \int_0^h dz\,\psi\frac{\partial\chi}{\partial r} \\ &= 2\pi\lambda h\frac{B_0^2}{\gamma_1^2}\int_0^a dr\, r J_1^2(\gamma_1 r) \\ &= \pi\lambda a^2 h\frac{B_0^2}{\gamma_1^2} J_0^2(x_{11}); \end{aligned} \tag{9.30}$$

here Eqs.(9.27) and (9.28) provide explicit representations for the fields, and Eq.(B.21)

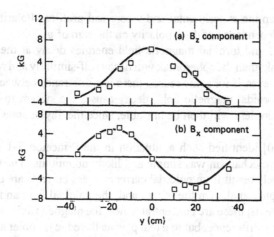

Fig.9.3 Measured poloidal and toroidal magnetic fields in Beta-II spheromak compared to predictions of Taylor relaxation. Coordinate system shown in figure above. (from Turner et al. [23])

has been used for the Bessel integrals. Using Eq.(4.12) the total magnetic energy is

$$W = \pi a^2 h \frac{\lambda^2}{\gamma_1^2} \frac{B_0^2}{2\mu_0} J_0^2(x_{11}). \tag{9.31}$$

The total magnetic energy is the sum of the toroidal and poloidal energies, i.e., $W = W_{tor} + W_{pol}$. Direct evaluation of the respective toroidal and poloidal magnetic energies

$$W_{tor} = \int \frac{B_{tor}^2}{2\mu_0} d^3r, \quad W_{pol} = \int \frac{B_{pol}^2}{2\mu_0} d^3r \tag{9.32}$$

shows that $W_{tor} = W_{pol}$, a specific example of the general result shown earlier in Sec.4.4.1.

9.2 Experimental measurements

Turner et al.[23] measured the Beta-II spheromak magnetic fields and showed that these fields were in good agreement with the Taylor state model. Figure 9.2 shows the Beta-II spheromak coordinate system (plasma is injected from right to left here) while Fig. 9.3 shows the measured poloidal field fitted to B_z from Eq.(9.12c) and the measured toroidal field (labeled B_x) fitted to B_ϕ from Eq. (9.12b); note the excellent agreement between the experimental and theoretical profiles. Polarity reversal of B_x results from the dependence of $\hat{\phi} \cdot \hat{x}$ polarity on the sign of y.

If the poloidal and toroidal magnetic field energies decay at the same rate in an isolated spheromak, then the spheromak will decay self-similarly and remain in a lowest energy state. However, if for some reason there is a difference between the decay rates of poloidal and toroidal magnetic field, decay causes the system to deviate from the Taylor state. The system will then be unstable, since the Taylor state is the minimum-energy state.

Ono et al.[110] identified such a situation in the Princeton S-1 spheromak. The differential decay mechanism was simple and likely ubiquitous. The toroidal field and its energy ultimately result from poloidal currents; these currents are concentrated near the edge of the spheromak. On the other hand, the poloidal field and its energy result from toroidal currents; these are concentrated near the magnetic axis. In an ideal plasma this would make no difference, but in a real plasma the edge is cooler and more resistive than the region near the magnetic axis. Thus, poloidal current typically decays faster than toroidal current resulting in a configuration which has $W_{pol} > W_{tor}$. This is a deviation from the minimum energy configuration and so instabilities will develop to relax the system back to a minimum energy state. Ono et al. observed a kink-like relaxation instability which equilibrated the poloidal and toroidal energies.

9.3 Safety factor

9.3.1 Limiting forms at magnetic axis, wall

From Eq.(2.47), the safety factor on the magnetic axis is

$$q_{axis} = \frac{\left(\sqrt{\psi_{rr}/\psi_{zz}} + \sqrt{\psi_{zz}/\psi_{rr}}\right)}{\lambda r_{axis}}. \tag{9.33}$$

The flux has the spatial dependence

$$\psi(r,z) \sim \gamma_1 r J_1(\gamma_1 r) \sin\left(\sqrt{\lambda^2 - \gamma_1^2}(z-h)\right) \tag{9.34}$$

so using Eq.(9.19) and $\gamma_1 r_{axis} = x_{01}$ it is seen that

$$\frac{\psi_{rr}}{\psi_{zz}} = \frac{\gamma_1^2}{\lambda^2 - \gamma_1^2}. \tag{9.35}$$

9.3 Safety factor

Using this ratio in Eq.(9.33) gives

$$q_{axis} = \frac{1}{x_{01}\sqrt{1 - \gamma_1^2/\lambda^2}}. \tag{9.36}$$

This functional form shows that q_{axis} decreases with increasing λ. Using Eq.(9.13) it is seen that $\lambda/\gamma_1 = \sqrt{1 + \pi^2 a^2/x_{11}^2 h^2}$ so

$$\begin{aligned}q_{axis} &= \frac{\lambda/\gamma_1}{x_{01}\sqrt{\lambda^2/\gamma_1 - 1}} \\ &= \frac{1}{x_{01}}\sqrt{1 + \frac{x_{11}^2 h^2}{\pi^2 a^2}}.\end{aligned} \tag{9.37}$$

The safety factor at the wall can be evaluated analytically using the argument presented earlier for spherical geometry (cf. p.85). On the geometric axis $B_r = 0$ and so the field line trajectory is given by $rd\phi/B_\phi = dz/B_z$. Because the field is nearly independent of z in the vicinity of the geometric axis, the force-free magnetic field in this region is characterized by the Lundquist solution, Eq.(1.2), giving $B_\phi/B_z = \lambda r/2$. Thus, on the geometric axis $d\phi = \lambda dz/2$ and so the increment in ϕ for a field line going up the geometric axis will be $\Delta\phi = \lambda h/2$. A field line going up the geometric axis becomes a field line adjacent to the wall, but at the wall, the toroidal field vanishes so there is no increment in toroidal angle as the field line moves along the wall. Thus the entire change in toroidal angle for a field line on the wall-facing flux surface will be $\Delta\phi = \lambda h/2$ and so the number of toroidal turns for one poloidal transit will be $\Delta\phi/2\pi = \lambda h/4\pi$. This is just the definition for the safety factor and so on the outermost flux surface

$$\begin{aligned}q_{wall} &= \frac{\lambda h}{4\pi} \\ &= \frac{1}{4}\sqrt{1 + \frac{x_{11}^2 h^2}{\pi^2 a^2}}.\end{aligned} \tag{9.38}$$

9.3.2 Evaluation of q profile

From Eqs.(9.15) and (9.16) the radial and toroidal magnetic fields can be expressed in terms of the poloidal flux function as

$$B_r = -\frac{1}{2\pi r}\frac{\partial\psi}{\partial z}, \quad B_\phi = \frac{\lambda}{2\pi r}\psi. \tag{9.39}$$

These fields can be used in Eq.(2.37) to give the safety factor

$$q = -\frac{\lambda\psi}{2\pi} \oint \frac{dr}{r\partial\psi/\partial z} \tag{9.40}$$

where the ψ in the numerator has been factored out of the integral because the integration is along a path of constant ψ. Since ψ appears both in the numerator and denominator, it is worthwhile to normalize the poloidal flux, Eq.(9.18), to its value on the magnetic axis, Eq.(9.21), by defining[71]

$$\bar{\psi}(r,z) = \frac{\psi(r,z)}{\psi_{axis}} = -\frac{\gamma_1 r J_1(\gamma_1 r)}{x_{01} J_1(x_{01})} \sin(k_1(z-h)). \tag{9.41}$$

Equation (9.40) can be written in terms of $\bar{\psi}$ as

$$q = \frac{\lambda\bar{\psi}(r,z)}{2\pi} \oint \frac{x_{01} J_1(x_{01}) dr}{k_1 r \; \gamma_1 r J_1(\gamma_1 r) \cos(k_1(z-h))} \tag{9.42}$$

and Eq.(9.41) can be recast as

$$\frac{\gamma_1 r J_1(\gamma_1 r)}{x_{01} J_1(x_{01})} \cos k_1(z-h) = \sqrt{\left(\frac{\gamma_1 r J_1(\gamma_1 r)}{x_{01} J_1(x_{01})}\right)^2 - \bar{\psi}^2}. \tag{9.43}$$

Equation (9.43) shows that on the midplane (i.e., $z = h/2$), the radial location of the flux surface must satisfy

$$\frac{\gamma_1 r J_1(\gamma_1 r)}{x_{01} J_1(x_{01})} = \bar{\psi}. \tag{9.44}$$

Recalling that the magnetic axis is at $r = ax_{01}/x_{11}$ (i.e., where $rJ_1(\gamma_1 r)$ is at a maximum), we see that Eq.(9.44) has two roots (which we label r_1 and r_2), one on either side of the magnetic axis. Defining $s = \gamma_1 r$, Eq.(9.42) becomes[71]

$$q(\bar{\psi}) = \frac{\lambda\bar{\psi}}{k_1 \pi} \int_{s_1}^{s_2} \frac{ds}{s \sqrt{\left(\frac{sJ_1(s)}{x_{01}J_1(x_{01})}\right)^2 - \bar{\psi}^2}} \tag{9.45}$$

or

$$q(\bar{\psi}) = \sqrt{1 + \frac{x_{11}^2 h^2}{\pi^2 a^2}} Q(\bar{\psi}) \tag{9.46}$$

where

$$Q(\bar{\psi}) = \frac{\bar{\psi}}{\pi} \int_{s_1}^{s_2} \frac{ds}{s\sqrt{\left(\frac{sJ_1(s)}{x_{01}J_1(x_{01})}\right)^2 - \bar{\psi}^2}} \qquad (9.47)$$

is a geometry-independent function of $\bar{\psi}$. Numerical evaluation of $Q(\bar{\psi})$ is straightforward. A plot of $Q(\bar{\psi})$ is shown in Fig.(9.4). Comparison with Eqs.(9.37) and (9.38) shows that $Q(1) = 1/x_{01} = .416$ (magnetic axis) and $Q(0) = 0.25$ (wall).

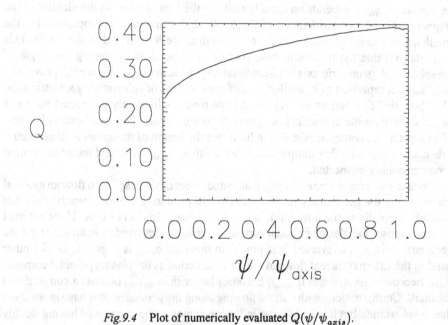

Fig.9.4 Plot of numerically evaluated $Q(\psi/\psi_{axis})$.

In the next chapter it will be shown that the spheromak is subject to a tilt instability when q approaches unity on the magnetic axis. This occurs when

$$\sqrt{1 + \frac{x_{11}^2 h^2}{\pi^2 a^2}} Q(1) = 1 \qquad (9.48)$$

which may be solved for h/a to give

$$\frac{h}{a} = \frac{\pi}{x_{11}}\sqrt{x_{01}^2 - 1} = 1.8 \qquad (9.49)$$

Since q is a monotonically increasing function of h/a, it is seen that if $h/a > 1.8$, q_{axis} will exceed unity.

9.3.3 Effect of flux conserver shape on q_{wall}

Equation (6.61) shows that the β limit is proportional to the square of the fractional shear, i.e., to $(q_{axis} - q_{wall})^2 / (q_{axis} + q_{wall})^2$. If the flux conserver has the cylindrical shape assumed here, then the ratio q_{axis}/q_{wall} is fixed and so there is no way to optimize the β limit. However, consideration of the derivation of Eq.(9.37) shows that for a given λ, q_{axis} depends on the ellipticity of the flux surfaces in the vicinity of the magnetic axis while consideration of Eq.(9.38) shows that q_{wall} is proportional to the length of the plasma geometric axis, i.e., to the distance h measured on the z axis. This suggests [67] that the fractional shear could be increased by shortening the length of the spheromak geometric axis while maintaining the same q_{axis}. Figure 9.5 provides a qualitative comparison of q profiles for different lengths of spheromak geometric axis. The "bow-tie" flux conserver [67] (third from top) is still simply connected but has a larger shear than the cylindrical or spherical configurations (top two figures) because of its shortened geometric axis. In the limit that the length of the spheromak geometric axis goes to zero, the flux dimples touch each other, $q_{wall} \to 0$, and the configuration becomes doubly connected.

Once the system becomes doubly connected, it becomes possible to flow an external current along the geometric axis; this current will produce a vacuum toroidal field that adds algebraically to the toroidal field due to plasma poloidal currents. If the external current opposes the plasma poloidal currents, it creates a reversed toroidal field near the geometric axis, i.e., a reversed field pinch. In this case q_{wall} is negative. On the other hand, if the external current flows in the same direction as the plasma poloidal currents, q_{wall} becomes elevated and if q_{wall} becomes larger than q_{axis}, one has a conventional tokamak. Configurations with current flowing along the geometric axis have more shear than a spheromak but this stabilizing influence comes at the expense of having doubly connected topology.

Uyama et al.[111] investigated the dependence of spheromak stability on geometric axis length by inserting a central conducting pole along the geometric axis of the CTCC-1 spheromak. By adjusting the pole's insertion depth, the various q profiles shown in Fig.9.5 could be obtained and by using different pole diameters the curvature of the q profile could be adjusted. These experiments showed that pole insertion reduced MHD instability, but this reduction did not significantly improve spheromak performance. Wright et al. [112] similarly did not see improvement in plasma performance due to insertion of a conductor on CTX; the failure to see improved performance was ascribed to increased radiation losses obscuring the beneficial change in q profile.

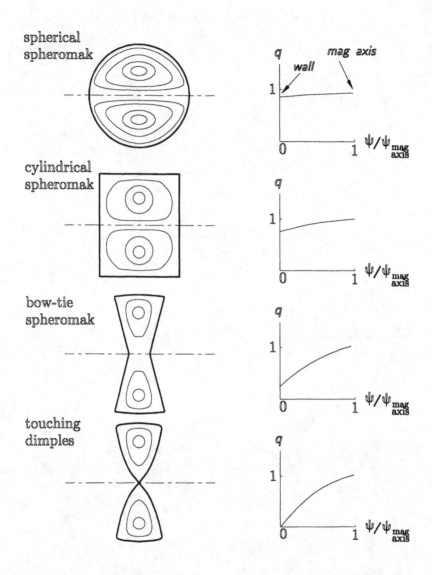

Fig.9.5 Qualitative dependence of q_{wall} on of dimpling of the flux conserver at the geometric axis. A spherical spheromak (top, negative dimpling) has the flattest q profile, while a fully dimpled spheromak (bottom) has the steepest q profile. Because the fully dimpled configuration is doubly connected, poloidal currents can be made to flow in the flux conserver, creating RFP or tokamak configurations and resulting in even larger q gradients.

CHAPTER 10
The Role of the Wall

We now examine the various ways a flux conserving wall influences an isolated spheromak. Most of these wall properties are relevant to driven spheromaks as well.

From a mathematical point of view, the energy minimization leading to the Taylor state required a flux-conserving wall to permit an integration by parts [cf. Eqs.(4.4,4.5)]. However, the Taylor state itself does not explicitly require a wall; it simply asserts that B satisfies a force-free equation having spatially uniform λ. Thus, a wall seems necessary for the system to relax to the minimum energy state but perhaps might not be necessary for the system to remain in this state. This subtle distinction has motivated consideration of spheromak designs that do not have a wall; so far these attempts have been unsuccessful, but as will be seen, they might eventually succeed. This chapter shows that the conducting wall performs three important functions: (i) it acts as an insulator for helicity, (ii) it provides equilibrium in the form of a steady uniform vertical field and (iii) it provides stability in the form of transient, non-axisymmetric stabilizing fields that provide a restoring force opposing the spheromak's predilection towards tilting. It appears straightforward to provide the equilibrium field by means other than a flux-conserving wall, but so far it has not been possible to provide the transient stabilizing fields by other means.

10.1 Helicity insulation

In Sec.5.2.2 [p.92, see especially Eq.(5.20)] it was shown that there cannot be helicity flux across a perfectly conducting wall that has no gaps. Such a wall prevents tangential electric fields from developing at the surface. These tangential electric fields would allow a helicity flux into or out of the system. If the system is not surrounded by a perfectly conducting wall and is not externally driven, then large tangential electric fields could develop with a polarity corresponding to a helicity flux out of the system. Such a situation could develop if the edge plasma is resistive, either because it is cold or because of a high rate of electron-neutral collisions. A conducting wall would short circuit the tangential electric fields associated with these resistive effects and reduce the helicity outflux.

10.2 Equilibrium

Let us now consider the wall contribution to the equilibrium of an isolated spheromak; this analysis is most easily performed in spherical geometry (the behavior in cylindrical geometry is qualitatively similar). The spherical wall is located at $r = a$ and is assumed to be a perfect conductor (flux conserver). There will be wall surface currents associated with the discontinuity in tangential magnetic field (which is finite in the plasma, but zero in the wall). As shown in Eq.(4.42) which we repeat here, the spheromak magnetic fields in spherical geometry are

$$B_r = 2B_0 \frac{a}{r} j_1(\lambda r) \cos \theta$$

$$B_\theta = -B_0 \frac{a}{r} \frac{\partial}{\partial r} (r j_1(\lambda r)) \sin \theta$$

$$B_\phi = \lambda a B_0 j_1(\lambda r) \sin \theta. \tag{10.1}$$

In order to have B_r vanish at the wall, λ must satisfy $j_1(\lambda a) = 0$. Since surface currents depend on the magnetic field discontinuity at the wall, these surface currents depend on the jump of B_θ at the wall, because B_θ is the only field which is non-zero at the wall.

The relevant component of Ampere's law is the toroidal component

$$\frac{1}{r} \frac{\partial}{\partial r} (r B_\theta) - \frac{1}{r} \frac{\partial B_r}{\partial \theta} = \mu_0 J_\phi \tag{10.2}$$

since this component involves the radial derivative of B_θ. Because B_θ is finite at $r = a_-$ (outer edge of the plasma) and zero at $r = a^+$ (inside the wall), the jump in B_θ indicates the existence of a wall surface current. This wall surface current density $I_\phi(\theta)$ will be in the toroidal direction and is

$$I_\phi(\theta) = \int_{a_-}^{a^+} dr J_\phi(r, \theta); \tag{10.3}$$

note that I_ϕ has dimensions of current per length.

Inserting Eq.(10.2) in (10.3) gives

$$\begin{aligned} I_\phi(\theta) &= \int_{a_-}^{a^+} dr \frac{1}{\mu_0 r} \frac{\partial}{\partial r} (r B_\theta) \\ &= \frac{B_0}{\mu_0} \sin \theta \left[\frac{\partial}{\partial r} (r j_1(\lambda r)) \right]_{r=a} . \end{aligned} \tag{10.4}$$

Using the Bessel identity[113]

$$\frac{1}{s} \frac{d}{ds} \left(s^2 j_1(s) \right) = s j_0(s) \tag{10.5}$$

10.2 Equilibrium

the surface current density can be expressed as

$$\mu_0 I_\phi(\theta) = s_{11} j_0(s_{11}) B_0 \sin\theta \tag{10.6}$$

where $s_{11} = 4.493$ is the first root of $j_1(s)$.

This wall surface current produces a vacuum magnetic field in the plasma region (i.e., $r < a$). It should be possible to replace the wall by external coils which provide the same vacuum magnetic field in the region $r < a$. If this could be done, then a spheromak equilibrium would be achieved without a nearby flux-conserving wall. This would be advantageous for fusion applications where it is generally desirable to minimize contact between the plasma and the wall.

Design of the external coil system requires knowledge of the vacuum field produced by the wall currents. The nature of the wall field becomes obvious when we recall[114] that the general form for vacuum magnet fields with $B_\theta \sim \sin\theta$ at the wall is

$$\begin{aligned}\mathbf{B}_{vac} &= c_{in} \nabla (r\cos\theta) \text{ for } r < a \\ &= c_{out} \nabla \left(r^{-1} \cos\theta\right) \text{ for } r > a.\end{aligned} \tag{10.7}$$

The coefficients c_{in}, c_{out} can be related to the wall surface current using the boundary conditions:

1. To maintain $\nabla \cdot \mathbf{B} = 0$, B_r must be continuous at the wall; this condition gives $c_{in} = -c_{out}/a^2$,
2. The jump in B_θ gives $(ac_{in} - c_{out}/a)\sin\theta = \mu_0 a I_\phi(\theta)$ so that the coefficients c_{in}, c_{out} can be determined.

Since $\nabla(r\cos\theta) = \hat{z}$ these two boundary conditions give the vacuum field produced by the wall surface current as

$$\mathbf{B}^{wall} = \begin{cases} +\frac{1}{2} s_{11} j_0(s_{11}) B_0 \, \hat{z} & \text{for } r < a \\ -\frac{1}{2} s_{11} j_0(s_{11}) B_0 \, a^2 \nabla\left(r^{-1}\cos\theta\right) & \text{for } r > a. \end{cases} \tag{10.8}$$

Since $j_0(s_{11})$ is negative, the magnetic field produced inside the plasma volume by the wall surface current is uniform and in the negative z direction, i.e., a so-called vertical field. The spheromak toroidal current $\mathbf{J}$ is in the positive ϕ direction so that $\mathbf{J} \times \mathbf{B}^{wall}$ is inward and, like the force due to the vertical field in a tokamak, balances the outward hoop force of the toroidal current loop.

On the midplane just inside the wall (i.e., at $r = a_-, \theta = \pi/2$), θ points downward so in this region, $B_z = -B_\theta$. Using Eq.(10.1) in this region gives

$$\begin{aligned} B_z &= \left[B_0 \frac{a}{r} \frac{\partial}{\partial r} (rj_1(\lambda r)) \sin\theta \right]_{r=a, \theta=\pi/2} \\ &= s_{11} j_0(s_{11}) B_0 . \end{aligned} \quad (10.9)$$

Since this field is the algebraic sum of the fields produced by the wall and by the plasma currents, it is clear that the vertical magnetic field due to the wall current equals the field due to the plasma internal currents. Thus at $r = a$, $\theta = \pi/2$, the z component of the magnetic field due to the plasma current adds to the wall field.

On the geometric axis the polar angle is $\theta = 0$ in the upper half plane and $\theta = \pi$ in the lower half plane. Both $B_\theta = B_\phi = 0$ on this axis and the radial field corresponds to the z-directed field. Thus, the z-directed spheromak field on the geometric axis is

$$\begin{aligned} B_z &= 2B_0 \lim_{r \to 0} \frac{a}{r} j_1(\lambda r) \\ &= \frac{2}{3} s_{11} B_0 \end{aligned} \quad (10.10)$$

which is upward and therefore opposite the direction of the field due to the wall image currents. Since this field is again the algebraic sum of the fields due to the plasma and the wall currents, the plasma toroidal current must be so strong that at the geometric axis it reverses the uniform B_z provided by the wall image currents. Thus, the spheromak can be considered as a form of field reversed configuration, although this name is normally reserved for systems having zero poloidal current.

Since the function of the wall image currents is to produce a uniform vertical field inside the plasma, one could conceivably eliminate the wall and instead use external coils to produce this vertical field. The spheromak could then be considered as a ball of plasma containing a toroidal current aligned to generate a poloidal magnetic field that aids the vacuum field in the midplane at $r = a$ and opposes the vacuum field on the geometric axis. Another possibility would be to have a segmented wall consisting of toroidal rings. These would conduct the toroidal surface current required for equilibrium, but would not allow current flow in the θ direction.

10.3 Tilt stability

The previous section showed that the spheromak behaves like a current loop immersed in an externally produced uniform field and oriented to oppose the external field on the axis of the loop. This means that the sense of the spheromak toroidal current is opposite that of the toroidal current flowing in the coils producing the external field. In fact, because anti-parallel currents repel each other, the opposing sense of the external coil (or wall) toroidal currents relative to the spheromak toroidal current is just the requirement for a vertical field to balance the hoop force.

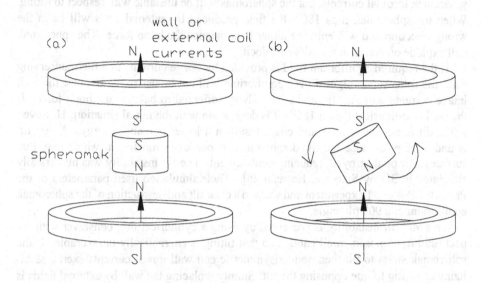

Fig.10.1 In order to balance the spheromak hoop force, the wall image toroidal current (or external coil current) flows in the opposite direction relative to the spheromak toroidal current (oppositely directed currents repel). However, this means that the wall or external coil currents are like magnet coils producing a field opposing the spheromak field (a). The spheromak is thus like a small magnet suspended between two coils producing an oppositely directed field and so is liable to tilt (b).

The spheromak toroidal current can be approximated as a magnetic dipole with magnetic moment $\mathbf{m} \approx I_\phi \pi r_{axis}^2 \hat{z}$ where r_{axis} is the radius of the magnetic axis and I_ϕ is now the spheromak toroidal current. The potential energy of a dipole in an external magnetic field is $U = -\mathbf{m} \cdot \mathbf{B}_{ext}$; since the B_z provided by the wall (or external coils) is in the negative z direction, the potential energy of the spheromak magnetic moment in the external field is at a maximum. As shown in Fig.10.1(a), the spheromak is like a small magnet between two large magnets (the wall currents or external coil currents) and this small magnet is oriented anti-parallel to the large magnets. Such a situation is unstable because the spheromak can lower its potential energy by flipping 180^0 as in Fig.10.1(b). Unfortunately, if the spheromak flips, the external field will be in the wrong direction for balancing the spheromak hoop force and equilibrium will be lost.

This reveals the flaw in replacing the wall by external equilibrium coils. So long as the spheromak does not tilt, the externally provided B_z balances the hoop force of the spheromak toroidal currents, but the spheromak will be unstable with respect to tilting. When the spheromak flips 180^0, the field produced by external coils will be in the wrong direction and will enhance rather than balance the hoop force. The spheromak will explode outwards at the Alfvén velocity.

If the required equilibrium field is provided by image currents in a flux-conserving wall, the spheromak might tilt but equilibrium will not be destroyed because the wall image currents will also tilt and so are always oriented to balance the hoop force. If the wall is cylindrical then a 180^0 tilt is degenerate with the initial situation. However, a 90^0 tilt is not degenerate and could result in a lower potential energy. A 90^0 tilt is undesirable, because the tilted spheromak is non-axisymmetric in which case flux surfaces will be destroyed, reducing confinement. The tilt instability was numerically simulated in 3D by Sato and Hayashi[28]. Their simulation used parameters of the Princeton Proto S-1 experiment and showed a clear tilt and destruction of the spheromak equilibrium at a 90^0 tilt angle.

This 90^0 tilt instability is prevented by using a cylindrical flux conserver with aspect ratio h/a chosen small enough so that tilting is energetically unfavorable. If the spheromak starts to tilt then non-axisymmetric end wall image currents exert a stabilizing restoring torque opposing the tilt. Simply replacing the wall by external fields is inadequate; although these coils can provide the required equilibrium field, they cannot produce the transient non-axisymmetric currents required to counteract tilt instability.

Mathematical analysis of the tilt instability threshold is difficult because of the non-axisymmetry of the tilted configuration and also because the boundary conditions ($B_r = 0$ on the sides, $B_z = 0$ on the end walls) on a cylindrical flux conserver lead to complicated relationships that the Helmholtz function χ must satisfy at the walls. The consequence of this complexity is that numerical techniques must be used to determine the instability threshold. This has been done independently by Finn, Manheimer and Ott [71] and by Bondeson et al.[91]. Both groups concluded that an axisymmetric spheromak in a cylindrical flux conserver is unstable with respect to tilting when $h/a > 1.67$ where h is the axial length of the flux conserver and a is its radius. The tilting analysis involved calculating λa as a function of h/a for axisymmetric and tilted configurations and then showing that above the instability threshold, the λa of a tilted configuration is less than that of an axisymmetric configuration. Since λ is a measure of the energy per helicity, the tilted configuration would have a lower energy for the same helicity. Similar conclusions were obtained earlier by Rosenbluth and Bussac who considered tilt stability in spherical geometry[19].

The discussion of $q(\psi)$ in the previous chapter leads to essentially the same conclusion, but with a slightly different numerical criterion. If, as shown in Fig. 10.2(a), one views tilt instability in a poloidal plane, it is seen that the spheromak bodily lifts its magnetic axis from the original midplane. The perturbation therefore has an $m = 1$ symmetry about the magnetic axis. If, as shown in Fig.10.2(b) one looks at tilt instability in the midplane ($z = 0$ plane), then a tilt instability involves upward motion at $\phi = 0$ and downward motion at $\phi = \pi$. The tilt instability therefore has $n = 1$ symmetry in the toroidal direction. To avoid deforming the internal spheromak field, a magnetic instability should satisfy $q = m/n = 1/1 = 1$, and so there ought to be a $q = 1$ surface at the magnetic axis. If $q < 1$ on the magnetic axis, then tilting would modify the internal field and this modification will typically require more energy than liberated by reducing $-\mathbf{m} \cdot \mathbf{B}$. It was found in the previous chapter that $q = 1$ appeared on the magnetic axis when $h/a > 1.8$; this suggests that it is necessary to have $h/a < 1.8$ in order to be tilt stable. This semi-quantitative result is not substantially different from what is obtained by Bondeson et al.'s[91] and Finn et al.'s[71] more careful analysis.

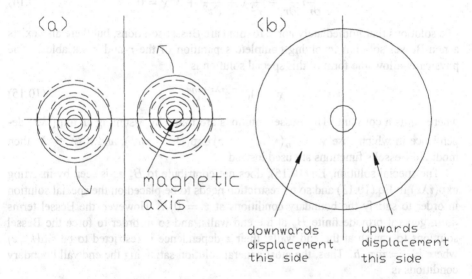

Fig.10.2 Tilt symmetry: (a) view of poloidal plane, showing tilt has $m = 1$ symmetry about the magnetic axis; (b) view of midplane showing tilt has $n = 1$ symmetry about z axis.

We now outline Bondeson et al.'s tilt stability analysis. Using Eqs. (9.2) and (9.3) the components of the force-free magnetic field in cylindrical geometry are

$$B_r = \frac{\lambda}{r}\frac{\partial \chi}{\partial \phi} + \frac{\partial^2 \chi}{\partial r \partial z} \tag{10.11}$$

$$B_\phi = -\lambda \frac{\partial \chi}{\partial r} + \frac{1}{r}\frac{\partial^2 \chi}{\partial z \partial \phi} \tag{10.12}$$

$$B_z = \frac{\partial^2 \chi}{\partial z^2} + \lambda^2 \chi. \tag{10.13}$$

Possible non-axisymmetric behavior is manifested by the ϕ dependence.

Without loss of generality we consider modes with $\chi \sim e^{in\phi}$ where n will later be set to unity. The configuration is assumed to be in a cylindrical flux conserver of length h and radius a with boundary conditions $B_r = 0$ at $r = a$, $B_z = 0$ at both $z = 0$ and $z = h$. The Helmholtz equation, Eq.(9.3) is

$$\frac{1}{r}\frac{\partial}{\partial r}r\frac{\partial \chi}{\partial r} - \frac{n^2}{r^2}\chi + \frac{\partial^2 \chi}{\partial z^2} + \lambda^2 \chi = 0. \tag{10.14}$$

The solutions that immediately come to mind are Bessel solutions, but there also exists a non-Bessel solution involving complete separation of the r and z variables. The physically allowable form of this special solution is

$$\chi = A_0 r^n e^{i\lambda z + in\phi} \tag{10.15}$$

where A_0 is a constant. The Bessel solutions are found by assuming an $\exp(ikz)$ dependence in which case $\chi \sim J_n(\sqrt{\lambda^2 - k^2}r)\exp(ikz + in\phi)$, and if $k^2 > \lambda^2$ then modified Bessel's functions are used instead.

The special solution, Eq.(10.15), does not contribute to B_z as is seen by inserting $\exp(i\lambda z)$ in Eq.(10.13) and so no restriction needs to be placed on the special solution in order to satisfy the boundary conditions at $z = 0, h$. However, the Bessel terms do in general provide finite B_z at the end walls, and so in order to force the Bessel solutions to vanish at the end walls, their z dependence is restricted to be $\sin(k_m z)$ where $k_m = m\pi/h$. Thus, the most general solution satisfying the endwall boundary conditions is

$$\chi = \left(A_0 r^n e^{i\lambda z} + \sum_{m=1}^{\infty} A_m J_n(\sqrt{\lambda^2 - k_m^2}\,r) \sin k_m z \right) e^{in\phi} \tag{10.16}$$

where I_n is used instead of J_n if $k_m^2 > \lambda^2$.

The radial boundary condition introduces real difficulty, because this boundary condition can only be satisfied by an infinite sum of χ eigenfunctions, and not by a single

10.3 Tilt stability

eigenfunction as in the axisymmetric case. Using Eq.(10.11) the condition $B_r = 0$ at $r = a$ gives

$$\begin{aligned}
0 = & \frac{in\lambda}{a}\left(A_0 a^n e^{i\lambda z} + \sum_{m=1}^{\infty} A_m J_n(\sqrt{\lambda^2 - k_m^2}\,a)\sin k_m z\right) \\
& + \frac{in\lambda}{a} A_0 a^n e^{i\lambda z} \\
& + \sum_{m=1}^{\infty} k_m\sqrt{\lambda^2 - k_m^2}\,A_m J_n'(\sqrt{\lambda^2 - k_m^2}\,a)\cos k_m z \quad (10.17)
\end{aligned}$$

which may be rearranged as

$$\begin{aligned}
0 = & 2in\lambda A_0 a^{n-1} e^{i\lambda z} \\
& + \sum_{m=1}^{\infty} A_m \left[\begin{array}{c} in\lambda a^{-1} J_n(\sqrt{\lambda^2-k_m^2}\,a)\sin k_m z \\ +k_m\sqrt{\lambda^2-k_m^2}\,J_n'(\sqrt{\lambda^2-k_m^2}\,a)\cos k_m z \end{array}\right]. \quad (10.18)
\end{aligned}$$

Equation (10.18) is now multiplied by $\cos(k_{m'}z)$ and integrated from $z = 0$ to $z = h$. The following three identities are used to develop a matrix equation for the A_m:

$$\begin{aligned}
\int_0^h dz\, e^{i\lambda z}\cos k_{m'}z &= i\lambda e^{i\lambda h/2}\frac{(e^{-i\lambda h/2} - e^{i\lambda h/2}\cos k_{m'}h)}{\lambda^2 - k_{m'}^2} \\
&= 2i\lambda e^{i\lambda h/2}\frac{\cos(\lambda h/2)}{\lambda^2 - k_{m'}^2} \text{ if } m' \text{ is odd} \\
&= 2\lambda e^{i\lambda h/2}\frac{\sin(\lambda h/2)}{\lambda^2 - k_{m'}^2} \text{ if } m' \text{ is even} \quad (10.19)
\end{aligned}$$

$$\begin{aligned}
\int_0^h dz\,\sin k_m z \cos k_{m'}z &= \frac{2k_m}{k_m^2 - k_{m'}^2} \text{ if } m \text{ and } m' \text{ have different parity} \\
&= 0 \text{ if } m \text{ and } m' \text{ have the same parity} \quad (10.20)
\end{aligned}$$

$$\int_0^h dz\,\cos k_m z\cos k_{m'}z = \frac{h}{2}\delta_{mm'}. \quad (10.21)$$

If m' is odd, then the cosine transform of Eq.(10.18) becomes

$$0 = -4n\lambda^2 A_0 a^{n-1} e^{i\lambda h/2} \frac{\cos(\lambda h/2)}{\lambda^2 - k_{m'}^2}$$
$$+ A_{m'} \frac{k_{m'} h}{2} \sqrt{\lambda^2 - k_{m'}^2} J_n'(\sqrt{\lambda^2 - k_{m'}^2} a)$$
$$+ \sum_{m=\text{even}}^{\infty} A_m \frac{in\lambda}{a} \frac{2k_m}{k_m^2 - k_{m'}^2} J_n(\sqrt{\lambda^2 - k_m^2} a) \qquad (10.22)$$

while if m' is even, then this transform becomes

$$0 = 4in\lambda^2 A_0 a^{n-1} e^{i\lambda h/2} \frac{\sin(\lambda h/2)}{\lambda^2 - k_{m'}^2}$$
$$+ A_{m'} \frac{k_{m'} h}{2} \sqrt{\lambda^2 - k_{m'}^2} J_n'(\sqrt{\lambda^2 - k_{m'}^2} a)$$
$$+ \sum_{m=\text{odd}}^{\infty} A_m \frac{in\lambda}{a} \frac{2k_m}{k_m^2 - k_{m'}^2} J_n(\sqrt{\lambda^2 - k_m^2} a). \qquad (10.23)$$

To illustrate how this infinite system of equations is solved, the system is truncated by discarding all A_m terms with $m > 2$. Using $m' = 1$ and $m' = 3$ Eq.(10.22) gives equations of the form

$$\alpha_1 A_0 + \beta_1 A_1 + \gamma_1 A_2 = 0 \qquad (10.24)$$
$$\alpha_3 A_0 + \gamma_3 A_2 = 0 \qquad (10.25)$$

while using $m' = 2$ Eq.(10.23) gives

$$\alpha_2 A_0 + \beta_2 A_1 + \delta_2 A_2 = 0. \qquad (10.26)$$

These constitute three homogeneous equations in the three unknowns A_0, A_1, and A_2. The determinant of this system gives an eigenvalue equation for λ as a function of h and a, and if all lengths are normalized to a, this eigenvalue relation becomes λa as a function of h/a. Bondeson et al. truncated the system at 10 terms to generate a system of 10 equations in 10 unknowns. By plotting the eigenvalue λa versus h/a for the $n = 1$ mode, it was found that the $n = 1$ eigenvalue is lower than the $n = 0$ eigenvalue when h/a exceeds 1.67, a result determined independently by Finn et al. and not too different from the $h > 1.8$ instability threshold determined semi-quantitatively from the $q = 1$ condition on the magnetic axis.

For fusion applications it would be desirable to eliminate a close-fitting wall and much effort has been expended trying to eliminate the wall. The Princeton group

10.3 Tilt stability

[115, 116, 117] experimented with non-axisymmetric figure-eight coils which would provide stabilization fields, but found that at best, these would merely slow down the tilt instability. Similar results were obtained using wall segments that were supposed to stabilize via line tying. The Princeton group[118] found that tilt stability could be achieved by inserting a funnel-shaped flux conserver that was essentially inwardly dimpled cylinder ends with no side walls. This is almost the same as a complete enclosing wall. It is possible that if much larger spheromaks are constructed, the tilt stability would become slower because of the larger moment of inertia and could then be easily stabilized by an active electronic feedback system.

By having close-fitting walls with $h/a < 1.8$ the Los Alamos CTX spheromak did not encounter difficulties with tilt instability. However, Barnes et al. [119] found that particle confinement could be improved by adding a small uniform magnetic field produced by external coils. This extra field pushed the spheromak separatrix slightly inwards from the wall and so reduced contact between the plasma and the wall. The result was a compromise between having no wall at all and having a wall right against the separatrix. The compromise apparently achieved the best of both worlds, i.e., (i) improved confinement by avoiding direct contact between the wall and closed flux surfaces, and (ii) tilt stability because of reasonably close proximity of a wall.

CHAPTER 11
Analysis of Driven Spheromaks: Strong Coupling

In this chapter we consider a spheromak tightly coupled to a helicity source having specified λ. This situation is analogous to the thermodynamic system in thermal contact with a constant temperature bath shown in Fig. 8.1(b). The thermodynamic system comes to the same temperature as the bath, has negligible temperature gradients, and negligible net heat flux in equilibrium. The MHD system is similarly assumed to equilibrate to the same λ as the source. However, we also permit the existence of a small λ gradient and corresponding small helicity flux to make up for a small distributed helicity dissipation. Existing spheromak experiments appear to be too lossy to be characterized by this strong coupling model and are more consistent with the weak coupling situation discussed in the next chapter; in other words, the assumption of nearly uniform λ is not a good characterization for existing experiments. It is nevertheless quite instructive to analyze the strong coupling case because it can be solved in closed form with modest effort and the solutions reveal important conceptual relationships. Furthermore, it would be desirable to arrange an experiment to be in the strong coupling regime if possible.

The analysis presented here differs in mathematical approach from the methods used by Turner[120] and by Jensen and Chu[39], but the essential result is the same. Turner solved the inhomogeneous force-free equation using a Green's function method; Jensen and Chu expressed the inhomogeneous solution as a summation of the eigenfunctions of the homogenous solution (this leads to the complication of a Gibbs phenomenon at the boundary[121]). We also note that Dixon et al.[122] examined the inhomogeneous problem in spherical geometry with non-axisymmetry taken into account. Here we solve the inhomogeneous force-free equation directly for arbitrary inhomogeneous boundary conditions assuming cylindrical geometry and axisymmetry.

In the strong coupling situation, λ is imposed by the combination of the gun bias flux and power supply current and it is assumed that the configuration settles down to have the same λ everywhere. Relative helicity, relative energy, and impedance are all dependent parameters. Because a small amount of helicity dissipation has been assumed and the system is stationary, one may imagine that after a steady state has been obtained, a great deal of helicity has been injected into the system and a great deal has been dissipated. The amount of helicity stored in the system self-adjusts so as to be consistent with the requirements of the stationary state. In contrast, for an isolated system, the helicity was set by initial conditions. Furthermore, impedance only has meaning for a driven system, since impedance is the ratio of the externally applied voltage to the current driven by this voltage.

Spheromak formation in a driven system occurs when the current flowing along the field lines attached to the gun electrodes exceeds a threshold determined by the bias flux and the geometry. When this threshold is exceeded, the field line topology changes discontinuously even though both the magnetic field and the magnetic flux change continuously. These concepts will now be discussed in detail.

11.1 Force-free equilibria with open field lines

We consider an axisymmetric plasma contained by a cylinder of length h and radius a; this arrangement is shown in Fig.11.1. The end wall at $z = 0$ corresponds to the coaxial gun and will be called the "gun end wall". The cylinder walls are flux conserving, but there is an annular gap at radius r_s in the gun end wall. Because of this gap, the gun end wall is segmented, consisting of a central disk surrounded by an outer annulus. This idealized model is topologically equivalent to coaxial gun spheromaks, and in particular, the central disk and outer annulus correspond respectively to the gun inner and outer electrode.

A power supply having open-circuit voltage V_{src} and internal resistance R_{src} is connected between the central disk and outer annulus. This power supply drives a current I_{gun} into the central disk and the return current flows back to the power supply from the outer annulus. Because of the finite internal resistance of the power supply, the voltage V_{gun} which appears across the gap differs from V_{src} by the voltage drop $I_{gun}R_{src}$.

An external coil creates a vacuum poloidal field $B_z^{gun}(r)$ which penetrates the gun end wall; the surface integral of this field gives the end wall flux function

$$\psi_g(r) = 2\pi \int_0^r B_z^{gun}(r) r dr. \qquad (11.1)$$

Any field line that enters the gun end wall also exits the gun end wall so there is no net flux through the gun end wall, i.e., $\psi_g(a) = 0$. The gun magnetic field $B_z^{gun}(r)$ reverses polarity at some radius r_s; for simplicity we assume that the gun has been designed so that the gap is located where this polarity reversal occurs. Thus all magnetic field lines that go from the central disk into the cylindrical volume exit through the outer annulus. The total flux intercepting the central disk is

$$\bar{\psi}_{gun} = \psi_g(r_s) = 2\pi \int_0^{r_s} B_z^{gun}(r) r dr; \qquad (11.2)$$

11.1 Force-free equilibria with open field lines

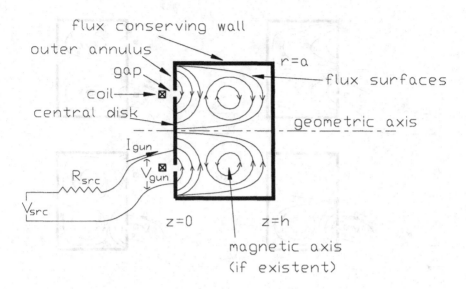

Fig.11.1 Experimental setup for a driven axisymmetric cylindrical spheromak. Poloidal flux surfaces and direction of poloidal magnetic field are shown. For clarity, the vacuum magnetic field outside of the plasma-filled cylinder is not shown.

this "gun flux" (also called bias flux) is equal in magnitude and opposite in polarity to the return flux intercepted by the outer annulus.

The system is assumed to be in an axisymmetric relaxed state and so will be described by the force-free equation with uniform λ, i.e.,

$$\nabla \times \mathbf{B} = \lambda \mathbf{B}. \tag{11.3}$$

The solutions to this equation can have various topologies as shown in Fig.11.2; boundary conditions determine which topology occurs. There are two distinct topologies for the field line projections shown in Fig.11.2, open and closed. Open field lines intercept the wall, closed field lines do not. It is seen that closed field lines may or may not exist.

The parameter λ can be expressed in terms of the gun end wall boundary conditions; this expression is determined by integrating the z component of Eq.(11.3) over the central disk giving

$$\lambda = \frac{\mu_0 I_{gun}}{\bar{\psi}_{gun}} \tag{11.4}$$

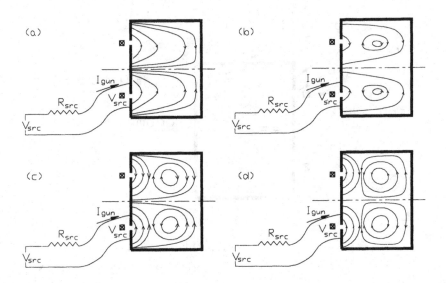

Fig.11.2 Sequence of poloidal flux toplogies as λ is increased from zero: (a) λ is zero or very small, so that the field is either a vacuum field or else very nearly a vacuum field, (b) the first flux extrema form giving closed field lines (o-points or equivalently magnetic axes), (c) the region of closed field lines nearly fills up the volume, (d) λ is slightly above the value producing a resonance so that a flipped spheromak is formed.

where

$$I_{gun} = 2\pi \int_0^{r_s} J_z(r) r dr \tag{11.5}$$

is the current from the power supply.

The general solution to Eq.(11.3) is

$$\mathbf{B} = \lambda \nabla \chi \times \nabla z + \nabla \times (\nabla \chi \times \nabla z) \tag{11.6}$$

where $\chi(r,z)$ is a scalar function satisfying the Helmholtz equation

$$\nabla^2 \chi + \lambda^2 \chi = 0; \tag{11.7}$$

this solution may be verified by substituting Eq.(11.6) into Eq.(11.3). By expressing Eq.(11.6) as

$$\mathbf{B} = \lambda \nabla \chi \times \hat{z} + \hat{z} \cdot \nabla \nabla \chi - \hat{z} \nabla^2 \chi$$

11.1 Force-free equilibria with open field lines

$$= \lambda \nabla \chi \times \hat{z} + \nabla \partial \chi / \partial z + \hat{z} \lambda^2 \chi \tag{11.8}$$

it is seen that

$$\lim_{\lambda \to 0} \mathbf{B} = \nabla \partial \chi / \partial z \tag{11.9}$$

so the $\lambda = 0$ limit is indeed a vacuum field.

In order to solve Eq.(11.6) using the method of separation of variables, we assume $\chi \sim \exp(\pm\sqrt{\gamma^2 - \lambda^2} z)$ in which case Eq.(11.7) becomes

$$\frac{1}{r} \frac{\partial}{\partial r} r \frac{\partial \chi}{\partial r} + \gamma^2 \chi = 0. \tag{11.10}$$

Equation (11.10) is a Bessel's equation of order zero and so χ has the form

$$\chi \sim J_0(\gamma r) e^{\pm \sqrt{\gamma^2 - \lambda^2} z}. \tag{11.11}$$

From Eq.(11.6) the components of the magnetic field are

$$B_r = \frac{\partial^2 \chi}{\partial r \partial z} \tag{11.12a}$$

$$B_\phi = -\lambda \frac{\partial \chi}{\partial r} \tag{11.12b}$$

$$B_z = \gamma^2 \chi. \tag{11.12c}$$

It is assumed that no magnetic field lines intercept either the cylinder side wall or the end wall at $z = h$; magnetic field lines intercept only the gun end wall. Since the walls are all assumed to be flux conserving, these initial conditions must hold at all times and so the boundary conditions which must be satisfied by solutions to Eq.(11.3) are:

1. $B_r(a, z) = 0$,
2. $B_z(r, h) = 0$,
3. $B_z(r, 0) = B_z^{gun}(r)$.

Application of radial boundary condition:

Boundary condition 1 shows that the radial magnetic field vanishes at $r = a$ for all z. From Eqs.(11.11) and (11.12a) it is seen that this requires $J_0'(\gamma a) = 0$. Since $J_0'(s) = -J_1(s)$ this means that γa is a root of J_1. Denoting the nth root of J_1 as x_{1n}, the allowed values of γ are $\gamma_n = x_{1n}/a$.

Application of axial boundary conditions:

Equation (11.12c) shows that χ must satisfy boundary conditions similar to B_z and, in particular, χ must vanish when B_z vanishes and χ will be finite when B_z is finite.

Since boundary conditions 2 and 3 show that B_z vanishes at $z = h$ but can be finite at $z = 0$, we find it convenient to introduce the axial function

$$S_n(z, \lambda) = -\frac{\sinh\left(\sqrt{\gamma_n^2 - \lambda^2}(z - h)\right)}{\sinh\left(\sqrt{\gamma_n^2 - \lambda^2}h\right)} \tag{11.13}$$

from which χ will be constructed. S_n has been defined to have the following properties:

1. S_n is a linear combination of $\exp(\pm\sqrt{\gamma_n^2 - \lambda^2}z)$ and so incorporates the assumed axial dependence of χ.
2. $S_n(h, \lambda) = 0$ and so incorporates the boundary condition $B_z = 0$ at $z = h$.
3. $S_n(0, \lambda) = 1$ in order to allow finite B_z at $z = 0$.
4. When $\lambda^2 > \gamma_n^2$ then $S_n(z)$ can be written in the equivalent form

$$S_n(z, \lambda) = -\frac{\sin\left(\sqrt{\lambda^2 - \gamma_n^2}(z - h)\right)}{\sin\left(\sqrt{\lambda^2 - \gamma_n^2}h\right)} \tag{11.14}$$

because $\sinh(ix) = i \sin x$.
5. If $\lambda^2 \simeq \gamma_n^2$ then

$$S_n(z, \lambda) \simeq -\frac{(z - h)}{h} \tag{11.15}$$

showing that the transition from sinh-like behavior to sine-like behavior is smooth and continuous as λ^2 increases from being smaller than γ_n^2 to being larger,
6. If $\lambda^2 > \gamma_n^2$, then S_n is sine-like and becomes resonant when $\sqrt{\lambda^2 - \gamma_n^2}h = m\pi$ where m is an integer, but if $\lambda^2 < \gamma_n^2$, then S_n is sinh-like, never resonant, and in fact is a monotonically decreasing function of z.

Thus the general form of χ satisfying the $r = a$ and $z = h$ homogeneous wall boundary conditions is

$$\chi = \frac{\bar{\psi}_{gun}}{2\pi} \sum_{n=1}^{\infty} c_n J_0(\gamma_n r) S_n(z, \lambda) \tag{11.16}$$

where the c_n are coefficients to be determined from the inhomogeneous boundary conditions at the $z = 0$ end wall. The constant factor $\bar{\psi}_{gun}/2\pi$ has been inserted to make the c_n dimensionless so that flux functions (to be calculated later) will be normalized to the gun flux. From Eq.(11.12a-11.12c) the magnetic field components are

$$B_r(r,z) = -\frac{\bar{\psi}_{gun}}{2\pi}\sum_{n=1}^{\infty} c_n \gamma_n J_1(\gamma_n r)\frac{\partial S_n(z,\lambda)}{\partial z}$$

$$B_\phi(r,z) = \lambda\frac{\bar{\psi}_{gun}}{2\pi}\sum_{n=1}^{\infty} c_n \gamma_n J_1(\gamma_n r) S_n(z,\lambda)$$

$$B_z(r,z) = \frac{\bar{\psi}_{gun}}{2\pi}\sum_{n=1}^{\infty} c_n \gamma_n^2 J_0(\gamma_n r) S_n(z,\lambda) \qquad (11.17)$$

so that, as required, B_r vanishes at $r = a$ and B_z vanishes at $z = h$. The c_n may now be determined using boundary condition (3) and Eq.(11.17) to give

$$B_z^{gun}(r) = \frac{\bar{\psi}_{gun}}{2\pi}\sum_{n=1}^{\infty} c_n \gamma_n^2 J_0(\gamma_n r). \qquad (11.18)$$

We now use the first of the Bessel orthogonality relations derived in Appendix 2 (see also [123]),

$$\int_0^a r J_0(x_{1,n'}r/a) J_0(x_{1,n}r/a) dr = \frac{a^2}{2}[J_0'(x_{1,n})]^2 \delta_{nn'}, \qquad (11.19)$$

to solve for c_n giving

$$c_n = \frac{4\pi}{[x_{1n}J_0'(x_{1n})]^2}\frac{\int_0^a B_z^{gun}(r) J_0(x_{1n}r/a) r dr}{\bar{\psi}_{gun}}. \qquad (11.20)$$

The solution to Eq.(11.3) is now completely determined and depends on just two quantities, $B_z^{gun}(r)$ and λ. The physical meaning of these quantities will now be discussed.

$B_z^{gun}(r)$ is established by the external coil shown schematically in Fig.11.1 and is constant in time so the c_n are also constant. We consider a time-dependent problem, but with the time-scale for changes in current to be so slow that the system is always in a force-free relaxed state. Thus, creating a spheromak involves ramping up the current I_{gun} on a time scale slow compared to any plasma instabilities, so that the plasma evolves through a sequence of relaxed states. Since $\bar{\psi}_{gun}$ is constant (because B_z^{gun} is constant), λ is proportional to I_{gun} and the slow increase of I_{gun} causes the plasma to evolve through a sequence of equilibria with successively larger λ. Each equilibrium is a distinct relaxed state described by Eq.(11.3). It must be emphasized that there is no presumption that the magnetic topology evolves smoothly as λ increases;

in fact, discontinuous bifurcations of field line topology can and do occur as λ increases. These discontinuous changes are analogous to the phase transformation which can occur in a thermodynamic system when temperature is changed. For example, thermodynamic equilibria can freeze or melt for slight changes in temperature; complex non-equilibrium dynamics produce these changes. The reconnection processes which create the closed flux surfaces of a relaxed MHD state correspond approximately to the non-equilibrium dynamics which produce a thermodynamic phase transformation.

When $\lambda = 0$, the plasma is current-free and $\mathbf{B}(r,z)$ is a vacuum field. As λ is increased above zero, currents start to flow in the plasma, and for small λ the magnetic field differs only slightly from the initial vacuum field. Thus for small λ, all field lines start at the central disk, enter the plasma volume, turn around inside, and then exit through the outer annulus. In other words, when λ is small, all field lines are open and deform continuously as λ is increased.

When λ increases to the point that it exceeds a particular γ_n, the associated S_n function changes from being sinh-like to being sine-like; at this point the magnetic topology associated with γ_n starts changing. These changes are best examined using the poloidal flux function introduced in the next section.

11.2 Flux surfaces

Because axisymmetry is assumed here, the magnetic field derived in the previous section has an associated flux function (if the configuration had been non-axisymmetric, then the solution to the Helmholtz equation would also be non-axisymmetric and typically, flux functions would not exist).

The axisymmetric solution to the force-free equation can be expressed in terms of flux surfaces as

$$\mathbf{B} = \frac{1}{2\pi} \left(\nabla \psi \times \nabla \phi + \lambda \psi \nabla \phi \right) \tag{11.21}$$

where ψ is the poloidal flux through a circle of radius r at axial location z (see Sec.2.6). As before, the toroidal component of $\mathbf{B}$ is specified to satisfy $\nabla \times \mathbf{B}_{tor} = \lambda \mathbf{B}_{pol}$.

Comparison of Eq.(11.12b) and the toroidal component of Eq.(11.21) shows that

$$\psi = -2\pi r \frac{\partial \chi}{\partial r} = \frac{2\pi r}{\lambda} B_\phi; \tag{11.22}$$

thus Eq.(11.16) immediately gives

$$\psi(r,z) = \bar{\psi}_{gun} \sum_{n=1}^{\infty} c_n \gamma_n r J_1(\gamma_n r) S_n(z,\lambda) \tag{11.23}$$

11.2 Flux surfaces

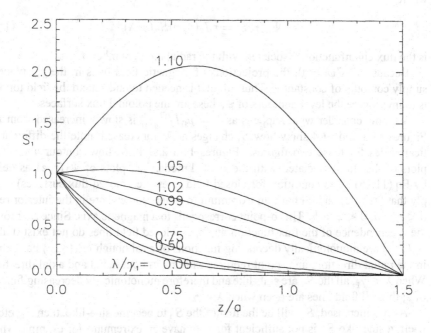

Fig.11.3 S_1 profiles for λ increasing from 0 to 1.10; note highly nonlinear dependence on λ. Here $h/a = 1.5$ has been assumed.

so it is unnecessary to make a separate investigation of the differential equation governing ψ. It is clear that ψ vanishes at both $r = 0$ and $r = a$. The vacuum flux satisfying the same boundary conditions as the actual flux is found by simply setting λ to zero; i.e.,

$$\psi_{vac}(r,z) = \bar{\psi}_{gun} \sum_{n=1}^{\infty} c_n \gamma_n r J_1(\gamma_n r) S_n(z,0). \quad (11.24)$$

For reference purposes, it is convenient to write Eq.(11.23) in the form

$$\psi(r,z) = \bar{\psi}_{gun} \sum_{n=1}^{\infty} c_n \gamma_n \Psi_n(r,z) \quad (11.25)$$

where

$$\Psi_n(r, z) = rJ_1(\gamma_n r)S_n(z, \lambda) \quad (11.26)$$

is the flux eigenfunction associated with the radial mode number n.

Because $\mathbf{B} \cdot \nabla \psi = 0$, the projections of magnetic field lines in the r, z plane are simply contours of constant ψ. Thus, all that is needed to understand the field topology is to investigate the level contours of ψ; these are the poloidal flux surfaces.

We now consider what happens as $\lambda = \mu_0 I/\bar{\psi}_{gun}$ is slowly increased from zero. Figures 11.3 and 11.4 show how S_1 changes as λ is increased; note the different vertical scales for these two figures. Figures 11.5 and 11.6 show contour and altitude plots of the flux associated with the $n = 1$ mode, i.e., plots of $\Psi_1(r, z)$ as defined by Eq.(11.26). The condition for closed field lines (i.e., closed flux surfaces) is simply that $\Psi_1(r, z)$ should have an extremum (o-point) somewhere in the interior region $0 < r < a, 0 < z < h$. This o-point corresponds to a magnetic axis. Since S_n provides the z-dependence of the flux function $\Psi_n(r, z)$, closed field lines do not exist if all the $S_n(z, \lambda)$ are monotonically decreasing functions of z (although $rJ_1(\gamma_n r)$ has extrema in the radial direction, an o-point requires extrema in both radial and axial directions). When $\lambda < \gamma_1$, all the S_n are sinh-like and therefore monotonically decreasing functions of z; thus all field lines are open when $\lambda < \gamma_1$.

As λ is increased, S_1 will be the first of the S_n to become sine-like, then S_2, etc. By itself, a sine-like S_n is not sufficient for ψ to have an extremum; for example when λ is just slightly larger than γ_n, S_n is sine-like but is nearly a straight line having a value of unity at $z = 1$ and a value of zero at $z = h$. As λ increases, S_n develops an upward bulge and only when $\partial S_n/\partial z = 0$ does S_n have a local maximum. From Eq.(11.14) it is seen that for each S_n, this first extremum occurs when

$$\lambda = \sqrt{\gamma_n^2 + \frac{\pi^2}{4h^2}}. \quad (11.27)$$

The extremum will be at $z = 0^+$ and will move towards larger z as λ is increased. The actual location of the extremum is

$$z = h - \frac{\pi}{2\sqrt{\lambda^2 - \gamma_n^2}}. \quad (11.28)$$

The extremum reaches the axial midpoint $z = h/2$ when

$$\lambda = \sqrt{\gamma_n^2 + \frac{\pi^2}{h^2}} \quad (11.29)$$

11.2 Flux surfaces

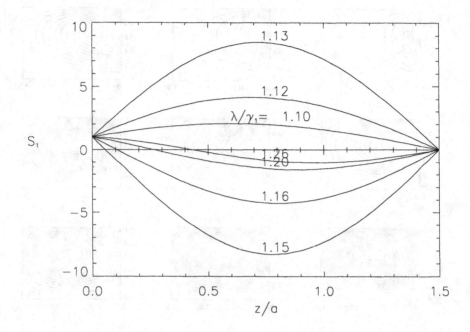

Fig.11.4 Continuation of dependence of S_1 on λ. When λ increases from 1.13 to 1.15, there is a resonance in S_1. When λ is above the critical value to give this resonance, S_1 is bipolar which corresponds to a flipped spheromak. Note the substantial change in vertical scale compared to the previous figure.

and when λ slightly exceeds this value, S_n passes through zero at $z = 0^+$. Because S_n was defined to be unity at $z = 0$, this forces S_n to become very large (resonant) when the extremum is near $h/2$. Resonance occurs because the denominator in Eq.(11.14) vanishes when λ satisfies Eq.(11.29).

Although this sequencing of S_n is true for all the n modes, resonant behavior happens first for the $n = 1$ mode. Thus, as λ increases from zero, the behavior of λ with respect to γ_1 is the dominant consideration. If $\lambda = \sqrt{\gamma_1^2 + \pi^2/h^2} - \varepsilon$ then the extremum will be large and positive, but if $\lambda = \sqrt{\gamma_1^2 + \pi^2/h^2} + \varepsilon$ the extremum will be large and negative and there will be a zero-crossing of S_n at $z = 0^+$. Since the poloidal flux is proportional to S_n it is clear that when there is an extremum (i.e., an o-point or, equivalently, a magnetic axis), the flux is larger at the extremum than at the gun.

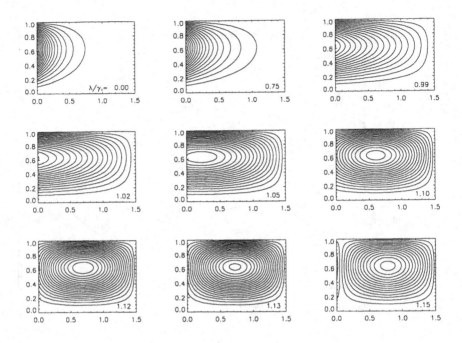

Fig.11.5 Contours of flux function $\Psi_1(r,z)$ for sequence of increasing λ; the cylinder has $h/a = 1.5$ and horizontal axis is z/a while the vertical axis is r/a. The lower right contour shows a flipped spheromak.

If the gun dimensions are small compared to a, i.e., $B_z^{gun}(r)$ is finite over a region small compared to a, then the Bessel transform of $B_z^{gun}(r)$ will contain a broad c_n spectrum and so the proportion of the spectral power contained in c_1 will be small. Nevertheless, as λ is increased from zero, the first S_n to become resonant will still be S_1. Thus, no matter what the specific radial dependence of $\psi^{gun}(r)$ happens to be, all guns will dominantly excite S_1 as λ is increased from zero provided there is the slightest amount of c_1. When λ is small (so that all the S_n are sinh-like) the S_n with large n decay rapidly with z and so the fields will be concentrated near the gun. Then, as λ is increased, the fields abruptly fill the volume as soon as $\lambda \simeq \sqrt{\gamma_1^2 + \pi^2/h^2}$. This sudden filling occurs because when λ approaches this critical value, S_1 has sine-like behavior and unlike the sinh function with its exponential behavior, the sine function fills up the volume. The behavior of the flux will be very sensitive in this region and as seen from Fig.(11.4) and from the denominator in Eq.(11.14), the bulk of the flux has reversed polarity for further small increase in λ. When this occurs, all fields also reverse polarity, but the helicity remains the same since helicity is related to the product of poloidal and toroidal fields.

This resonant buildup of flux and then flipping is shown in Figs. 11.5 and 11.6; the former is a contour plot of flux, while the latter is a surface plot of the same flux. In each of these figures, λ/γ_1 is increased through a sequence of nine values and the corresponding flux profile is plotted. The critical value for flipping is $\lambda/\gamma_1 = \sqrt{1 + \pi^2 a^2/x_{11}^2 h^2} = 1.14$ for the ratio $h/a = 1.5$ which has been used here.

Specifically, for $\lambda/\gamma_1 = \sqrt{1 + \pi^2 a^2/x_{11}^2 h^2} - \varepsilon$, the topology will be closed flux surfaces surrounded by a few open flux surfaces (cf. Fig.11.2(c) and lower middle contour in Fig.11.5), while for $\lambda/\gamma_1 = \sqrt{1 + \pi^2 a^2/x_{11}^2 h^2} + \varepsilon$ the topology will be open flux surfaces separated from the gun by a separatrix at the z location where S_1 vanishes (cf. Fig.11.2(d) and lower right contour in Fig.11.5). The former case corresponds to a spheromak surrounded by open field lines (i.e., the $\lambda/\gamma_1 = 1.13$ case in Figs. 11.5 and 11.6), while the latter case corresponds to a flipped spheromak (i.e., the $\lambda/\gamma_1 = 1.15$ case in Figs. 11.5 and 11.6). Both flipped and unflipped spheromaks were observed[62] in the CTX spheromak, but it should be stated that the flipped spheromaks were not interpreted in terms of the model presented here.

Yee and Bellan[124] have made high speed photographs of the plasma emerging from a magnetized coaxial plasma gun. These photos can only be compared qualitatively to the model presented here because (i) there was no flux conserver for these photographs, so the bulging flux tubes had no wall to push against and (ii) the system was not filled with plasma, so that λ was non-uniform and maximum in the vicinity of the gun. Example photographs shown in Fig.11.7 indicate the beginnings of a toroidal twist to the poloidal flux tubes emanating from the center electrode and bending back to the outer electrode.

11.3 Safety factor variation with lambda

The safety factor $q(\psi)$ only has meaning for closed flux surfaces and so it only makes sense to consider q when λ has exceeded the threshold for the first closed flux surfaces to form. In Sec.9.3 it was shown that for a cylindrical flux $\psi(r,z) \sim \gamma_1 r J_1(\gamma_1 r) \sin\left(\sqrt{\lambda^2 - \gamma_1^2}(z-h)\right)$ the safety factor has the form

$$q_{axis} = \frac{1}{x_{01}\sqrt{1 - \gamma_1^2/\lambda^2}}. \tag{11.30}$$

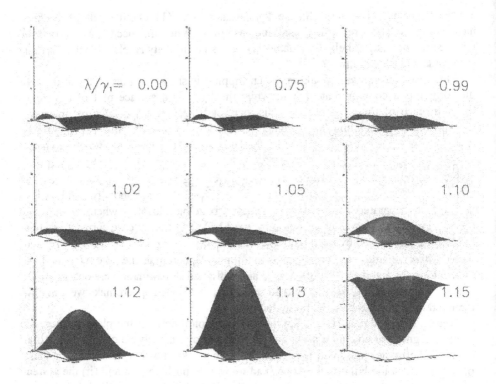

Fig.11.6 Contours of flux function $\Psi_1(r,z)$ for sequence of increasing λ; the cylinder has $h/a = 1.5$ and horizontal axis is z/a while the vertical axis is r/a. The lower right contour shows a flipped spheromak.

This functional form shows that q_{axis} decreases with increasing λ. Equations (11.27) and (11.29) show that for closed flux surfaces λ/γ_1 lies in the range

$$1 + \frac{\pi^2 a^2}{4x_{11}^2 h^2} < \frac{\lambda^2}{\gamma_1^2} < 1 + \frac{\pi^2 a^2}{x_{11}^2 h^2} \qquad (11.31)$$

where the lower limit corresponds to the formation of the first closed flux surface and the upper limit corresponds to approaching the first resonance. This range is parameterized by the ratio h/a and for larger h/a the extent of the range is smaller. Figure (11.8) plots $q_{axis}(\lambda/\gamma_1)$ from Eq.(11.30) for four different values of h/a and shows how q_{axis} decreases as λ increases and also that for h/a smaller than the tilt stability condition, q_{axis} passes through unity as λ is increased. Thus, one would expect to see

11.3 Safety factor variation with lambda

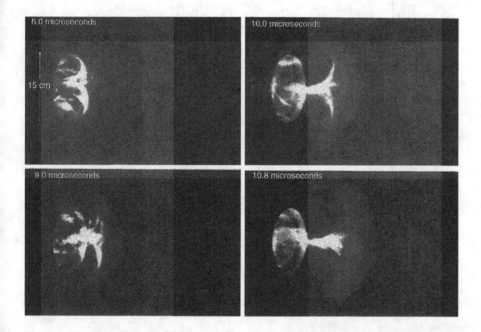

Fig.11.7 Poloidal flux bulging outwards from a coaxial magnetized plasma gun as gun current increases with time (plasma breakdown at $t = 0$); the flux tubes start to swirl because of the toroidal magnetic field produced by the increasing poloidal current (from Yee and Bellan[124]).

$m = 1$, $n = 1$ mode activity develop on the magnetic axis as λ is increased (here m, n are the mode numbers of an MHD fluctuation).

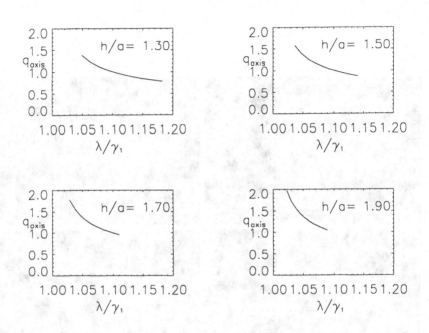

Fig.11.8 Dependence of q_{axis} on λ.

11.4 Flux amplification

A curious feature (first observed experimentally by Alfvén et al.[8] and by Lindberg et al.[9]) is that the poloidal flux in the flux conserver can exceed the poloidal flux imposed by the external coil. While this seems to violate the concept of flux conservation, it is not a cause of concern because no promise of *continuous* evolution of topology has been made. Flux amplification is immediately evident in Fig.11.6 from the resonant behavior of the peak of ψ as λ approaches $\sqrt{\gamma_1^2 + \pi^2/h^2}$. In fact, in order to have closed field lines, it is essential for the poloidal flux inside the volume to exceed its value at the gun (i.e., at $z = 0$). Flux amplification begins when λ exceeds $\sqrt{\gamma_1^2 + \pi^2/4h^2}$, the condition for a local flux maximum to first occur. If there is no maximum then the flux decays monotonically from the gun.

11.5 Relative helicity

Because some field lines are open, the relative helicity defined in Eq.(3.34) must be used, i.e., to avoid gauge ambiguity, the helicity accounting must refer to a vacuum magnetic field satisfying the same boundary conditions as the actual field. The vacuum solution with the prescribed boundary conditions is just the $\lambda = 0$ version of Eq.(11.21), namely

$$\mathbf{B}_{vac} = \frac{1}{2\pi} \nabla \psi_{vac} \times \nabla \phi = \frac{1}{2\pi} \nabla \times (\psi_{vac} \nabla \phi) \qquad (11.32)$$

where ψ_{vac} is the poloidal flux produced by the external coil.

Using Eq.(11.8) the vacuum magnetic field can also be expressed as

$$\mathbf{B}_{vac} = \nabla \frac{\partial \chi_{vac}}{\partial z} \qquad (11.33)$$

while using Eq.(9.29) the vacuum vector potential is

$$\mathbf{A}_{vac} = \frac{1}{2\pi} \psi_{vac} \nabla \phi + \nabla g \qquad (11.34)$$

where g is an arbitrary gauge function.

The relative helicity defined by Eq.(3.34) may be computed using Eq.(11.21) to give $\mathbf{B}$ and $\mathbf{B}_{vac}$ and Eq.(9.29) to give $\mathbf{A}$ and $\mathbf{A}_{vac}$ so that

$$\begin{aligned} K_{rel} &= \int d^3r \, (\mathbf{B} - \mathbf{B}_{vac}) \cdot (\mathbf{A} + \mathbf{A}_{vac}) \\ &= \int d^3r \left\{ \left[\nabla \left(\frac{\psi - \psi_{vac}}{2\pi} \right) \times \nabla \phi + \lambda \frac{\psi}{2\pi} \nabla \phi \right] \right. \\ &\qquad \left. \cdot \left[\frac{(\psi + \psi_{vac})}{2\pi} \nabla \phi + \lambda \chi \nabla z + \nabla (f + g) \right] \right\} \end{aligned} \qquad (11.35)$$

where f and g are gauge potentials for the actual and vacuum fields respectively. The term involving the gauge potentials is of the form $\int d^3r \, (\mathbf{B} - \mathbf{B}_{vac}) \cdot \nabla (f + g) = \int_S d\mathbf{s} \cdot [(\mathbf{B} - \mathbf{B}_{vac}) (f + g)]$ and so vanishes because by assumption $d\mathbf{s} \cdot (\mathbf{B} - \mathbf{B}_{vac}) = 0$. Multiplying out the remaining terms gives the relative helicity as

$$K_{rel} = \frac{\lambda}{2\pi} \int d^3r \left(\frac{\chi}{r} \frac{\partial}{\partial r} (\psi - \psi_{vac}) + \frac{\psi(\psi + \psi_{vac})}{2\pi r^2} \right). \qquad (11.36)$$

The first term in the integral can be integrated by parts and simplified using Eq.(11.22),

$$\int d^3r \frac{\chi}{r} \frac{\partial}{\partial r} (\psi - \psi_{vac}) = 2\pi \int_0^h dz \int_0^a dr \chi \frac{\partial}{\partial r} (\psi - \psi_{vac})$$

$$= -2\pi \int_0^h dz \int_0^a dr \frac{\partial \chi}{\partial r} (\psi - \psi_{vac})$$

$$= \int d^3r \frac{\psi}{2\pi r^2} (\psi - \psi_{vac}) \qquad (11.37)$$

so that Eq.(11.36) becomes

$$K_{rel} = 2\lambda \int d^3r \left(\frac{\psi}{2\pi r}\right)^2$$

$$= 2\lambda \left(\frac{\bar\psi_{gun}}{2\pi}\right)^2 \sum_{n,m} c_n c_m \gamma_n \gamma_m \times$$

$$\int d^3r J_1(\gamma_n r) J_1(\gamma_m r) S_n(z,\lambda) S_m(z,\lambda). \qquad (11.38)$$

The radial integral may be evaluated using the second Bessel orthogonality relation derived in Appendix B, namely

$$\int_0^a r J_1(x_{1n}r/a) J_1(x_{1m}r/a) dr = \frac{a^2}{2} [J_0(x_{1n})]^2 \delta_{nm}. \qquad (11.39)$$

Thus, the relative helicity is

$$K_{rel} = \lambda \frac{\bar\psi_{gun}^2}{2\pi} \sum_{n=1}^{\infty} [c_n x_{1n} J_0(x_{1n})]^2 \int_0^h dz S_n^2(z,\lambda)$$

$$= \frac{\bar\psi_{gun}^2}{4\pi} \sum_{n=1}^{\infty} [c_n x_{1n} J_0(x_{1n})]^2 F(\gamma_n h, \lambda h) \qquad (11.40)$$

where

$$F(\Gamma, \Lambda) = \Lambda \left[\frac{\cosh\left(\sqrt{\Gamma^2 - \Lambda^2}\right)}{\sqrt{\Gamma^2 - \Lambda^2} \sinh\left(\sqrt{\Gamma^2 - \Lambda^2}\right)} - \frac{1}{\sinh^2\left(\sqrt{\Gamma^2 - \Lambda^2}\right)} \right] \qquad (11.41)$$

gives the dependence of the relative helicity on $\Lambda = \lambda h$ and $\Gamma = \gamma h$. If $\Lambda > \Gamma$, then F can also be expressed using the relations $\sinh(ix) = i \sin x$, $\cosh(ix) = \cos x$ in the

11.5 Relative helicity

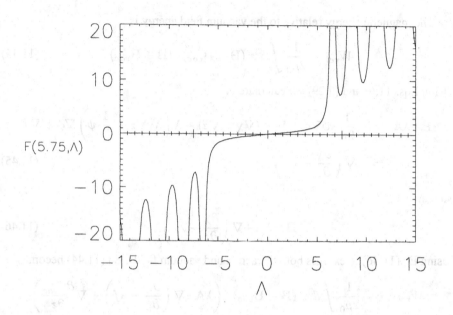

Fig. 11.9 Plot of $F(5.75, \Lambda)$ showing how helicity resonantly depends on λ.

equivalent form

$$F(\Gamma, \Lambda) = \Lambda \left[\frac{1}{\sin^2\left(\sqrt{\Lambda^2 - \Gamma^2}\right)} - \frac{\cos\left(\sqrt{\Lambda^2 - \Gamma^2}\right)}{\sqrt{\Lambda^2 - \Gamma^2} \sin\left(\sqrt{\Lambda^2 - \Gamma^2}\right)} \right]. \qquad (11.42)$$

From Eq.(11.42) it is seen that $F(\Gamma, \Lambda)$ has resonances whenever

$$\Lambda^2 = \Gamma^2 + m^2 \pi^2. \qquad (11.43)$$

The first and physically most important of these resonances occurs when $\lambda^2 = \gamma_1^2 + \pi^2/h^2$; this resonance is at the λ of the Taylor state. Figure 11.9 plots $F(5.75, \Lambda)$; the choice of $\Gamma = 5.75$ is relevant to the physically interesting situation of $h/a = 1.5$ and $\gamma_1 = 3.83/a$. The resonances are clearly visible; also it is seen that F is antisymmetric with respect to Λ and so K_{rel} is similarly antisymmetric.

11.6 Relative energy

The magnetic energy relative to the vacuum field energy is

$$W_{rel} = \frac{1}{2\mu_0} \int d^3r \, (\mathbf{B} - \mathbf{B}_{vac}) \cdot (\mathbf{B} + \mathbf{B}_{vac}). \tag{11.44}$$

Using Eqs.(11.6) and (9.29) we calculate

$$\begin{aligned}\mathbf{B} - \lambda\mathbf{A} &= \frac{1}{2\pi}\lambda\psi\nabla\phi + \nabla\times(\nabla\chi\times\nabla z) - \lambda\left(\lambda\chi\nabla z + \frac{1}{2\pi}\psi\right)\nabla\phi + \nabla f \\ &= \nabla\left(\frac{\partial\chi}{\partial z} - \lambda f\right)\end{aligned} \tag{11.45}$$

so that

$$\mathbf{B} = \lambda\mathbf{A} + \nabla\left(\frac{\partial\chi}{\partial z} - \lambda f\right). \tag{11.46}$$

Using Eq.(11.46) to express both the actual and vacuum fields, Eq.(11.44) becomes

$$\begin{aligned}W_{rel} &= \frac{1}{2\mu_0}\int d^3r\,(\mathbf{B}-\mathbf{B}_{vac})\cdot\left(\lambda\mathbf{A}+\nabla\left(\frac{\partial\chi}{\partial z}-\lambda f\right)+\nabla\frac{\partial\chi_{vac}}{\partial z}\right) \\ &= \frac{\lambda}{2\mu_0}\int d^3r\,(\mathbf{B}-\mathbf{B}_{vac})\cdot\mathbf{A} \\ &\quad +\frac{1}{2\mu_0}\int d\mathbf{s}\cdot\left[(\mathbf{B}-\mathbf{B}_{vac})\left(\frac{\partial\chi}{\partial z}-\lambda f+\frac{\partial\chi_{vac}}{\partial z}\right)\right] \\ &= \frac{\lambda}{2\mu_0}\int d^3r\,(\mathbf{B}-\mathbf{B}_{vac})\cdot\mathbf{A}.\end{aligned} \tag{11.47}$$

The surface integral vanishes because $(\mathbf{B}-\mathbf{B}_{vac})\cdot d\mathbf{s} = 0$ on all bounding surfaces.

The relative energy can be related to the relative helicity using Eq.(11.47) to give

$$\begin{aligned}W_{rel} &= \frac{\lambda}{2\mu_0}\left[K_{rel} - \int d^3r\,(\mathbf{B}-\mathbf{B}_{vac})\cdot\mathbf{A}_{vac}\right] \\ &= \frac{\lambda}{2\mu_0}\left[K_{rel} - \int d^3r\,\frac{\lambda\psi\psi_{vac}}{(2\pi r)^2}\right] \\ &= \frac{\lambda}{2\mu_0}\left[K_{rel} - \frac{\bar{\psi}_{gun}^2}{4\pi}\sum_{n=1}^{\infty}[c_n x_{1n}J_0(x_{1n})]^2\lambda\int_0^h dz\, S_n(z,\lambda)S_n(z,0)\right].\end{aligned} \tag{11.48}$$

The z integral can be evaluated to give

$$\lambda \int_0^h dz S_n(z,\lambda) S_n(z,0) = \frac{\Gamma \coth \Gamma - \sqrt{\Gamma^2 - \Lambda^2} \coth \sqrt{\Gamma^2 - \Lambda^2}}{\Lambda}. \quad (11.49)$$

Using Eqs.(11.40) and 11.41) the relative energy can be written as

$$W_{rel} = \frac{\bar{\psi}_{gun}^2}{8\pi h \mu_0} \sum_{n=1}^{\infty} [c_n x_{1n} J_0(x_{1n})]^2 G(\gamma_n h, \lambda h) \quad (11.50)$$

where

$$G(\Gamma, \Lambda) = \frac{\Gamma^2}{\sqrt{\Gamma^2 - \Lambda^2}} \coth \sqrt{\Gamma^2 - \Lambda^2} - \frac{\Lambda^2}{\sinh^2\left(\sqrt{\Gamma^2 - \Lambda^2}\right)} - \Gamma \coth \Gamma \quad (11.51)$$

if $\Gamma > \Lambda$ and

$$G(\Gamma, \Lambda) = -\frac{\Gamma^2}{\sqrt{\Lambda^2 - \Gamma^2}} \cot \sqrt{\Lambda^2 - \Gamma^2} + \frac{\Lambda^2}{\sin^2\left(\sqrt{\Lambda^2 - \Gamma^2}\right)} - \Gamma \coth \Gamma \quad (11.52)$$

if $\Lambda > \Gamma$.

Figure 11.10 plots $G(\Gamma, \Lambda)$ as a function of Λ for $\Gamma = 5.75$; this plot can be compared with Fig. 11.9. It is seen that W_{rel} is an even function of Λ and has resonances when $\Lambda = \sqrt{\Gamma^2 + m^2 \pi^2}$ in a fashion similar to K_{rel}. Each n mode will have its own set of resonances, but the resonance with the lowest λ will be the one corresponding to x_{11}. Since dramatic behavior occurs when λ approaches this first resonance, the first resonance is of most interest. The quantity $\Lambda = \lambda h$ is an independent variable determined by the gun current, gun flux, and the physical dimensions. If a very large λ is imposed by the gun, then there will result a very large W_{rel} and only a moderately large K_{rel} because W_{rel} is approximately a factor of λ larger than K_{rel}.

11.7 Gun efficiency

The external circuit can be changed by short circuiting (crowbarring) the gap so that the inner disk is connected to the outer annulus. In this case there is no longer external control over I_{gun} and λ becomes a free parameter. The system will then relax to the lowest allowed energy state, namely the state with the smallest allowed λ for the given helicity. This is akin to the isolated configuration described in Chapter 9, but differs slightly because here there are still open field lines intercepting the end wall, whereas the completely isolated configuration does not have open field lines. If the gun had driven the system to a λ exceeding the lowest allowed λ for the given helicity

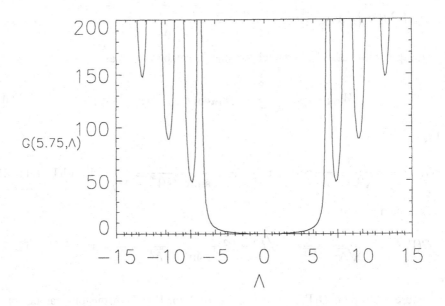

Fig.11.10 Plot of $G(5.75, \Lambda)$ showing how magnetic energy resonantly depends on λ.

then substantial energy would have to be dissipated for the system to relax to its lowest energy state.

Because helicity is conserved, the plasma retains whatever K_{rel} it had at the instant the gun was crowbarred; because K_{rel} has resonances, λ can be considered as a multi-valued function of K_{rel}. When the gun is crowbarred there will be a set of allowed λ's for the given K_{rel} and the system will relax to the state with the lowest λ in this set. The lowest λ will always be below the first resonance of K_{rel} and will be the Taylor state; i.e. the lowest energy for the helicity.

If the gun is designed to produce a driven configuration very similar to the Taylor state (i.e., c_1 is the predominant radial coefficient and λ is close to the first resonance), then very little energy will be dissipated when the gun is crowbarred and λ adjusts itself to the Taylor value. If not, then substantial energy will have to be dissipated when the gun is crowbarred. In particular, if the driven λ is such as to excite a resonance associated with a radial mode c_n with $n \neq 1$ then all the energy associated with this higher order mode will have to be dissipated in order to relax to the lowest energy mode. This indicates that guns with radial dimensions much smaller a will be inefficient, because in such a case most of the gun energy will go into driving modes with high n.

In reality, poorly matched guns will tend to drive large λ gradients, much like a single small, hot fire in a large house will produce much larger temperature gradients than a distributed, lower temperature heat source.

11.8 Gun impedance and load line

Equation (5.26) showed that the plasma impedance depended on both the total magnetic energy and on the open flux. Thus the impedance of a coaxial gun with uniform resistivity is

$$Z = \frac{2\mu_0 \eta}{\bar{\psi}_{gun}^2}(W_{rel} + W_{vac}). \qquad (11.53)$$

When λ is small, current flows along the vacuum field lines and the impedance is controlled by the geometry of the vacuum field. However, when $\lambda \to \sqrt{\gamma_1^2 + \pi^2/h^2}$ the impedance becomes resonant because W_{rel} becomes resonant.

In the driven system $\lambda = \mu_0 I_{gun}/\bar{\psi}_{gun}$ is determined by the external power supply via I_{gun} and by the flux boundary condition imposed at the gun end wall via $\bar{\psi}_{gun}$. The manner in which I_{gun} is determined depends on whether the external power supply acts as a current source or a voltage source. Because of the variable impedance of the spheromak, the same power supply will behave either as a current source or as a voltage source depending on the ratio of the spheromak impedance to the power supply impedance. Typically, the power supply impedance R_{src} is designed so that at small λ, the power supply impedance is much larger than the plasma impedance. In this case the power supply behaves as a constant current source so that $V_{gun} = I_{gun}Z$ is proportional to the plasma impedance. However, as λ approaches $\sqrt{\gamma_1^2 + \pi^2/h^2}$ so that Z starts to become resonant and hence larger than R_{src}, the power supply can no longer function as a current source. Now λ will adjust to a value consistent with both the power supply impedance and plasma impedance. There will be voltage spikes when λ is in the vicinity of a resonance because small changes in current result in large changes in impedance.

In order to make a quantitative determination of the impedance we now evaluate W_{vac} the energy associated with the vacuum field. We use Eq.(11.33) and the property $\nabla^2 \chi_{vac} = 0$ to express the vacuum energy as

$$W_{vac} = \frac{1}{2\mu_0} \int d^3r \nabla \frac{\partial \chi_{vac}}{\partial z} \cdot \nabla \frac{\partial \chi_{vac}}{\partial z}$$

$$= \frac{1}{2\mu_0} \int d^3r \nabla \cdot \left(\frac{\partial \chi_{vac}}{\partial z} \nabla \frac{\partial \chi_{vac}}{\partial z} \right)$$
$$= \frac{1}{2\mu_0} \int ds \cdot \left(\frac{\partial \chi_{vac}}{\partial z} \nabla \frac{\partial \chi_{vac}}{\partial z} \right). \quad (11.54)$$

Equation (11.16) shows that the vacuum Helmholtz function is

$$\chi_{vac} = \frac{\bar{\psi}_{gun}}{2\pi} \sum_{n=1}^{\infty} c_n J_0(\gamma_n r) S_n(z,0)$$
$$= -\frac{\bar{\psi}_{gun}}{2\pi} \sum_{n=1}^{\infty} c_n J_0(\gamma_n r) \frac{\sinh \gamma_n (z-h)}{\sinh \gamma_n h}. \quad (11.55)$$

The surface integral in Eq.(11.54) can be decomposed into the contributions from the side wall, the end wall at $z = h$, and end wall at $z = 0$ (i.e., the gun end wall). The side wall contribution contains a factor of the form $\hat{r} \cdot \nabla \partial \chi_{vac}/\partial z$ which is proportional to $J_0'(\gamma_n a) = -J_1(\gamma_n a) = 0$; thus the side wall terms all vanish. Bearing in mind that ds points in the negative z direction at the gun end wall and in the positive z direction at the $z = h$ end wall, the surface integral can be written as

$$W_{vac} = \frac{\pi}{\mu_0} \int_0^a rdr \left[\frac{\partial \chi_{vac}}{\partial z} \frac{\partial^2 \chi_{vac}}{\partial z^2} \right]_{z=0}^{z=h}. \quad (11.56)$$

However, $\partial^2 \chi_{vac}/\partial z^2$ vanishes at $z = h$ so only the contribution from the gun end wall survives. After evaluating the z-derivatives of χ at the gun end wall it is seen that

$$W_{vac} = \frac{\bar{\psi}_{gun}^2}{4\pi\mu_0} \sum_{m,n=1}^{\infty} c_n c_m \gamma_n \gamma_m^2 \coth \gamma_n h \int_0^a rdr J_0(\gamma_n r) J_0(\gamma_m r)$$
$$= \frac{\bar{\psi}_{gun}^2}{8\pi h \mu_0} \sum_{n=1}^{\infty} [c_n x_{1n} J_0(x_{1,n})]^2 \Gamma_n \coth \Gamma_n \quad (11.57)$$

where $\Gamma_n = \gamma_n h$ and Eq.(11.19) has been used.

Combining Eqs.(11.50) and (11.57) it is seen that the total magnetic energy is

$$W_{tot} = W_{rel} + W_{vac}$$
$$= \frac{\bar{\psi}_{gun}^2}{8\pi\mu_0 h} \sum_{n=1}^{\infty} [c_n x_{1n} J_0(x_{1n})]^2 \left(\frac{\Gamma_n^2 \coth \sqrt{\Gamma_n^2 - \Lambda^2}}{\sqrt{\Gamma_n^2 - \Lambda^2}} - \frac{\Lambda^2}{\sinh^2 \sqrt{\Gamma_n^2 - \Lambda^2}} \right)$$
$$\quad (11.58)$$

11.8 Gun impedance and load line

and thus the impedance is

$$Z = \frac{\eta}{4\pi h} \sum_{n=1}^{\infty} [c_n x_{1n} J_0(x_{1n})]^2 \left(\frac{\Gamma_n^2 \coth \sqrt{\Gamma_n^2 - \Lambda^2}}{\sqrt{\Gamma_n^2 - \Lambda^2}} - \frac{\Lambda^2}{\sinh^2 \sqrt{\Gamma_n^2 - \Lambda^2}} \right).$$
(11.59)

The operating point of a given experiment is determined by the relative magnitudes of the plasma impedance Z and the power supply impedance R_{src} (cf. Fig.11.1). From the point of view of the power supply circuit, the voltage applied to the spheromak will be the source voltage V_{src} less the drop across the series resistor, i.e.,

$$V_{gun} = V_{src} - I_{gun} R_{src}.$$
(11.60)

This provides a load-line description for the steady state gun current, since we also have the plasma impedance relation

$$V_{gun} = I_{gun} Z(I_{gun}).$$
(11.61)

The actual steady-state operating point for I_{gun} is where the V_{gun} provided by the power supply circuit matches the value required by the spheromak. This is found graphically by overlaying plots of V_{gun} v. I_{gun} calculated from both Eqs.(11.60) and (11.61) and identifying points of intersection; these will be the places where both equations give the same voltage and current. A convenient dimensionless parameter to use is

$$\Lambda = \lambda h = \mu_0 I_{gun} h / \bar{\psi}_{gun}$$
(11.62)

so the source circuit relation can be written as

$$V_{gun} = V_{src} - V_{drop} \Lambda$$
(11.63)

where

$$V_{drop} = \frac{\bar{\psi}_{gun} R_{src}}{h \mu_0}$$
(11.64)

is the voltage drop across the power supply series resistance when $\Lambda = 1$. The gun voltage can be written using Eq.(11.53) as

$$V_{gun} = V_{ref} \frac{\Lambda}{4\pi} \sum_{n=1}^{\infty} [c_n x_{1n} J_0(x_{1n})]^2 \left(\frac{\Gamma_n^2 \coth \sqrt{\Gamma_n^2 - \Lambda^2}}{\sqrt{\Gamma_n^2 - \Lambda^2}} - \frac{\Lambda^2}{\sinh^2 \sqrt{\Gamma_n^2 - \Lambda^2}} \right)$$
(11.65)

where

$$V_{ref} = \frac{\eta \bar{\psi}_{gun}}{\mu_0 h^2}. \qquad (11.66)$$

It is useful to normalize all voltages to V_{ref} so that the gun voltage as determined by the source circuit will be

$$\bar{V}_{gun} = \bar{V}_{src} - \bar{V}_{drop}\Lambda \qquad (11.67)$$

where a bar means that the voltage is normalized to V_{ref}. Similarly, the normalized gun voltage as determined by the spheromak physics is

$$\bar{V}_{gun} = \frac{\Lambda}{4\pi} \sum_{n=1}^{\infty} [c_n x_{1n} J_0(x_{1n})]^2 \left(\frac{\Gamma_n^2 \coth \sqrt{\Gamma_n^2 - \Lambda^2}}{\sqrt{\Gamma_n^2 - \Lambda^2}} - \frac{\Lambda^2}{\sinh^2 \sqrt{\Gamma_n^2 - \Lambda^2}} \right). \qquad (11.68)$$

The normalized voltage drop across the power supply series resistor is

$$\bar{V}_{drop} = \frac{V_{drop}}{V_{ref}} = \frac{R_{src} h}{\eta}; \qquad (11.69)$$

this is approximately the ratio of the source impedance to the nominal resistance of a simple unmagnetized plasma with linear dimensions of order h. If this ratio is very large, then the source acts very much like a current source, but if this ratio is very small then the power supply acts like a voltage source (often called a "stiff" source, since the output voltage is virtually unaffected by the load). For a given c_n spectrum, the non-dimensional operating point Λ is a function of $\Gamma_n = \gamma_n h$, $\bar{V}_{drop}$, and $\bar{V}_{src}$. Since the location of the first resonance is usually the determining factor, the operating point is just a function of three dimensionless parameters, namely Γ_1, $\bar{V}_{drop}$, and $\bar{V}_{src}$.

This impedance model can be compared to extensive spheromak impedance measurements made by Barnes et al.[125] (who proposed a different model based on the gun being loaded by accelerated nozzle-like flows and thus not being in a force-free equilibrium). In particular, for one set of measurements (labeled the 1984c measurements in Ref.[125]), the relevant independent parameters were:

$$\begin{aligned}
\eta &= 5 \times 10^{-3} \quad \Omega\text{-m} \\
h &= 0.6 \quad \text{m} \quad \text{(nominal)} \\
a &= 0.7 \quad \text{m} \\
V_s &= 6 \quad \text{kV} \\
\bar{\psi}_g &= 20 \quad \text{mWb} \\
R_{src} &= 6 \quad \text{m}\Omega
\end{aligned}$$

11.8 Gun impedance and load line

We will also assume that the gun design was such that the c_1 coefficient was dominantly excited. Thus, Eq.(11.20) becomes simply

$$\psi(r,z) = \bar{\psi}_{gun} c_1 \gamma_1 r J_1(\gamma_1 r) S_1(z,\lambda). \tag{11.70}$$

Since $\bar{\psi}_{gun}$ is the value of $\psi(r,z)$ at $z=0, r=r_s$, it is seen that

$$c_1 = \frac{1}{[xJ_1(x)]_{\max}} \simeq 0.8 \tag{11.71}$$

where $[xJ_1(x)]_{\max}$ is where $xJ_1(x)$ is at its first maximum (this is where B_z reverses sign on the gun end wall).

Using these parameters gives $V_{ref} = 220$ Volts so that the relevant non-dimensional quantities are

$$\bar{V}_{src} = 27$$
$$\bar{V}_{drop} = 0.7$$
$$\gamma_1 h = 3.3$$

Figure (11.11) plots $\bar{V}_{gun}$ given respectively by Eqs.(11.67) and (11.68). It is seen that the source load-line (straight line, Eq.(11.67)) crosses the spheromak value Eq.(11.68) at more than one point. The left-most intersection would be a spheromak surrounded by open field lines, while the next two intersections would be flipped spheromaks, separated from the gun by a separatrix located at a small finite z. Flipped spheromaks may not always be possible, since the second and third intersections will disappear if the source load line becomes steeper as would happen if $\bar{V}_{src}$ becomes very large.

The left-most intersection gives $\Lambda \approx 4$ and so reverting to dimensioned variables, this predicts an operating point current of

$$I_{gun} = \frac{\bar{\psi}_{gun}}{\mu_0 h} \Lambda = 100\,\text{kA} \tag{11.72}$$

a factor of two lower than the measured current of 200 kA. The reason for this discrepancy is most likely the inadequacy of the constant λ model. In reality the configuration has a λ gradient resulting in much more complex physics. This situation is discussed in the next chapter.

Analysis of Driven Spheromaks: Strong Coupling

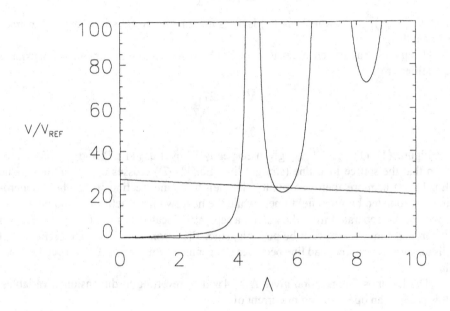

Fig.11.11 Plot of normalized gun voltage as determined by intersection of load-line from external circuit (downward sloping straight line) and plasma voltage (curve with resonances). Horizontal axis is normalized current $\Lambda = \lambda h$, vertical axis is $\bar{V} = V/V_{ref}$.

CHAPTER 12
Helicity Flow and Dynamos

Chapter 8 showed that there is an analogy between λ and temperature, and in particular, that λ gradients drive helicity transport much like temperature gradients drive heat transport. These λ gradients will be discussed in the present chapter and will be shown to be closely related to dynamo concepts and helicity flow.

12.1 Downhill flow of helicity

The parameter λ can be interpreted in several different ways depending on the context. For example, we have already seen that λ can be interpreted as an eigenvalue, as a ratio of current to flux, as a ratio of energy to helicity, as a criterion for tilt stability, and as an imperfect analog to temperature; it is conceivable that there are still more interpretations. The juxtaposition of these various interpretations can give a good sense of spheromak phenomenology. Section 11.1 showed that in cylindrical geometry with inhomogeneous boundary conditions (i.e., driven spheromaks), closed flux surfaces start forming when λ exceeds $\sqrt{\gamma_1^2 + \pi^2/4h^2}$ where $\gamma_1 = x_{11}/a$ and $x_{11} = 3.83$ is the first zero of J_1. These closed flux surfaces initially occupy a small region of space but progressively fill up the entire volume as λ increases to $\sqrt{\gamma_1^2 + \pi^2/h^2}$. The closed flux surfaces are surrounded by outer open field lines going to the gun. The drive becomes resonant as λ approaches $\sqrt{\gamma_1^2 + \pi^2/h^2}$ and if λ exceeds $\sqrt{\gamma_1^2 + \pi^2/h^2}$ then the volume is filled by a flipped configuration with the same helicity, but all fields reversed. This flipped configuration is completely detached from the gun; i.e., is not surrounded by open field lines from the gun. The existence of flipped solutions has been dismissed as a mathematical artifact by some authors on the grounds that the configuration must first pass through a resonance (infinite energy state) before reaching the flipped state. On the other hand, flipped configurations were observed in CTX by Barnes et al. [62], but were not interpreted as the result of having $\lambda > \sqrt{\gamma_1^2 + \pi^2/h^2}$. In the author's opinion, the resonance need not be insurmountable, because it simply describes the energy of a driven force-free equilibrium having λ corresponding to the eigenvalue of the homogeneous solution. If the configuration is not in a force-free equilibrium at the instant λ traverses the forbidden value but is instead dynamically accelerating, the configuration might be able to access the regime to the right of the resonance. Thus, the configuration could cross the forbidden value, but not remain at this value.

From another point of view, comparison between driven and isolated spheromaks showed that λ is an independent parameter in the driven situation but is a dependent parameter in the isolated situation. The gun current and flux fix λ in the driven situation, whereas for the isolated situation, λ self-adjusts to give the lowest energy state; i.e., the smallest λ consistent with the boundary conditions.

The analysis of tilt stability in Sec.10.3 also involved consideration of λ. This analysis showed that an isolated cylindrical spheromak is tilt-unstable when $h/a > 1.7$ because in this case the λ of a tilted spheromak is smaller than the λ of an untilted spheromak.

The discussion in the above paragraphs was based on the assumption of uniform λ. In actual experiments, λ is usually observed to be non-uniform, indicating that uniformity of λ is an idealization. Later in this chapter we will show that non-uniformity of λ is associated with helicity dissipation, a process explicitly excluded from the Taylor relaxation model. In the real world, helicity is invariably dissipated to some extent, and as will be shown, a λ gradient is required to drive a helicity flux which replenishes the helicity lost due to dissipation. Kitson and Browning[126] have shown that the resonant response discussed in Chapter 11 still occurs when there is a λ gradient, but a weighted average of λ must be used rather than the local value (Kanki et al.[127] have also investigated equilibria with λ gradients).

12.1.1 An isolated spheromak acting as a helicity source

The interpretation of λ depends on whether one is considering driven situations, isolated situations, or tilt stability. Jarboe et al. [128] and Fernandez et al.[105] suggested using the symbol λ to describe both the source properties and the geometrical properties of the volume, distinguishing the two λ's by subscripts. Thus, what we have called just λ until now becomes the gun or source λ, i.e.,

$$\lambda_{src} = \frac{\mu_0 I}{\psi_{gun}} \qquad (12.1)$$

while the eigenvalue associated with an isolated minimum-energy state for an axisymmetric cylinder having $h/a < 1.7$ is

$$\lambda_{me} = \sqrt{\gamma_1^2 + \pi^2/h^2} \,. \qquad (12.2)$$

This seems pedantic so far, but this concept could be used to describe the propagation of helicity through a series of variously shaped volumes linked by orifices of various sizes[105, 128]. Each volume would have a characteristic λ_{me} determined by the geometrical features of that volume. The first volume would be connected to a gun which produces λ_{src} as given by Eq.(12.1). If λ_{src} exceeded λ_{me1}, a spheromak would be formed in volume 1. If this spheromak becomes decoupled from the gun (e.g., by the gun turning off or by a separatrix developing between gun and spheromak), then the spheromak should relax to the lowest energy state of an isolated configuration, i.e., to a state with λ_{me1}. The current per flux of this isolated configuration would be given by $\lambda_{me,1}$ and now, the key point, is that this isolated spheromak could be considered as the effective "gun" sourcing the next volume in the configuration. To see this, let us call the isolated spheromak in region 1 by the name spheromak #1. Suppose that some of the field lines (and hence flux and current) from spheromak #1 poke slightly into region 2. Since the ratio of current to flux for spheromak #1 is given by λ_{me1} then λ_{me1} must be the ratio of current to flux on the field lines poking into region 2. From the point of view of an observer in region #2, there is a gun with λ_{me1} acting on region #2. If λ_{me1} exceeds λ_{me2} the spheromak in volume 1 would generate a spheromak in volume 2. This argument can be repeated for each successive volume.

The net result is that helicity will be transported from one volume to the next only if it can move to a region with smaller λ_{me}. In particular, a spheromak can pass through an orifice only if its λ exceeds the inverse of the characteristic dimension of the orifice. Since λ is also the ratio of energy to helicity, and since helicity is assumed to be conserved, configurations will only move to the next volume if their energy can be reduced in so doing. If one considers the motion as a function of the local λ, then the configuration always moves in the direction of minimizing λ, hence the idea that helicity runs "downhill" towards regions of ever decreasing λ. Furthermore, because λ is inversely proportional to the physical dimensions, the configuration always moves in the direction associated with a geometrical expansion if helicity is conserved.

12.2 Dynamos and relaxation mechanisms

12.2.1 Paradox of the driven axisymmetric spheromak

The driven axisymmetric spheromak is paradoxical because it appears to violate basic electric circuit concepts. On the one hand, the steady-state driven spheromak has currents flowing through plasma-facing electrodes and voltage drops across these electrodes; this gives a net injection of helicity and power to balance the helicity and energy dissipation in the plasma. On the other hand since $\mathbf{J} = \lambda \mathbf{B}$, the path of current flow is along magnetic field lines allowing one to visualize individual magnetic field lines as acting like insulated wires. The paradox results because when λ exceeds $\sqrt{x_{11}^2/a^2 + \pi^2/4h^2}$, closed flux surfaces form. The field lines forming these closed flux surfaces are not connected to the source and yet the source somehow drives steady-state DC current along these disconnected closed field lines. This violates Cowling's theorem because a steady-state axisymmetric equilibrium is being supported in a dissipative medium. It is also a violation of common sense because in effect a battery (the source) is sustaining a DC current in a resistive closed loop conductor which is not electrically connected to the battery.

The resolution of this paradox lies in the dynamics. The Taylor relaxation argument avoids consideration of the dynamics and simply specifies the final state towards which the system evolves after dynamic activity is complete. Thus, the Taylor relaxation argument does not preclude non-axisymmetric dynamics, it merely states that all dynamics lead to certain minimum energy meta-equilibria some of which happen to be axisymmetric. Hence, if non-axisymmetric dynamic processes lead to axisymmetric equilibria then both Cowling's theorem and the Taylor relaxation principle can be satisfied. We therefore expect that the evolution of closed flux surfaces necessarily involves non-axisymmetric dynamics. It is if one periodically spliced the battery into the resistive closed loop conductor, ramped up the current, and then removed the splice.

Experiments bear out this conjecture and indicate that the actual dynamo process always involves non-axisymmetric behavior. Systems do not evolve smoothly and axisymmetrically from one force-free equilibrium to another. Instead, systems jump from one state to another via transient non-axisymmetric processes. Relaxation may involve substantial energy dissipation and the transient non-axisymmetric process must provide this dissipation.

It is probable that no universal dynamo mechanism exists because the Taylor argument merely predicts the final state for a given helicity and geometry. There is no explicit time-scale in Taylor relaxation and it is simply presumed that the relevant time scale will be the reconnection time scale. Plasmas with different physical characteristics (e.g., resistive or collisionless, radiation-dominated or not) may have different dynamics which would nevertheless lead to the same final state. Developing a detailed understanding of how the dynamo works in a specific experiment does not necessarily mean that the dynamo problem has been solved for all situations. Furthermore, dynamos involve electric fields parallel to the magnetic field and often electrostatic potential drops along field lines; these situations are poorly described by MHD. Although parallel electric fields are allowed by resistive MHD and are an important feature of resistive tearing, the observed reconnection rate in most experimental situations is much faster than predicted by resistive MHD. This discrepancy with resistive MHD suggests that the actual reconnection process involves non-MHD effects which allow collisionless parallel electric fields and which give a correct prescription of electrostatic potentials varying along field lines.

Any dynamo mechanism must involve non-linear processes because a dynamo which is both linear and non-axisymmetric could not provide a toroidal electromotive force (EMF). Non-axisymmetry means that the dynamo fields vary as $\exp(in\phi)$ where $n \neq 0$; clearly the toroidal average of a linear field of this form must vanish. If $n = 0$, then a

12.2 Dynamos and relaxation mechanisms

linear field could provide a net toroidal EMF, but this would violate Cowling's theorem. The electric field $\mathbf{E}'$ in a frame moving with velocity $\mathbf{U}$ is

$$\mathbf{E}' = \mathbf{E} + \mathbf{U} \times \mathbf{B} \tag{12.3}$$

and so if a plasma moves with an oscillatory velocity $\tilde{\mathbf{U}}$ and the magnetic field has an oscillatory component $\tilde{\mathbf{B}}$, there will be a time-averaged non-linear effective electric field resulting from the cross-product term. Thus, the average electric field in the plasma frame will be

$$<\mathbf{E}'> = <\mathbf{E}> + <\tilde{\mathbf{U}} \times \tilde{\mathbf{B}}>. \tag{12.4}$$

The term $<\tilde{\mathbf{U}} \times \tilde{\mathbf{B}}>$ is called the MHD dynamo term and can drive steady state current in the plasma frame. Since current is a frame-independent quantity, this means that the dynamo also drives current in the lab frame. The dynamo involves $\tilde{\mathbf{U}}$ and $\tilde{\mathbf{B}}$ components which are transverse to the product $<\tilde{\mathbf{U}} \times \tilde{\mathbf{B}}>$ and can be in any direction.

The dynamo process is not exclusive to spheromaks. It is important to RFP's and is also the probable cause of the anomalously fast current penetration observed [129] during the early phase of the discharge in large tokamaks.

12.2.2 Physical constraints on coefficients of Fourier expansions in cylindrical coordinates

Regardless of the physical model or governing equations, the fields $\tilde{\mathbf{U}}$, $\tilde{\mathbf{B}}$ are tightly constrained [130] to a specific symmetry that can be easily related to experimental measurements. These constraints are a fundamental mathematical property and result from the requirement that physical quantities must always be continuous and infinitely differentiable (i.e., mathematically regular) in the region of interest. Quantities expressed in cylindrical coordinates r, ϕ, z tend not to be regular unless so constrained. For example, if f is a scalar quantity and we make the simple assumption that $f = r$, then f is not regular at $r = 0$. This is apparent by transforming to Cartesian coordinates so that $f = \sqrt{x^2 + y^2}$ which does not have a continuous derivative at $x = 0$, $y = 0$. Thus, no physical quantity could have the dependence $f = r$. When solving equations one typically finds mathematically allowed solutions that are not regular in the region of interest and must be discarded; for example the Bessel $Y_m(r)$ function is discarded from physical problems including the origin because $Y_m(r)$ is not regular at $r = 0$.

Lewis and Bellan derived [130] constraints for vectors based on the condition that the vector be regular at $r = 0$; the constraints are independent of the set of equations under consideration. The derivation involved transforming a physical vector from cylindrical coordinates to Cartesian coordinates and then requiring regularity in Cartesian coordinates. The form of the constraints can be seen by decomposing a vector $\mathbf{V}$

representing some arbitrary physical quantity into azimuthal Fourier modes, i.e., writing **V** as

$$\mathbf{V}(r,\theta) = \sum_{m=0}^{\infty} \mathbf{V}_m(r) e^{im\theta}. \tag{12.5}$$

The form of the Fourier coefficients $\mathbf{V}_m(r)$ must be constrained in order to ensure regularity of **V** at $r = 0$. In particular, $\mathbf{V}_m(r)$ must have the following form for $m \neq 0$

$$\begin{aligned}
V_{rm} &= \alpha_m r^{|m|-1} + r^{|m|+1} g_m(r^2) \\
V_{\theta m} &= i\,\text{sign}(m)\, \alpha_m r^{|m|-1} + r^{|m|+1} h_m(r^2) \\
V_{zm} &= r^{|m|} f_m(r^2)
\end{aligned} \tag{12.6}$$

while for $m = 0$, **V** must have the form

$$V_{r0} = rg_0(r^2), \quad V_{\theta 0} = rh_0(r^2), \quad V_{z0} = f_0(r^2) \tag{12.7}$$

where $f_m(r^2) = f_0^{(m)} + f_2^{(m)} r^2 + f_4^{(m)} r^4 + ...$ and similarly for g_m and h_m. Thus, near $r = 0$ the only finite Fourier components are $V_{r1}, V_{\theta 1}$, and V_{z0}. Hence, any non-linear product involving r and θ components on axis can only involve $|m| = 1$ terms since all other terms vanish at $r = 0$. This is true no matter what mathematical model is invoked.

There are two axes for a spheromak, the geometric axis and the magnetic axis. The angle about the geometric axis is the toroidal angle ϕ and the angle about the magnetic axis is the poloidal angle θ. The regularity constraint applies directly to quantities expressed in cylindrical coordinates r, ϕ, z defined with respect to the geometric axis. It can also be applied in a limiting fashion to the region near the magnetic axis if a large aspect ratio expansion is made so as to create a cylindrical coordinate system r, θ, z defined with respect to the magnetic axis.

12.2.2.1 Dynamo field along the geometric axis

Suppose that there is a dynamo providing an effective electric field along the geometric axis

$$E_{dyn}^{ga} = \hat{z} \cdot <\tilde{\mathbf{U}} \times \tilde{\mathbf{B}}>. \tag{12.8}$$

Thus, the vector components $\tilde{U}_r, \tilde{B}_\phi$ and/or $\tilde{U}_\phi, \tilde{B}_r$ would have to be finite on the geometric axis. The regularity constraint shows that the angular dependence of these quantities must be $\exp(in\phi)$ with $|n| = 1$. This dynamo would drive poloidal current down the geometric axis. Depending on its polarity, this poloidal current would either increase or decrease the toroidal magnetic field.

12.2.2.2 Dynamo along the magnetic axis

A dynamo along the magnetic axis would drive toroidal current on the magnetic axis by providing an effective electric field

$$E_{dyn}^{ma} = \hat{\phi} \cdot < \tilde{\mathbf{U}} \times \tilde{\mathbf{B}} > . \quad (12.9)$$

If a large aspect ratio approximation is used for the region near the magnetic axis and a local cylindrical coordinate system (r, θ, z) is constructed where r is now the minor radius of the spheromak and z is now in the ϕ direction, it is seen that $\tilde{U}_r$, $\tilde{B}_\theta$ and/or $\tilde{U}_\theta$, $\tilde{B}_r$ would have to be finite on the geometric axis. The regularity constraint shows that the poloidal angular dependence of these quantities must be $\exp(im\theta)$ where $|m| = 1$ is the poloidal mode number. This dynamo would drive toroidal current which would either increase or decrease the poloidal magnetic field. In effect, this is a more constrained prescription than Cowling's theorem. Cowling's theorem shows that the dynamo driving current on the magnetic axis cannot be axisymmetric, i.e., the dynamo must have a toroidal dependence $\exp(in\phi)$ where n is a finite integer. Furthermore, in order to be finite on the magnetic axis the dynamo fields must have a poloidal dependence $\exp(im\theta)$ where $|m| = 1$. Thus, dynamo fields driving current along the magnetic axis must vary as $\exp(\pm i\theta + in\phi)$ where n is a non-zero integer.

The relationship between the geometric axis dynamo and the magnetic axis dynamo can be explained qualitatively in terms of kink stability [45]. In a straight cylinder, kinks are destabilized by having too much axial current and are stabilized by having large axial magnetic fields. The axial magnetic field is a vacuum magnetic field (i.e., is produced by currents located elsewhere) and therefore any perturbation of this field requires work. Thus, the axial magnetic field can be construed as a rigid backbone which stabilizes the system. If there is excess current flowing along the geometric axis, then the geometric axis will become kink unstable. However, the current flowing along the geometric axis is poloidal and so creates the toroidal magnetic field on the magnetic axis. Hence, a strong current on the geometric axis tends to drive kinks on the geometric axis and simultaneously makes the magnetic axis rigid and immune to kinks. Conversely, a strong toroidal current along the magnetic axis tends to destabilize kinks on the magnetic axis and simultaneously makes the geometric axis rigid and stable.

12.3 Observations of dynamo behavior

Fluctuations consistent with dynamo behavior have been clearly identified in the S-1 spheromak (Janos et al.[131, 115], Ono et al.[118]), in the CTX spheromak (Knox et

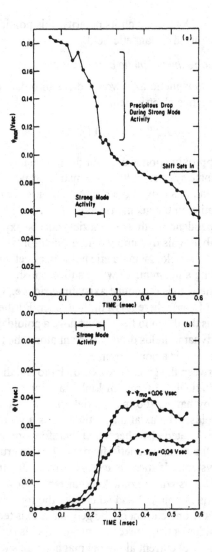

Fig. 12.1 Data from Janos et. al [115] showing conversion of poloidal flux into toroidal flux during spheromak formation. Plot (a) shows the time evolution of ψ_{max}, the poloidal flux on the magnetic axis. Plot (b) shows the time evolution of the toroidal flux included in a closed poloidal flux surface a distance $\psi - \psi_{max}$ from the magnetic axis for two choices of $\psi - \psi_{max}$. There is a correlation between strong MHD mode activity and a large transfer of poloidal to toroidal flux.

al.[132], Wright et al.[112]), in the CTCC spheromak (Uyama et al.[111]), in the FACT spheromak (Nagata et al.[133]), and in the SPHEX spheromak (al-Karkhy et al.[69], Duck et al.[44]). These fluctuations are usually quite coherent and reproducible indicating that the dynamo does not depend on randomness, but is instead deterministic. Although the dynamo mechanism is not well understood, the observed coherence and reproducibility suggests that it should be possible to develop a complete theoretical understanding. The coherence and reproducibility permit measurement of the symmetry of the dynamo fluctuations and provides useful input for theoretical models.

It is important to note that there can be a difference in behavior depending on whether the spheromak is sustained by a coaxial plasma gun (CTX method) or is sustained inductively (S-1 method). The coaxial gun drives current on outer poloidal field lines and so is effectively creating toroidal flux; a generous poloidal current on these outer field lines means that the spheromak will be toroidal flux-rich and relaxation will tend to convert toroidal flux into poloidal flux (i.e., create toroidal current flowing on the magnetic axis). To the extent that $q(\psi) = d\Phi/d\psi$ is proportional to Φ/ψ, the excess of toroidal flux relative to poloidal flux means that the q will start out high and relaxation will tend to decrease q. The situation is reversed with inductive drive. Here, the transformer action induces a toroidal current and so creates poloidal flux. Thus, a generous transformer drive of toroidal current will create a configuration that is rich in poloidal flux compared to the Taylor state and relaxation will convert ψ into Φ. The richness in ψ means that the q will start out low and relaxation will tend to increase q. Because the inductive system has the flexibility of also inducing poloidal currents, it can also be run in a toroidal flux rich mode and then behave in a manner similar to the coaxial plasma gun.

Janos et al.[115, 131] observed that dynamo-instigated conversion of poloidal flux into toroidal flux was an integral part of the formation sequence of the S-1 spheromak. This conversion, shown in Fig.12.1, was correlated with strong MHD modes having spatial dependence $\exp(im\theta + in\phi)$ where θ is the poloidal angle and ϕ is the toroidal angle. The upper curve in Fig.12.1 shows the poloidal flux on the magnetic axis as a function of time while the lower two curves show the toroidal flux within two closed poloidal flux surfaces having a fixed amount of poloidal flux, in one case 0.04 V-s and in the other 0.06 V-s. There is a precipitous drop in poloidal flux and a corresponding rise in toroidal flux during the time interval from 0.15 ms to 0.25 ms. During this time the profile of $q = d\Phi/d\psi$ passed through the m/n sequence $1/5, 1/4, 1/3, 1/2$ as shown by the right hand scale of Fig.12.2. This occurred in discrete steps each of which was associated with a sequence of coherent MHD modes measured by an array of magnetic probes. The modes were quite well defined, had $m = 1$ and evolved in a sequence $n = 5, 4, 3, 2$ corresponding to the $q = m/n$ sequence $1/5, 1/4, 1/3, 1/2$. The progression from $n = 3$ to $n = 1$ is shown in Fig.12.3. These observations

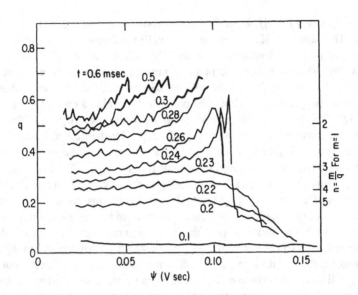

Fig.12.2 $q(\psi)$ as a function of time from Ref.[115] showing increase of q with time as toroidal field builds up and poloidal field decreases. The horizontal extent of the curves shrinks with time as the maximum poloidal field decreases. Right hand scale shows the n value for $q = m/n$ where $m = 1$. The evolution of q is consistent with the progression of toroidal modes through decreasing mode numbers $n = 5, 4, 3, 2, 1$.

indicate that conversion of poloidal to toroidal flux is associated with non-axisymmetric MHD mode activity. The $m = 1$ mode structure was determined from a soft x-ray measurement and is consistent with the regularity requirement that only an $m = \pm 1$ dynamo can exist on the magnetic axis. This data confirms that in order for a plasma to relax to have the proportion of poloidal to toroidal flux corresponding to a Taylor state, non-axisymmetric dynamics must inevitably occur.

Uyama et al.[111] observed a step-wise relaxation phenomenon in the CTCC spheromak in which non-uniformity of resistivity caused a peaking of toroidal current near the magnetic axis and hence a deviation from the Taylor state. Both magnetic energy and magnetic helicity decay, but because the system deviates from the Taylor state W/K increases (i.e., λ is not the lowest possible value). At a certain time the system would abruptly go unstable and decay back into a Taylor state (minimum $\lambda = 2\mu_0 W/K$).

Ono et al.[110] investigated relaxation events in a sustained S-1 spheromak plasma and also noted significant dynamo activity. As shown in Fig.12.4(a) Ono et al. ob-

12.3 Observations of dynamo behavior

served that the ratio ψ/Φ tended to increase monotonically and then suddenly drop; this would repeat in a relaxation cycle. The monotonic increase of ψ/Φ was ascribed to a peaking of toroidal current on the magnetic axis due to core electron heating and a preferential decay of Φ due to higher edge resistivity dissipating poloidal currents (similar to Uyama et al.[111]). The increase of ψ/Φ would cause the configuration to deviate from the Taylor state and when q_{axis} decreased from 0.55 to 0.50 and below as shown in Fig.12.4(c), an MHD instability with $m = 1$, $n = 2$ would develop which would restore the ratio ψ/Φ back to the Taylor state. Ono et al interpreted the sharp drop in ψ/Φ at 370 μs as a conversion from poloidal flux to toroidal flux, i.e. $\dot{\psi} < 0$ and $\dot{\Phi} > 0$. Changing the poloidal flux on the magnetic axis would require a non-dissipative reduction in the on-axis toroidal current, i.e., an anti-dynamo on the magnetic axis opposing the toroidal current flow. Such a dynamo would require $m = 1$ symmetry. Nagata et al. made similar measurements of dynamo activity on the Flux Amplification Compact Toroid (FACT) and observed a quasi-regular relaxation cycle [133].

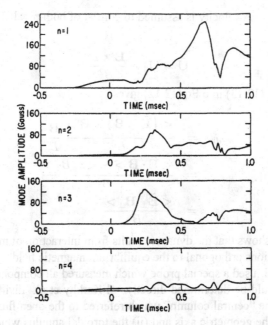

Fig.12.3 Progression of toroidal mode numbers from $n = 3$ to $n = 1$ observed by Janos et al. [115] during formation of the S-1 spheromak.

al-Karkhy et al.[69] investigated how non-axisymmetric modes sustained the SPHEX coaxial gun spheromak and characterized the dynamo by a nonlinear resistive MHD Ohm's law

$$\tilde{\mathbf{E}} + \tilde{\mathbf{U}} \times (\mathbf{B} + \tilde{\mathbf{B}}) = \eta \left(\mathbf{J} + \tilde{\mathbf{J}}\right) \tag{12.10}$$

where tilde means a fluctuating quantity with zero time-average, and no tilde denotes a steady-state quantity. Time averaging of Eq.(12.10) gives

$$<\tilde{\mathbf{U}} \times \tilde{\mathbf{B}}> = \eta \mathbf{J}. \tag{12.11}$$

Since the Taylor state has **J** parallel to **B**, the dynamo term $<\tilde{\mathbf{U}} \times \tilde{\mathbf{B}}>$ must also be parallel to **B** and so can be considered as an effective time-averaged parallel electric field $E_{d\parallel}$; this can be written as

$$E_{d\parallel} \equiv <\tilde{\mathbf{U}} \times \tilde{\mathbf{B}}> \cdot \frac{\mathbf{B}}{B} = \eta J_\parallel. \tag{12.12}$$

The oscillating plasma velocity is assumed to consist of both parallel motion and $E \times B$ drift, i.e.,

$$\tilde{\mathbf{U}} = \tilde{\mathbf{U}}_\parallel + \frac{\tilde{\mathbf{E}} \times \mathbf{B}}{B^2}. \tag{12.13}$$

Substitution of Eq.(12.13) into Eq.(12.12) gives

$$\begin{aligned} E_{d\parallel} &= \frac{<\left(\tilde{\mathbf{E}} \times \mathbf{B}\right) \times \tilde{\mathbf{B}}>}{B^2} \cdot \frac{\mathbf{B}}{B} \\ &= \frac{<\tilde{\mathbf{E}} \cdot \tilde{\mathbf{B}}> - <\tilde{E}_\parallel \tilde{B}_\parallel>}{B} \\ &= \frac{<\tilde{\mathbf{E}}_\perp \cdot \tilde{\mathbf{B}}_\perp>}{B}. \end{aligned} \tag{12.14}$$

Equation (12.14) shows that the dynamo results from interaction of magnetic and electric field perturbations orthogonal to the equilibrium magnetic field.

al-Karkhy et al. used a special probe which measured all components of $\tilde{\mathbf{E}}, \tilde{\mathbf{B}}$, and **B** so that $E_{d\parallel}$ could be determined directly. al-Karkhy et al. divided the plasma in SPHEX into (i) the "central column" which referred to the open flux leaving the gun and going along the geometric axis and (ii) the toroidal annulus which referred to the closed flux surfaces about the magnetic axis. Thus, as shown in Fig.12.5, the open field lines formed a central column about which the closed field lines were wrapped. The central column had coherent $n = 1$ oscillations and the $E_{d\parallel}$ dynamo electric field deduced from these oscillations was oriented to drive a current opposing the current flowing out from the gun center electrode. Thus, the central column contained an antidynamo which acted as a sink for the power coming from the gun. On the other hand, in the torus (annulus) comprising the closed field lines, measurement of $E_{d\parallel}$ indicated that the magnetic axis dynamo was aligned to drive the toroidal current on the magnetic axis and had a magnitude suitable for sustaining this current. The magnetic axis dynamo did not have $n = 1$ symmetry and appeared to be turbulent.

Duck et al.[44] showed that this central column in SPHEX has a helical shape and that rotation of this helix accounts for the observed $n = 1$ mode behavior. Furthermore, the current along the central column deviates from being force-free and in fact most of the current returns to the flux conserver wall in regions well away from the gun (below the midplane in Fig. 12.5).

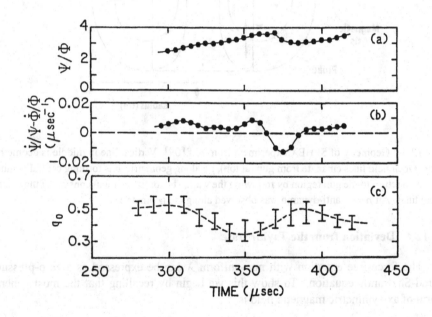

Fig. 12.4 Time evolution data from Ref. [110]; (a) ψ/Φ (poloidal flux/toroidal flux), (b) $\dot\psi/\psi - \dot\Phi/\Phi$, (c) q at the magnetic axis.

These experiments provide compelling evidence supporting the assertion that non-axisymmetric dynamo activity must occur in order to relax to a Taylor state.

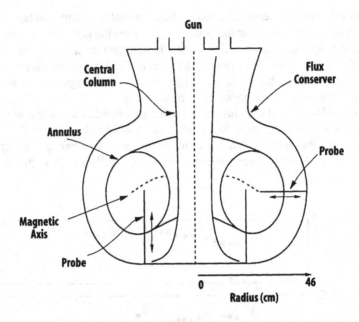

Fig.12.5 Geometry of SPHEX experiment from Ref.[69]. Vertical line in middle is geometric axis. Open field lines emanate from gun (at top), go along geometric axis forming central column and then return to the gun region by following the wall. The torus (annulus) consists of the closed field lines. An $n = 1$ anti-dynamo was observed along the geometric axis.

12.4 Deviation from the Taylor state

The force-free equation with non-uniform λ can be expressed as a zero-pressure Grad-Shafranov equation. To show this we begin by recalling that the most general form of axisymmetric magnetic field is

$$\mathbf{B} = \frac{1}{2\pi} (\nabla \psi \times \nabla \phi + \mu_0 I \nabla \phi) \qquad (12.15)$$

where I and ψ are axisymmetric functions representing poloidal flux and poloidal current. Taking the divergence of the force-free equation $\nabla \times \mathbf{B} = \lambda \mathbf{B}$ gives

$$\mathbf{B} \cdot \nabla \lambda = 0 \qquad (12.16)$$

12.4 Deviation from the Taylor state

and combining Eqs.(12.15) and (12.16) gives

$$(\nabla\psi \times \nabla\phi + \mu_0 I \nabla\phi) \cdot \nabla\lambda = 0. \tag{12.17}$$

Since all quantities are assumed axisymmetric, $\nabla\phi \cdot \nabla\lambda = 0$ leaving $\nabla\psi \times \nabla\phi \cdot \nabla\lambda = 0$ which implies $\lambda = \lambda(\psi)$. Thus, general axisymmetric force-free states must have λ constant on a flux surface, but unlike the Taylor state may have different values on different flux surfaces.

Knox et al.[132] conducted a series of experiments on the CTX spheromak to investigate λ non-uniformity, i.e., the extent to which the plasma deviated from a Taylor state. Their measurements were reasonably modeled by a linear dependence of λ on ψ; i.e., the simplest non-trivial dependence. Knox et al. found it convenient to express $\lambda(\psi)$ in terms of $\bar\lambda$, the flux-averaged λ, giving

$$\lambda(\psi) = \bar\lambda \left[1 + \alpha \left(2\psi/\psi_{\max} - 1\right)\right] \tag{12.18}$$

which satisfies the flux-averaged condition

$$\frac{1}{\psi_{\max}} \int_0^{\psi_{\max}} \lambda(\psi) d\psi = \bar\lambda. \tag{12.19}$$

This simple representation of the dependence of λ on ψ invokes α as a single-parameter measure of the deviation from the Taylor state. If α is positive, then $\lambda(\psi)$ increases with ψ so that λ is larger at the magnetic axis then at the edge. Conversely, if α is negative, then $\lambda(\psi)$ decreases with ψ so that λ is largest at the edge (in this discussion λ is assumed positive).

The determination of α is obtained using the Grad-Shafranov equation in the limit of zero pressure. To derive this equation consider the axisymmetric current

$$\mathbf{J} = \frac{1}{2\pi\mu_0} \nabla \times (\nabla\psi \times \nabla\phi + \mu_0 I \nabla\phi) \tag{12.20}$$

which has toroidal current component

$$\mathbf{J}_{tor} = -\frac{r^2 \nabla\phi}{2\pi\mu_0} \nabla \cdot \left(\frac{1}{r^2} \nabla\psi\right) \tag{12.21}$$

and poloidal component

$$\mathbf{J}_{pol} = \frac{1}{2\pi} \nabla I \times \nabla\phi. \tag{12.22}$$

Using the force-free condition $\mathbf{J}_{pol} \times \mathbf{B}_{tor} + \mathbf{J}_{tor} \times \mathbf{B}_{pol} = 0$ with $\mathbf{B}$ determined by Eq.(12.15) gives $I = I(\psi)$ and

$$r^2 \nabla \cdot \left(\frac{1}{r^2} \nabla \psi \right) + \mu_0^2 I \frac{dI}{d\psi} = 0; \quad (12.23)$$

this is the zero-pressure Grad-Shafranov equation. The combination of $\mu_0 \mathbf{J}_{pol} = \lambda \mathbf{B}_{pol}$, Eq.(12.15), and Eq.(12.22) show that

$$\mu_0 \nabla I \times \nabla \phi = \lambda(\psi) \nabla \psi \times \nabla \phi. \quad (12.24)$$

However, because $I = I(\psi)$, this gives

$$\mu_0 \frac{dI}{d\psi} = \lambda(\psi). \quad (12.25)$$

This may be integrated using Eq.(12.18) and the boundary condition $I(0) = 0$ to give

$$\mu_0 I(\psi) = \left[1 + \alpha \left(\frac{\psi}{\psi_{max}} - 1 \right) \right] \bar{\lambda} \psi \quad (12.26)$$

so that the Grad-Shafranov equation becomes

$$r^2 \nabla \cdot \left(\frac{1}{r^2} \nabla \psi \right) + \left[1 + \alpha \left(\frac{\psi}{\psi_{max}} - 1 \right) \right] \left[1 + \alpha \left(2 \frac{\psi}{\psi_{max}} - 1 \right) \right] \bar{\lambda}^2 \psi = 0 \quad (12.27)$$

or if we define the normalized flux

$$\tilde{\psi} = \frac{\psi}{\psi_{max}}$$

then it is seen that the Grad-Shafranov equation actually describes $\tilde{\psi}$, i.e.,

$$r^2 \nabla \cdot \left(\frac{1}{r^2} \nabla \tilde{\psi} \right) + \left[1 + \alpha \left(\tilde{\psi} - 1 \right) \right] \left[1 + \alpha \left(2\tilde{\psi} - 1 \right) \right] \bar{\lambda}^2 \tilde{\psi} = 0. \quad (12.28)$$

The boundary conditions are

1. $\tilde{\psi} = 0$ on the side walls (i.e., $r = a$)
2. $\tilde{\psi} = 0$ on the end wall opposite the gun (i.e., $z = h$)
3. $\tilde{\psi}$ is finite on the gun end wall (i.e., $z = 0$), but if we assume that the system is driven close to resonance, we may approximate $\tilde{\psi} \approx 0$ at the gun end wall as well (see for example bottom middle plot of Fig.11.6)
4. $\tilde{\psi} = 1$ on the magnetic axis (normalization assumption).

12.5 MHD dynamo, helicity flux, and lambda gradient

If it is also assumed that $\bar{\lambda}$ in Eq.(12.18) is not too different from the Taylor value, then $\bar{\lambda}$ can be estimated from the flux conserver geometry using Eq.(9.13). Equation (12.27) may then be solved numerically [132] for various α using the boundary conditions listed above. The solution gives $\nabla\tilde{\psi}$ at the wall and, in particular, the poloidal dependence of $\nabla\tilde{\psi}$ at the wall. This poloidal dependence is normalized to its mean value so that it becomes a dimensionless quantity.

The calculated normalized poloidal dependence can now be compared to an experimental quantity which is easy to measure. Since the poloidal field is $\mathbf{B}_{pol} = (2\pi)^{-1}\nabla\psi \times \nabla\phi$, measurement of the poloidal field at the wall directly gives $\nabla\psi$ at the wall. The discontinuity in poloidal field across the wall corresponds to the wall toroidal current so a measure of wall toroidal current can equivalently be used to determine $\nabla\psi$ at the wall (this was done in Ref.[132]). The measured $\nabla\psi$ at the wall can be normalized to its mean poloidal value so that it also becomes a dimensionless quantity dependent on poloidal position at the wall. The parameter α is then adjusted to give best fit between the measured and calculated normalized $\nabla\psi$'s at the wall. Once α has been determined, the normalization ψ_{max} is adjusted to match the calculated $\nabla\tilde{\psi}$ at the wall with the measured $\nabla\psi$ at the wall.

By repeating this prescription for a sequence of times in the discharge, Knox et al. determined α as a function of time. They observed that α was negative during the formation phase when the gun current was driving the spheromak; this corresponds to $\lambda(\psi)$ peaking at the spheromak edge. In contrast, during the decay phase when the spheromak was isolated and the gun turned off, they observed that α was positive which corresponds to $\lambda(\psi)$ peaked on the magnetic axis. This is consistent with the concept that helicity flows in the direction of decreasing λ. When the gun is on, it injects helicity into the spheromak whereas when the spheromak is decaying, helicity flows outwards from the magnetic axis to the wall.

One can make a simple analogy of this behavior to the familiar situation of a roast being heated in the oven. When the oven is turned on and heating the roast, the outside of the roast is hotter than the inside and there is a heat flux from outside to inside. If the oven is turned off after the roast has been heated, then the outside of the roast cools off first. Now the temperature gradient is reversed (inside hot, outside cold) and so the heat flux is from inside to outside.

12.5 MHD dynamo, helicity flux, and lambda gradient

12.5.1 Growth rate

We now combine the concepts presented earlier in this chapter to develop a simple

MHD dynamo theory. The MHD energy principle gave an expression for δW, the change in potential energy associated with a perturbation. Equation (6.38) in Section 6.2 showed that in the $\beta = 0$ limit,

$$\delta W = \frac{1}{2\mu_0} \int_V d^3r \left(B_{1\perp}^2 - \mathbf{A}_1 \cdot \mathbf{B}_{1\perp} \lambda \right) \quad (12.29)$$

and that the term involving $\mathbf{A}_1 \cdot \mathbf{B}_{1\perp}$ integrated to zero if λ were uniform. Using Eq.(12.18) to give the spatial dependence of λ it is seen that δW will have the form

$$\delta W = \frac{1}{2\mu_0} \int_V d^3r \, B_{1\perp}^2 - \frac{\bar{\lambda}\alpha}{\mu_0 \psi_{\max}} \int_V d^3r \, \mathbf{A}_1 \cdot \mathbf{B}_{1\perp} \psi \, . \quad (12.30)$$

The stabilizing term $B_{1\perp}^2$ vanishes on rational surfaces [i.e. where $q = m/n$, see Eq.(6.50)] and so we expect instabilities to occur on rational surfaces and the growth rate of these instabilities (i.e., the extent to which δW is negative) should be proportional to the magnitude of α.

12.5.1.1 Effect on flux function and toroidal current

Using Eq.(12.18) it is seen that the gradient of λ is related to α by

$$\frac{1}{2\bar{\lambda}} \nabla \lambda = \frac{\alpha}{\psi_{\max}} \nabla \psi \quad (12.31)$$

or since $\lambda = \lambda(\psi)$

$$\frac{d\lambda}{d\psi} = \frac{2\bar{\lambda}}{\psi_{\max}} \alpha. \quad (12.32)$$

We can formally solve for the functional dependence of $\psi(r, z)$ in the presence of a small λ gradient by making the expansion

$$\psi(r, z) = \psi^{(0)}(r, z) + \alpha \psi^{(1)}(r, z) \quad (12.33)$$

where $\psi^{(0)}$ is the uniform λ solution given by Eq.(11.23). Since $\psi^{(0)}$ satisfies the prescribed inhomogeneous boundary condition at the wall (i.e., $\psi^{(0)}$ equals the imposed gun flux at the electrodes), we have the boundary condition $\psi^{(1)} = 0$ at the wall. Substituting for ψ in Eq.(12.27) using Eq.(12.33) and then linearizing using α as a small

12.5 MHD dynamo, helicity flux, and lambda gradient

parameter gives

$$r^2 \nabla \cdot \left(\frac{1}{r^2}\nabla\psi^{(1)}\right) + \bar{\lambda}^2 \psi^{(1)} = -\left(3\frac{\psi^{(0)}}{\psi_{\max}} - 2\right)\bar{\lambda}^2 \psi^{(0)}. \qquad (12.34)$$

This is a well-posed problem since $\psi^{(0)}$ is given by Eq.(11.23) and there is a boundary condition $\psi^{(1)} = 0$ at the wall. To solve this inhomogeneous system consider the homogenous equation

$$r^2 \nabla \cdot \left(\frac{1}{r^2}\nabla\psi\right) + \sigma^2 \psi = 0 \qquad (12.35)$$

with homogeneous boundary conditions; here σ is an eigenvalue to be determined. The solutions of Eq.(12.35) are just the flux functions of an isolated configuration, namely

$$\psi_{mn} = r J_1(x_{1n}r/a) \sin\left(\sqrt{\sigma_{mn}^2 - x_{1n}^2/a^2}\,(z-h)\right) \qquad (12.36)$$

where

$$\sigma_{mn} = \sqrt{\frac{m^2\pi^2}{h^2} + \frac{x_{1n}^2}{a^2}}. \qquad (12.37)$$

The solution to Eq.(12.34) is expressed as a sum of the homogeneous solutions, i.e.,

$$\psi^{(1)}(r,z) = \sum_{m,n} a_{mn} \psi_{mn}(r,z) \qquad (12.38)$$

which clearly satisfies the boundary condition $\psi^{(1)} = 0$ at the wall. Equation (12.38) is inserted into Eq.(12.34) to give

$$\sum_{m,n} a_{mn}\left(\bar{\lambda}^2 - \sigma_{mn}^2\right)\psi_{mn}(r,z) = -\left(3\frac{\psi^{(0)}}{\psi_{\max}} - 2\right)\bar{\lambda}^2 \psi^{(0)} \qquad (12.39)$$

and so the coefficients a_{mn} can be found by taking advantage of the orthogonality of the ψ_{mn}, i.e.

$$a_{mn} = -\frac{\bar{\lambda}^2}{\bar{\lambda}^2 - \sigma_{mn}^2} \frac{\int d^3r \left(3\frac{\psi^{(0)}}{\psi_{\max}} - 2\right) \psi^{(0)} \psi_{mn}}{\int d^3r\, \psi_{mn}^2} \qquad (12.40)$$

The coefficient a_{mn} diverges if $\bar{\lambda}^2 \to \sigma_{mn}^2$; such divergence would invalidate the presumed expansion Eq.(12.33) which was based on the assumption that $\psi^{(1)}$ is finite. Assuming that $\bar{\lambda}$ is slowly increased from zero as the spheromak is established, it is seen that in order to ensure that none of the a_{mn} diverge, we must require $\bar{\lambda}^2$ to be smaller than the smallest of the σ_{mn}^2; this means that $|\bar{\lambda}|$ must be smaller than the uniform $|\lambda|$ corresponding to the isolated Taylor state; i.e., using Eq.(9.13) this means we require

$$|\bar{\lambda}| < \lambda_{Taylor} \equiv \sqrt{\frac{x_{11}^2}{a^2} + \frac{\pi^2}{h^2}}. \tag{12.41}$$

This restriction has implications for α, for consider Eq.(12.18) evaluated at the wall (i.e., where $\psi = 0$),

$$\lambda_{wall} = \bar{\lambda}(1 - \alpha). \tag{12.42}$$

If λ_{wall} is increased, then α has to adjust to satisfy Eq.(12.41) for

$$\bar{\lambda} = \frac{\lambda_{wall}}{1 - \alpha} < \lambda_{Taylor}. \tag{12.43}$$

This can be solved for α to give

$$\alpha < 1 - \frac{\lambda_{wall}}{\lambda_{Taylor}} \tag{12.44}$$

so that if $\lambda_{wall} > \lambda_{Taylor}$ then α will be negative. If α becomes so negative as to make λ go through zero, then the linear theory fails and a nonlinear dependence of λ on ψ should be used.

This result can be stated: resonances such as occurred in the strongly coupled (i.e., uniform λ) case are avoided providing $\bar{\lambda}$ (the flux averaged lambda) is smaller than the Taylor λ of an isolated state[126]. If λ_{wall} is increased, then λ falls off more quickly so as to keep $\bar{\lambda}$ below λ_{Taylor}.

12.5.2 Relationship between fluctuations, dissipation, lambda gradient, and helicity flux

The dynamo electric field must be parallel to **B** in order to sustain force-free current. Assuming an MHD dynamo, the effective dynamo electric field will be

$$\mathbf{E}_{dyn} = \frac{<\tilde{\mathbf{U}} \times \tilde{\mathbf{B}}> \cdot \mathbf{B}}{B^2}\mathbf{B} \tag{12.45}$$

12.5 MHD dynamo, helicity flux, and lambda gradient

where <> denotes time averaging. It is further assumed that the ideal MHD Ohm's law is a reasonable first approximation for the perturbed field, i.e., $\tilde{\mathbf{E}} + \tilde{\mathbf{U}} \times \mathbf{B} \approx 0$ so that Eq.(12.45) can be rearranged as

$$\begin{aligned}
\mathbf{E}_{dyn} &= <\tilde{\mathbf{E}} \cdot \tilde{\mathbf{B}}> \mathbf{B}/B^2 \\
&= -<\left(\nabla\tilde{\phi} + \frac{\partial \tilde{\mathbf{A}}}{\partial t}\right) \cdot \tilde{\mathbf{B}}> \mathbf{B}/B^2 \\
&= -<\nabla\cdot\left(\tilde{\phi}\tilde{\mathbf{B}}\right) + \frac{\partial \tilde{\mathbf{A}}}{\partial t}\cdot\nabla\times\tilde{\mathbf{A}}> \mathbf{B}/B^2;
\end{aligned} \quad (12.46)$$

this expression is gauge invariant, because the electric field is gauge invariant. Since the perturbations are assumed to be periodic, an integration by parts with respect to time shows that

$$\left\langle \frac{\partial \tilde{\mathbf{A}}}{\partial t}\cdot\nabla\times\tilde{\mathbf{A}} \right\rangle = -\left\langle \tilde{\mathbf{A}}\cdot\nabla\times\frac{\partial \tilde{\mathbf{A}}}{\partial t} \right\rangle \quad (12.47)$$

in which case

$$\begin{aligned}
\left\langle \frac{\partial \tilde{\mathbf{A}}}{\partial t}\cdot\nabla\times\tilde{\mathbf{A}} \right\rangle &= \frac{1}{2}\left\langle \frac{\partial \tilde{\mathbf{A}}}{\partial t}\cdot\nabla\times\tilde{\mathbf{A}} - \tilde{\mathbf{A}}\cdot\nabla\times\frac{\partial \tilde{\mathbf{A}}}{\partial t} \right\rangle \\
&= \frac{1}{2}\nabla\cdot\left\langle \tilde{\mathbf{A}}\times\frac{\partial \tilde{\mathbf{A}}}{\partial t} \right\rangle.
\end{aligned} \quad (12.48)$$

The dynamo electric field in Eq.(12.46) can therefore be written as

$$\mathbf{E}_{dyn} = \frac{\mathbf{B}}{2B^2}\nabla\cdot\mathbf{h} \quad (12.49)$$

where

$$\mathbf{h} = -\left\langle 2\tilde{\phi}\tilde{\mathbf{B}} + \tilde{\mathbf{A}}\times\frac{\partial \tilde{\mathbf{A}}}{\partial t} \right\rangle \quad (12.50)$$

is the MHD dynamo helicity flux *into* a volume (the minus sign corresponds to the definition of inwards flow into a volume, in contrast to the usual definition of flux as being out of a volume). Equation (12.49) is in the canonical form postulated by Boozer [134] for a mean-field dynamo.

In steady state, the dynamo electric field sustains the Ohmic voltage drop, i.e.,

$$\mathbf{E}_{dyn} = \eta \mathbf{J}. \qquad (12.51)$$

We now multiply Eq.(12.51) by **J** and integrate over a thin toroidal shell volume between two adjacent equilibrium flux surfaces. This gives

$$\int_V d^3r \mathbf{E}_{dyn} \cdot \mathbf{J} = \int_V d^3r \eta J^2 \qquad (12.52)$$

showing that dynamo power balances Ohmic dissipation. We now substitute for $\mathbf{E}_{dyn}$ to obtain

$$\frac{1}{2\mu_0} \int_V d^3r \, \lambda \nabla \cdot \mathbf{h} = \int_V d^3r \eta J^2. \qquad (12.53)$$

Integrating by parts on the left hand side gives[134]

$$\frac{1}{2\mu_0} \left(\int_{S_{out}} d\mathbf{s} \cdot \mathbf{h} \lambda + \int_{S_{in}} d\mathbf{s} \cdot \mathbf{h} \lambda - \int_V d^3r \, \mathbf{h} \cdot \nabla \lambda \right) = \int_V d^3r \eta J^2. \qquad (12.54)$$

where S_{out} and S_{in} are respectively the outside and inside surfaces of the thin toroidal shell. Since $d\mathbf{s}$ points in opposite directions on these inside and outside surfaces, the two surface terms cancel in the limit of infinitesimal shell thickness. Thus, the dynamo can be expressed as

$$-\frac{1}{2\mu_0} \int_V d^3r \, \mathbf{h} \cdot \nabla \lambda = \int_V d^3r \eta J^2. \qquad (12.55)$$

Since this holds locally on each thin toroidal shell, the integrands on the left and right hand sides must be equal[75], i.e.,

$$-\frac{1}{2\mu_0} \mathbf{h} \cdot \nabla \lambda = \eta J^2. \qquad (12.56)$$

Equation (12.56) shows three important properties of the dynamo:

1. There must be finite $\nabla \lambda$ everywhere if the dynamo is to balance Ohmic dissipation locally.
2. The direction of $\nabla \lambda$ must be opposite the direction of the helicity flux **h**.
3. There must be fluctuating fields everywhere inside V in order for **h** to be finite everywhere.

12.5 MHD dynamo, helicity flux, and lambda gradient

Property #2 shows that $\nabla \lambda$ must reverse direction when h reverses direction. Hence a spheromak that is driven by an external source will have an inward h and therefore λ will decrease going from the source to the spheromak. On the other hand, an isolated decaying spheromak will have an outward h and so λ will decrease going from the center outwards. This is in agreement with Knox et al.'s CTX measurements [132].

Since h depends on the existence of MHD fluctuations, and since the energy principle shows that these fluctuations can only grow if λ has a gradient, it should be possible to construct a quasi-linear dynamo model. In fact, using Eqs.(8.6) and (12.56), it is seen that this model would be of the form

$$\frac{D_K}{2\mu_0} |\nabla \lambda|^2 = \eta J^2. \tag{12.57}$$

where the helicity diffusion coefficient D_K would depend on the fluctuation amplitudes. Dynamo models of this sort have been discussed by Bhattacharjee and Hameiri [135] and also by Strauss[136]. One can conclude from the form of Eqs.(12.56) and (12.57) that the lambda gradient and helicity flux will be controlled by the way D_K depends on η. If $D_K \sim \eta^p$ where p is positive then for a given current Eq.(12.57) shows that $|\nabla \lambda| \sim \eta^{(1-p)/2}$ and Eqs.(8.6) shows that $|h| \sim \eta^{(1+p)/2}$. Thus, as the resistivity decreases the fluctuation level $\sim |h|$ will always decrease. However, $|\nabla \lambda|$ only decreases with decreasing resistivity if $p < 1$.

Although this simple model identifies several issues and relationships intrinsic to a dynamo, it does not describe reconnection physics or mode dynamics. It is likely that the ideal MHD description for the modes is only approximate. The actual modes most certainly will have to involve non-ideal behavior[135], either from resistivity or from non-MHD physics. It is also clear that quality of confinement is highly coupled to the dynamo problem because the dynamo involves fluctuations which tend to destroy confinement and yet the dynamo creates and sustains the closed flux surfaces which produce confinement. If confinement is good, then only a small dynamo is required. This intricate relationship highlights the importance of reducing all other forms of confinement degradation as much as possible. The next chapter discusses confinement in spheromaks and describes methods that have been used to improve confinement.

Finally, it should be noted that there is an interaction between the λ gradient and the q profile[103]. In Sec.11.3 it was shown that the safety factor on the magnetic axis, q_{axis}, varies approximately inversely with λ while Eq.(9.38) showed that the safety factor at the separatrix (wall or near wall), q_{sep}, is proportional to λ. Thus, a driven spheromak (λ peaked at wall) will tend to have both q_{axis} and q_{sep} increase relative to the corresponding Taylor state value while a decaying spheromak (λ peaked on magnetic axis) will have both q_{axis} and q_{sep} depressed relative to the Taylor state values. The net effect is approximately that of simply adding a constant to the q profile, the constant being positive for a driven spheromak and negative for a decaying spheromak. This shifting of the pedestal of the entire q profile was observed experimentally by Knox et al. [132] and also calculated numerically by Mayo and Marklin [102]. The shifting of the pedestal means that spheromaks sustained by coaxial guns have $q \approx 1$ while decaying spheromaks have q somewhat less than unity. The MHD modes for the sustained CTX experiment had $n = 1$ while the modes for the decaying spheromak had $n = 2$ or $n = 3$. The effect changes q but not q' and so decaying spheromaks will have larger q'/q than sustained spheromaks.

CHAPTER 13
Confinement and Transport in Spheromaks

13.1 Overview

The spheromak configuration is attractive because it offers a potentially low cost method for confining plasma. While confinement in spheromaks to date has been modest, it is not known whether this is a consequence of specific shortcomings of existing designs or else a fundamental defect of the spheromak concept. There is evidence that the former is the case suggesting that spheromak confinement could be much improved by understanding and then removing the shortcomings of present-day experiments.

The essence of toroidal magnetic confinement is the existence of closed flux surfaces: if flux surfaces do not exist or are not closed, then confinement will be severely degraded or lost entirely. The issue of whether flux surfaces are closed is related to the issue of symmetry and the Lagrangian concept of conservation of canonical momentum. In cylindrical coordinates r, ϕ, z the canonical angular momentum of a particle with mass m and charge q is

$$P_\phi = mrv_\phi + qrA_\phi. \tag{13.1}$$

The quantity rA_ϕ is proportional to the poloidal flux function, i.e., $\psi = 2\pi rA_\phi$ so the canonical angular momentum is just

$$P_\phi = mrv_\phi + q\psi/2\pi. \tag{13.2}$$

For a configuration symmetric in the ϕ direction, Lagrangian mechanics shows that P_ϕ is a constant of the motion for each particle. If a particle has zero mass, then the constancy of P_ϕ would mean that the particle must move in such a way that ψ remains constant, i.e., zero mass particles would be confined to surfaces of constant ψ. However, if the mass is finite, then the particle trajectory can deviate from a constant ψ surface; this deviation is estimated by Taylor expanding about the zero-mass trajectory,

$$\begin{aligned} \psi &= \psi^{\text{zero mass}} + \delta r \frac{\partial \psi}{\partial r} + \delta z \frac{\partial \psi}{\partial z} \\ &= 2\pi \left(q^{-1} P_\phi + \delta r \, rB_z - \delta z \, rB_r \right). \end{aligned} \tag{13.3}$$

If P_ϕ is a constant of the motion, then we may use Eqs.(13.1) and (13.3) to write

$$\begin{aligned}
0 &= P_\phi(r+\delta r, z+\delta z, v_\phi + \delta v_\phi) - P_\phi(r,z,v_\phi) \\
&= m\delta(rv_\phi) + q\delta\psi/2\pi \\
&= mr\delta v_\phi + mv_\phi \delta r + q\left(\delta r\, rB_z - \delta z\, rB_r\right).
\end{aligned} \qquad (13.4)$$

Both v_ϕ and δv_ϕ are of the order of the thermal velocity v_T, and typically $\delta r \ll r$; therefore Eq. (13.4) can be approximated as

$$0 = m\, v_T + q\left(\delta r\, B_z - \delta z\, B_r\right). \qquad (13.5)$$

The quantities B_r and B_z are the components of the poloidal field, while δr and δz are components of the displacement in the poloidal direction. Thus, $\delta r\, B_z - \delta z\, B_r \sim \Delta_{pol} B_{pol}$ where $\Delta_{pol} = \sqrt{(\delta r)^2 + (\delta z)^2}$ and $B_{pol} = \sqrt{B_r^2 + B_z^2}$. The deviation of a particle trajectory from the zero-mass flux surface is therefore the poloidal Larmor radius

$$\Delta_{pol} = \frac{m\, v_T}{qB_{pol}} \equiv r_{pol}; \qquad (13.6)$$

i.e., the Larmor radius determined using the poloidal magnetic field. Equation (13.5) shows that if there is symmetry in the ϕ direction each particle must remain within a poloidal Larmor radius of the flux surface associated with a zero-mass particle having the same canonical momentum. Because this argument applies to each particle, it shows that axisymmetry should produce perfect confinement – no particle will be able to deviate more than a poloidal Larmor radius from a flux surface. Interparticle collisions alter this picture because the force associated with a collision is localized and therefore not axisymmetric. In effect, collisions knock particles off flux surfaces. The toroidal and poloidal magnetic fields of a spheromak are comparable, so there is no significant distinction between the order of magnitude of the poloidal Larmor radius and the toroidal Larmor radius.

There is a fundamental conflict between the axisymmetry required for good confinement and the non-axisymmetry required for the dynamo that creates and sustains the spheromak. An essential issue for spheromaks is whether a middle ground exists where a modest non-axisymmetry can provide the dynamo needed to sustain the spheromak and yet not be so large as to adversely affect confinement.

Spheromak confinement is complicated and at the present time cannot be derived from first principles. Instead, a practical understanding of spheromak confinement can be pieced together from carefully designed experiments. While the actual confinement in spheromaks to date has not been competitive with tokamaks, there have been tantalizing hints that much better confinement may be achieved by eliminating specific, identified problems. For example, the energy confinement of early spheromaks was dominated by impurity radiation, a problem not intrinsic to the spheromak concept. Using wall conditioning techniques, impurity radiation has been reduced sufficiently so as to be unimportant compared to other loss mechanisms.

13.2 Confinement times

Since spheromak confinement is not well understood at the present time, the most important issue is identification of the dominant processes. Order of magnitude accuracy suffices for this purpose and there is little point in expending substantial effort to reconcile discrepancies that are of the order of factor of two or so.

The confinement properties of any physical quantity X can be characterized by a rate equation

$$\frac{dX}{dt} = -\frac{X}{\tau_X} + s_X \tag{13.7}$$

where s_X is a constant source function and τ_X is a decay time (also called confinement time). In the limit of large τ_X, X increases as $X(t) = X(0) + ts_X$, while in the limit of no source X decays as $X(t) = X(0)\exp(-t/\tau_X)$. Equation (13.7) can be solved for the confinement time

$$\tau_X = \frac{-X}{dX/dt - s_X} \tag{13.8}$$

so that determination of the confinement time requires measurement of X, dX/dt, and s_X. This determination is easiest in the special situations where s_X is small enough to be ignored or where dX/dt is nearly zero.

The physical quantities of greatest interest are magnetic helicity, magnetic energy, electron and ion thermal energy, and particle density. The relevant confinement times are denoted as the helicity decay time τ_K, the magnetic energy decay time τ_B, the thermal energy confinement time τ_E, and the particle confinement time τ_p. The magnetic energy decay time can be further sub-categorized into a toroidal field energy decay time τ_{Bt} and a poloidal field energy decay time τ_{Bp}. Sometimes the plasma is modeled as a short-circuited inductor L in series with a resistor R, a system which has an energy decay time $L/2R$; this rate is called the resistive time scale and is the rate at which global processes cause magnetic energy and helicity to decay if the system behaves as a rigid conductor with finite resistivity.

13.3 Survey of transport mechanisms

The thermal energy confinement time (usually called the energy confinement time) ought to be related to the confinement properties of the magnetic configuration; i.e., to the quality of the flux surfaces. Much is known about energy confinement in tokamaks and stellarators, less is known about energy confinement in reversed field pinches and very little is known about energy confinement in spheromaks. Energy transport can be caused by diffusion, conduction, or radiation.

13.3.1 Diffusive processes

In a diffusive process the outward flux Γ_f of a quantity f is related to the gradient of f by

$$\Gamma_f = -D\nabla f \qquad (13.9)$$

where D is called the diffusion coefficient. Combining Eq.(13.9) with the continuity equation for f gives the diffusion equation

$$\frac{\partial f}{\partial t} = D\nabla^2 f. \qquad (13.10)$$

Let us consider diffusion of f in cylindrical geometry with the boundary condition that $f = 0$ at $r = a$. This boundary condition can be satisfied by having $f \sim J_0(rx_{01}/a)$. If we also assume that f has a temporal dependence $\sim \exp(-\gamma t)$ then substitution into Eq.(13.10) gives $\gamma = Dx_{01}^2/a^2$. Thus, the solution to the diffusion equation in cylindrical geometry with boundary condition $f = 0$ at $r = a$ decays self-similarly as

$$f = J_0(rx_{01}/a) \exp\left(-Dtx_{01}^2/a^2\right) \qquad (13.11)$$

and so the e-folding time will be

$$\tau = \frac{a^2}{Dx_{01}^2} \simeq \frac{a^2}{6D}. \qquad (13.12)$$

Diffusion is a random-walk process; if the size of the random step is Δx and the time between successive random steps is Δt, then the diffusion coefficient scales as $D \sim (\Delta x)^2/\Delta t$.

Bohm diffusion

This diffusion has more of a historical than a physical basis, but is still used as a benchmark because it often gives a reasonable description of poorly confined plasmas. Bohm diffusion is given by

$$D_{Bohm} = \frac{1}{16}\frac{T_e}{B} \text{ m}^2/\text{s} \qquad (13.13)$$

where T_e is in eV, B is in Tesla, and the coefficient 1/16 was originally chosen to fit early measurements. It is the simplest dimensionally correct diffusion coefficient that

13.3 Survey of transport mechanisms

depends on both T and on B and which prescribes improved confinement with increasing magnetic field, as would be expected for a magnetic confinement system.

Classical/neoclassical diffusion

The best conceivable energy confinement would be classical/neo-classical where thermal energy is transported out of the plasma because of occasional Coulomb collisions between gyrating particles.

In a strongly magnetized plasma only collisions between unlike particles can produce particle transport [137]. The classical particle diffusion rate is intrinsically ambipolar and can be determined by eliminating **J** from the steady state MHD Ohm's law

$$\mathbf{U} \times \mathbf{B} = \eta_\perp \mathbf{J} \tag{13.14}$$

using the MHD force balance

$$\mathbf{J} \times \mathbf{B} = \nabla P \tag{13.15}$$

to obtain

$$\mathbf{U}_\perp = -\frac{\eta_\perp}{B^2} \nabla_\perp P \tag{13.16}$$

so that the outward particle flux for constant temperature is

$$\Gamma_\perp = -\frac{\eta_\perp n \kappa (T_e + T_i)}{B^2} \nabla_\perp n. \tag{13.17}$$

The classical particle diffusion coefficient is therefore

$$\begin{aligned}
D_{cl}^{particle} &= \frac{\eta_\perp n \kappa (T_e + T_i)}{B^2} \\
&= \frac{m_e \nu_{ei}(T_e + T_i)}{e^2 B^2} = 2\nu_{ei} r_{Le}^2
\end{aligned} \tag{13.18}$$

where ν_{ei} is the electron-ion collision frequency, r_{Le} is the electron Larmor radius, and $T_e = T_i$ has been assumed.

The classical transport rate has a spatial step size corresponding to the Larmor radius and a time step corresponding to the electron-ion collision period (i.e., a collision between unlike particles). This diffusion coefficient could also have been expressed in terms of ion quantities since $m_e \nu_{ei} = m_i \nu_{ie}$. Using $\eta_\perp = 10^{-4} Z \ln \Lambda / T_e^{3/2}$, the classical particle diffusion coefficient can be expressed numerically as

$$D_{cl}^{particle} = 3.2 \times 10^{-23} \frac{n\, Z \ln \Lambda}{T_e^{1/2} B^2}\ \mathrm{m^2/s} \tag{13.19}$$

assuming $T_e = T_i$.

While like-particle collisions do not provide particle transport, like-particle collisions do transport energy. Since $v_{ei} \sim v_{ee}$ and $v_{ii} \sim (m_i/m_e)^{1/2} v_{ie}$, ion-ion collisions provide the dominant perpendicular energy transport and the diffusion coefficient for this ion-ion energy transport is approximately $(m_i/m_e)^{1/2}$ larger than the particle transport coefficient. Thus, the classical energy diffusion rate will be

$$D_{Ecl}^{ion} = \left(\frac{m_i}{m_e}\right)^{1/2} \frac{\eta_\perp n \kappa T_i}{B^2}. \tag{13.20}$$

Neoclassical diffusion takes into account toroidal geometry effects and in accordance with Eq.(13.6) the step size for diffusion becomes the poloidal Larmor radius $\sim T^{1/2}/B_{pol}$; i.e., the poloidal field is used instead of the toroidal field. For spheromaks, there is no significant distinction between classical and neoclassical transport because B_{pol} is of the order of B_{tor}.

Drift wave turbulence diffusion

Confinement in tokamaks has been demonstrated during the 1980's and 1990's to be governed largely by drift wave turbulence. The transport is still diffusive but the random step size is now $2\pi/k_\perp$ where $k_\perp$ is the nominal wavenumber of the turbulence in the poloidal cross-section. The random step time is the life-time $2\pi/\omega^*$ of a turbulence cell where $\omega^* = k_\perp T/aB$ is the characteristic frequency of the turbulence. The k-spectrum of drift turbulence is such that $k_\perp \rho_s \sim 1$ where $\rho_s = (T_e/T_i)^{1/2} r_{Li}$ is an effective Larmor radius. Assuming equal electron and ion temperatures, the diffusion associated with drift wave turbulence will have the scaling

$$D_{turb} \approx 2\pi \omega^*/k_\perp^2 = 2\pi\, \omega^* \rho_s^2 = 2\pi \frac{T}{B} \frac{r_{Li}}{a} \approx 32\pi \frac{r_{Li}}{a} D_{Bohm}. \tag{13.21}$$

This scaling is sometimes called gyro-Bohm because it is Bohm diffusion times the factor r_{Li}/a. Turbulent diffusion in tokamaks has been substantially reduced by decreasing the magnitude of the drift wave turbulence; this is accomplished by shearing the plasma rotational velocity to stretch and thin out the turbulence cells. The energy confinement time for this drift wave diffusion scales as $B^2 T^{-3/2}$, so that the temperature scaling is less favorable than for classical/neoclassical diffusion. If $r_{Li}/a \sim 10^{-2}$ then it is diffi-

13.3 Survey of transport mechanisms

cult to distinguish gyro-Bohm scaling from Bohm scaling; such a situation often occurs because the minor radius of a device is typically $\sim 10^2$ ion Larmor radii.

Transport along open field lines

In the worst case, the configuration is permeated by open field lines so that particles in the central core simply travel along field lines to the wall. If the collisional mean free path is shorter than the path length along the open field line, this process will be diffusive; otherwise it will be a free flow. In both cases the outward particle flux will have to be ambipolar to maintain charge neutrality. For comparable electron and ion temperatures, the ions will not move as fast as the electrons and so an ambipolar electric field will be established which will retard the electrons and accelerate the ions until there are equal ion and electron fluxes. In the free flow case, the plasma convects out at the ion acoustic velocity $c_s = \sqrt{\kappa T_e/m_i}$ while in the diffusive case, the ambipolar parallel diffusion coefficient will be[138] $D_{amb} = 2v_{T_i}^2/\nu_{ii}$. If l is the length of the open field line, the parallel confinement time will be $\tau_\parallel = l/c_s$ for convective flow and will be $\tau_\parallel = l^2/4D_{amb}$ for diffusive flow. Open field lines could be either permanent or transient. Open field lines occur when there is loss of axisymmetry so that flux surfaces fail to exist. Loss of axisymmetry can also result in the field lines becoming stochastic so that the trajectory of a field line fills up a volume rather than traces out a surface. Parallel transport will dominate perpendicular transport if $\tau_\parallel < \tau_\perp$.

The Taylor state $\nabla \times \mathbf{B} = \lambda \mathbf{B}$ allows equilibria with both closed flux surfaces (axisymmetric solutions) and stochastic open field lines. This can be seen from Eq.(12.16) which shows that λ is constant along a field line. Since the Taylor state has λ constant throughout the entire volume, it is consistent with the Taylor state to have one stochastic field line fill up the entire volume. While this is allowed, it need not be the case, because as shown in Chapter. 9, axisymmetric isolated Taylor states have associated flux surfaces. Furthermore, usually there is a λ gradient, so the field lines are not degenerate with respect to λ and a single field line could not fill up the entire volume.

13.3.2 Non-diffusive processes

Line radiation

Energy confinement in plasmas with electron temperatures below about 50 eV is typically dominated by optical line radiation from impurities, especially carbon and oxygen. These impurities originate from the wall and can be reduced by wall conditioning. Radiation energy loss is a particularly serious problem in small experiments because the ratio of wall surface area to plasma volume is relatively large. As the plasma is heated to higher temperatures, the impurities become multiply-ionized and do not radiate so strongly.

Charge exchange

When a moderate energy ion collides with a cold neutral there is a possibility for the ion to capture an electron from the neutral; the hot ion is neutralized becoming a hot neutral and the original neutral is ionized becoming a cold ion. This process, called charge exchange, has a cross-section comparable to the ionization cross-section. If there is a substantial neutral density, as often occurs at the plasma edge, then charge exchange can constitute a significant ion energy loss mechanism.

13.3.3 Magnetic energy and magnetic helicity

Magnetic energy dissipates rapidly as the configuration relaxes to a Taylor state (see Sec.4.2), but once in a Taylor state, there are no longer fine-scale dissipative processes, the circuit topology stays constant, and the magnetic energy now decays at the same slow, resistive rate as helicity. Let us consider the entire helical current in a spheromak as a bundle of N helical conductors each carrying current I_i so that $I = \sum_{i=1}^{N} I_i$. Suppose that the subdivision into individual conductors has been arranged so that all the I_i are equal and further suppose that these N conductors are connected in series so as to have a total inductance L. If Φ_i is the flux linked by the ith conductor with contour C_i and cross-section S_i, the inductive energy of this system is

$$\begin{aligned}
\frac{1}{2}LI^2 &= \frac{1}{2}\sum_{i=1}^{N}\Phi_i I \\
&= \frac{1}{2}\sum_{i=1}^{N}\oint_{C_i} d\mathbf{l}\cdot\mathbf{A}\int d\mathbf{s}\cdot\mathbf{J} \\
&= \frac{1}{2}\int d^3r\,\mathbf{A}\cdot\mathbf{J} \\
&= \frac{1}{2\mu_0}\int d^3r\,\mathbf{A}\cdot\nabla\times\mathbf{B} \\
&= \frac{1}{2\mu_0}\int d^3r\left(\nabla\cdot(\mathbf{B}\times\mathbf{A})+B^2\right).
\end{aligned} \quad (13.22)$$

Using the integral of the force-free state, i.e., $\mathbf{B} = \lambda\mathbf{A}+\nabla f$, the term $\nabla\cdot(\mathbf{B}\times\mathbf{A})$ vanishes and it is seen that

$$\int \frac{B^2}{2\mu_0}d^3r = \frac{1}{2}LI^2. \quad (13.23)$$

Since the power to sustain the magnetic energy is $P = IV$ where V is the voltage

13.3 Survey of transport mechanisms

driving the current I, the magnetic energy decay time is

$$\tau_B = \frac{\int \frac{B^2}{2\mu_0} d^3r}{P} = \frac{\frac{1}{2}LI^2}{IV} = \frac{L}{2R} \tag{13.24}$$

where R is the series resistance of the inductor. Hence, the magnetic energy of a Taylor state decays at the resistive rate.

An alternate way to express this decay time is to invoke the force-free relation $\mathbf{J} = \lambda \mathbf{B}/\mu_0$ to obtain

$$\tau_B = \frac{\int \frac{B^2}{2\mu_0} d^3r}{\int \eta_\parallel J^2 d^3r} = \frac{\mu_0}{2\lambda^2} \frac{\int B^2 d^3r}{\int \eta_\parallel B^2 d^3r} \tag{13.25}$$

where we have used the parallel resistivity because $\mathbf{J} = \lambda \mathbf{B}$. If the resistivity is uniform, then

$$\tau_B = \mu_0 / 2\eta_\parallel \lambda^2, \tag{13.26}$$

while if the resistivity is non-uniform, the decay rate will be dominated by the regions with high resistivity. The inverse dependence on λ^2 immediately shows that large configurations should last longer. Equation (4.12) showed that the helicity for an isolated spheromak is $K = 2\mu_0 W_B/\lambda$ so the helicity decay rate of a Taylor state is the same as the magnetic energy decay rate, i.e., $\tau_K = \tau_B$. This is to be contrasted to the process of relaxation to a Taylor state where magnetic energy decays rate much faster than magnetic helicity because of fine-scale reconnection.

13.3.4 Relationship between magnetic energy and thermal energy decay times

The magnetic energy density is $W_B = B^2/2\mu_0$ and the thermal energy is $W_{th} = \frac{3}{2}(n_i \kappa T_i + n_e \kappa T_e)$. Charge neutrality gives $Zn_i = n_e$ where Z is the charge on the ions. The ratio of pressure to magnetic energy is defined as β, i.e.,

$$\beta \equiv \frac{n_e(Z^{-1}\kappa T_i + \kappa T_e)}{B^2/2\mu_0} = \frac{2}{3}\frac{W_{th}}{W_B}. \tag{13.27}$$

In a decaying spheromak with no external power input, the power lost by the decaying magnetic field acts as a source which heats the particles. Thus, the magnetic rate equation is

$$\frac{dW_B}{dt} = -\frac{W_B}{\tau_B} \tag{13.28}$$

and the power going into heating the particles is just $-dW_B/dt$, i.e., magnetic energy is converted into thermal energy. Using Eqs.(13.26) and (13.27) the thermal rate equation becomes

$$\begin{aligned}\frac{dW_{th}}{dt} &= -\frac{W_{th}}{\tau_E} + \frac{W_B}{\tau_B} \\ &= -\left(\frac{D_E x_{01}^2}{a^2} - \frac{4}{3}\frac{\eta_\| x_{11}^2}{\beta a^2 \mu_0}\right) W_{th}.\end{aligned} \tag{13.29}$$

If the heating from magnetic energy dissipation suffices to balance energy transport, the thermal energy will remain in steady state; such a balancing would require an energy diffusion

$$\frac{D_E}{D_{Ecl}^{ion}} = 7\frac{\eta_\|}{\eta_\perp}\frac{\sqrt{m_e/m_i}}{\beta^2} \tag{13.30}$$

where D_{Ecl}^{ion} is given by Eq.(13.20). For a hydrogen plasma with $\eta_\perp = 2\eta_\|$ this gives $D_E/D_{Ecl}^{ion} \approx 0.1/\beta^2$ showing that the energy transport would have to be much faster than classical in a low β plasma for a steady state to result. In such a case the relation between thermal and magnetic confinement times would be

$$\tau_E = \frac{3}{2}\beta\tau_B. \tag{13.31}$$

The thermal energy confinement time τ_E is smaller than the magnetic energy confinement time τ_B because τ_B is a measure of the lifetime of the confinement system while τ_E is a measure of the effectiveness of the confinement system.

13.3.5 Dissipation in a single flux tube

If a system is force-free, but not in a Taylor state, then Eq.(12.16) indicates that λ is constant along a field line rather than uniform throughout the entire volume. Thus, if flux surfaces exist, λ is a surface quantity, i.e., $\lambda = \lambda(\psi)$. If there are no sources and the configuration is enclosed by a flux conserving wall, the respective energy and

13.3 Survey of transport mechanisms

helicity decay equations for a given flux tube will be[139]

$$\frac{dW_B}{dt} = -\int_{FT} d^3r\, \mathbf{E}\cdot\mathbf{J},$$
$$\frac{dK}{dt} = -2\int_{FT} d^3r\, \mathbf{E}\cdot\mathbf{B} \qquad (13.32)$$

where the integrals are over the volume of the flux tube. The volume elements making up a flux tube can be expressed as $d^3r = d\mathbf{s}\cdot d\mathbf{l}$ where $d\mathbf{s}$ is the cross-section of the flux tube and $d\mathbf{l}$ is the length along the flux tube. The axial current carried by the flux tube is $I = \mathbf{J}\cdot d\mathbf{s}$ and the flux is $\Phi = \mathbf{B}\cdot d\mathbf{s}$. Thus, the decay equations become

$$\frac{dW_B}{dt} = -I\int d\mathbf{l}\cdot\mathbf{E} = IV,$$
$$\frac{dK}{dt} = -2\Phi\int d\mathbf{l}\cdot\mathbf{E} = 2\Phi V \qquad (13.33)$$

where $V = -\int \mathbf{E}\cdot d\mathbf{l}$ is the voltage drop along the length of the flux tube. One can define an effective average electric field for the flux tube as[139]

$$E_{eff} = \frac{\int d\mathbf{l}\cdot\mathbf{E}}{\int dl} \qquad (13.34)$$

where $\int dl$ is the length of the flux tube. The helicity decay equation can be solved to give

$$E_{eff} = -\frac{dK/dt}{2\Phi\int dl} \qquad (13.35)$$

a useful expression, because all quantities on the right hand side are measurable; Fernandez et al.[139] determined E_{eff} for the outer flux tubes using this method. The outer flux tubes are nearly completely poloidal and in close proximity to the wall. Fernandez et al. measured E_{eff} in CTX experiments configured with an open-mesh flux conserver and found the relation between E_{eff} and the neutral pressure was reminiscent of the Paschen curves characterizing breakdown of a neutral gas (see Sec.14.1). Fernandez et al. interpreted this correlation as a regulation by electron-neutral collisions of the effective electric field along the flux tube. This interpretation implies that collisions with neutrals dominated other loss mechanisms for the outer field lines of the open flux conserver configuration. This is an undesirable situation and indicates the importance of reducing the edge neutral density so that neutral collisions become less important than Coulomb collisions.

13.4 Experiments on transport in spheromaks

Because there are so many possible transport mechanisms, the dominant transport mechanism can vary from one device to another and from one operating regime to another. Transport must be investigated on each device via careful experimental measurements and once the dominant mechanism has been identified, steps can be taken to reduce it. We now discuss transport measurements that have been made on several devices.

13.4.1 S-1 measurements

Levinton et al.[140] measured the magnetic and energy confinement times in the Princeton S-1 spheromak. The main result was that β in the central core region was proportional to J^2 over the entire operating range whereas the volume-averaged β scaled as J (the central core region was defined as the region $r < a/3$ where r is the minor radius, and a is the minor radius of the outermost closed flux surface). This discrepancy between core and volume-averaged scaling indicates confinement in the core is better than the volume-averaged confinement time. Also since $J \sim B$ these measurements can also be interpreted as giving the scaling of β with B.

The S-1 group also determined magnetic energy decay times and found that the ratio of τ_{Bp} to τ_{Bt} remained at approximately 1.4 indicating that the toroidal field was more short-lived than the poloidal field. This is consistent with the edge having higher resistivity than the region near the magnetic axis (core region) because the currents which generate the toroidal field are poloidal currents flowing near the plasma edge whereas the currents generating the poloidal field are toroidal currents flowing near the magnetic axis. This difference in rates indicates that a decaying spheromak will cease to be in a Taylor state because decay results in excess poloidal field energy compared to toroidal field energy. A new relaxation will have to occur to equalize the toroidal and poloidal field energies.

The S-1 magnetic energy decay times were measured to be in the range $100 - 400$ μs and thermal energy decay times in the range 6-12 μs.

13.4.2 S-1 particle confinement measurement

Mayo et al.[141] measured the local carbon diffusion in the S-1 spheromak by following the diffusion of locally injected carbon atoms using spectroscopic methods.

13.4 Experiments on transport in spheromaks

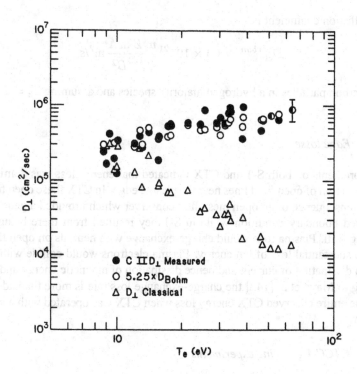

Fig.13.1 Experimentally measured carbon II diffusion coefficient $D_\perp$ on the S-1 spheromak from Mayo et al. [141]. The measured diffusion (solid circles) is in good agreement with $5.25 D_{Bohm}$ (open circles), and is much larger than classical diffusion (triangles).

They determined that the carbon diffusion coefficient was consistent with Bohm scaling, but had a magnitude about five times higher. The measured diffusion coefficient, shown in Fig.13.1, was as much as two orders of magnitude higher than the classical diffusion coefficient calculated assuming the carbon test particles were dominantly colliding with the hydrogen majority species. The carbon-hydrogen collision process means that the carbon-proton collision frequency ν_{CH} should be used instead of the electron-proton collision frequency which has the effect of increasing the diffusion coefficient by a factor $\sim (m_p/m_e)^{1/2}$ since the reduced mass of a carbon-proton system is used instead of the reduced mass of an electron-proton system. Thus, the classical carbon diffusion coefficient is

$$D_{cl}^{carbon} = 1.4 \times 10^{-21} \frac{n\, Z \ln \Lambda}{T_e^{1/2} B^2} \text{ m}^2/\text{s} \qquad (13.36)$$

for test carbon particles in a hydrogen majority species and assuming $T_e = T_i$.

13.4.3 Edge losses

Measurements on both S-1 and CTX indicated that energy loss was dominated by the large fraction of open field lines near the plasma edge. In CTX these open field lines were the consequence of an open mesh flux conserver which provided a geometrically complicated boundary condition while in S-1 they resulted from there being no flux conserving wall. Plasma ions would charge-exchange with neutrals on open field lines leading to substantial loss of ion energy. Plasma electrons would collide with neutrals leading to dissipation of current and hence dissipation of magnetic energy and helicity. According to Mayo et al.[142] the charge-exchange loss rate is more than adequate to explain the entire observed CTX energy loss when CTX was operated with a mesh flux conserver.

13.4.4 CTCC-1 gettering experiments

Uyama et al.[111] investigated how impurity radiation affected spheromak performance and showed that desorption of oxygen and carbon from the flux conserver wall was the principal mechanism causing poor confinement in their experiment. They addressed the following questions: (1) does ablation of the gun electrodes lead to injection of high Z metal impurities into the spheromak, (2) are low Z impurities injected into the spheromak from the gun electrodes or from the flux conserver wall, (3) is it possible to control impurities by wall conditioning, and (4) does reduction of impurities improve spheromak performance?

Uyama et al. measured the Doppler shift of impurity radiation along the line of sight from the gun electrode into the flux conserver; a substantial blue shift would indicate a directed velocity of the emitting impurity ion from the gun into the flux conserver. They found that the Doppler blue shift was negligible for the Cr II line (chromium) compared to that of the OIII (oxygen) line. This indicated that the heavy chromium impurity (a constituent of the steel alloy in the gun electrode) was not transported into the main discharge, whereas the much lighter oxygen impurity was indeed convected into the main discharge. Thus, ablation of heavy metal impurities from the gun electrode did not appear to be a problem.

The CTCC group then examined the effectiveness of titanium gettering of the flux conserver wall. This standard vacuum technique involves electrically heating a small titanium ball to a high temperature so that it sublimates and deposits a thin layer of titanium on all exposed surfaces. A retractable titanium ball was inserted into the interior of the flux conserver and operated for 60 hours, depositing many monolayers of titanium. The ball was retracted at intermediate time intervals of 15, 30, and 45 hours so that spheromak discharges could be made at these times to determine the dependence of impurity line emission on titanium gettering. The results were quite definitive and showed a monotonic decrease of the O II, C II and C III emission lines such that after 60 hours these lines became negligible. On the other hand, the O V line started to increase after about 30 hours of gettering. Since the O V line only exists when the electrons are hotter than 10 eV, this increase of O V indicated that the electron temperature increased substantially due to gettering. When the gas puff valve timing was adjusted to fuel only the region near the gun, Uyama et al. noted that the O III, O IV, and O V signals developed sequentially in the spheromak. This was interpreted as indicating that highly ionized oxygen was not being directly transported from the gun and instead the oxygen was being desorbed from the wall of the flux conserver and then evolved through a sequence of increasing ionization states in the spheromak. These measurements indicate that desorption of carbon and oxygen impurities from the flux conserver wall is a critical problem which must be resolved in order to improve spheromak performance.

13.4.5 The CTX gettered, solid flux conserver experiment

The preceding discussion showed that leading causes for transport in spheromaks are:
1. line radiation from wall-originated impurities,
2. excess neutrals causing drag on the current-carrying electrons and charge exchange with ions,
3. open field lines due to the error fields of an imperfect flux conserver,
4. transient open field lines associated with dynamo action.

It is important to minimize all these effects in order to improve confinement. The CTX group, realizing that their open-mesh flux conserver produced large field errors (i.e., open field lines) rebuilt CTX to have a solid, highly symmetric flux conserver. Furthermore, the inside wall of this solid flux conserver was titanium gettered as in CTCC. Experiments were performed using a quiescent decaying spheromak so that the spheromak would be axisymmetric, have no MHD activity, and therefore have good flux surfaces. Gas puffing rather than static fill was used to avoid excessive neutral density. The result[66] of these carefully optimized methods was that both τ_E and τ_B increased by an order of magnitude. In particular, τ_E increased to 0.2 ms while τ_B increased to 2 ms. These were achieved in spheromaks having $T_e = 100$ eV. Later measurements in a much smaller flux conserver[67] indicated that T_e could be as high as 400 eV, but with a reduced confinement time.

13.4.6 Evidence for good confinement: hard X-rays on CTX

The evidence presented so far suggests that confinement in early spheromaks was poor, but could be substantially improved by identifying and then minimizing the various causes of excessive transport. There also exists some intriguing evidence suggesting that, under the right circumstances, confinement in spheromaks can be rather good. This evidence is the observation by Chrien et al.[143] of hard x-ray bursts during the decay phase of the CTX spheromak. These bursts lasted 20-100μs and usually occurred during the quiescent decay of an isolated spheromak with short-circuited (crowbarred) gun power supply. The x-rays were measured with plastic scintillators and also with germanium detectors. The x-rays penetrated 12.7 mm thick stainless steel with about a 40% attenuation, suggesting an x-ray energy of 1-3 MeV.

The x-rays were presumably generated by 1-3 MeV electrons striking the wall. Such a high electron energy indicates the existence of excellent flux surfaces because the only electric field available to accelerate electrons would be the electric field generated by the decaying magnetic flux of the isolated spheromak. Chrien et al. estimated this electric field to be 200 V/m on the magnetic axis and 100 V/m at the plasma edge. Because the circumference of the magnetic axis was about 1 m, an electron would have to make approximately 5000 toroidal transits or 10,000 poloidal transits in order to be accelerated to 1 MeV. Equivalently, the electron would have to travel 5-10 km before being lost. This demonstrates that excellent confinement occurs at least some of the time in a decaying spheromak. The acceleration time required to reach these energies is $20 - 50$ μs, indicating that the good confinement must last this long.

13.5 Anomalous ion heating

Because spheromaks typically have both a large current density J and a low electron temperature T_e, it was assumed for many years that the dominant heating mechanism in spheromaks would be classical Ohmic heating, a mechanism which heats electrons but not ions. The relatively high density in spheromak experiments also indicated an electron-ion energy equilibration rate shorter than the experimental duration so that it was assumed that the ions would be in thermal equilibrium with the electrons, i.e., $T_i = T_e$.

However, measurements of the Doppler width of impurity line optical emission con-

13.5 Anomalous ion heating

sistently showed that impurity ions were two to six times hotter than the electrons (a similar situation is observed in reversed field pinches). The density of the various impurities is very small compared to the majority species (typically hydrogen) and it was not clear whether these high impurity temperatures were representative of the majority species temperature or simply the result of a heating mechanism specific to impurities. This question was settled by Mayo et al. [144] who measured the majority species temperature directly using charge exchange neutral particle energy analysis and demonstrated that the impurity ion temperature was indeed representative of the majority species ion temperature.

Determining ion temperature from impurity temperature is not completely straightforward because as the temperature increases, additional electrons are stripped from the impurity ions resulting in different spectral characteristics. However, the mere existence of highly ionized ions indicates high ion temperatures. Mayo, Hurlburt, and Fernandez[145] deconvolved the impurity ion temperatures in the CTX and S-1 spheromaks to calculate ion temperature profiles and found, for CTX configured with a solid flux conserver, that the ion temperature was approximately twice the electron temperature while for the S-1 spheromak the ion temperature was approximately six times the electron temperature. Several cases were studied for each device. For CTX two specific cases involved core electron temperatures of 100 eV and 400 eV which had respective core ion temperatures of 250 and 1000 eV respectively. For S-1 two cases involved electron core temperatures of 20 and 50 eV which had respective core ion temperatures of 300 and 400 eV.

These $T_i \gg T_e$ situations indicate that contrary to classical assumptions:

1. electrons cannot be heating ions (it must be the other way around),
2. there must exist a powerful anomalous ion heating mechanism,
3. most of the thermal energy is in the ions.

Considerable additional insight into anomalous ion heating has been provided by Ono et al.[146] who observed strong ion heating in an experiment where two spheromaks with opposite helicity polarities were merged[46]. These spheromaks had oppositely directed toroidal fields and the same sense of poloidal field. Thus the toroidal currents were parallel causing the two spheromaks to attract each other. It is important to realize that the toroidal currents of two adjacent coaxial spheromaks will exert forces on each other, but the corresponding poloidal currents will not; this is because the toroidal current creates poloidal fields which extend outside the spheromak, whereas the poloidal current creates toroidal fields which exist only within the spheromak.

The merging of two spheromaks with opposite initial helicities[46] resulted in a plasma having zero net helicity (i.e., a field reversed configuration); this resulted from the annihilation of the oppositely directed toroidal magnetic fields of the original spheromaks. Annihilation of the toroidal fields meant that the field lines transform from being helical to being purely poloidal, in other words the field lines were partially straightened. This transformation means that a field line on the large radius side of the configuration would have a positive toroidal velocity while a field line on the small radius side would develop a negative toroidal velocity.

Ono et al. used Doppler shift measurements to determine time and space-resolved ion flow in the reconnection regions and found ion flow velocities consistent with these toroidal velocities of the magnetic field. This transformation of initially helical magnetic field lines into purely poloidal field lines was first recognized by Yamada et al.[46] who called this a "slingshot" process because it resembles the straightening out of a stretched rubber band. The direction of ion motion was consistent with the plasma being frozen into the magnetic field, but the measured plasma velocities were somewhat smaller than the velocity of the moving, straightening magnetic field lines. Ono et al. observed strong ion heating at the time of maximum velocity shear and in particular the ion temperature increased from 10 eV to 200 eV in about 20 μs. The cause of this very strong ion heating was postulated to be ion viscosity resulting from the strong velocity shear (however, a quantitative estimate of the ion viscosity and the viscous heating rate was not provided). An energy budget showed that about 80% of the energy in the initial, oppositely directed toroidal fields was transformed into ion thermal energy. A 3D MHD simulation of two merging flux tubes illustrated the main qualitative features of the slingshot mechanism in a manner reasonably consistent with the experiment. Kornack et al.[147] observed similar behavior in a merging spheromak experiment at Swarthmore and in particular, measured very energetic ion flows produced by reconnection.

Ono et al.'s experiment is probably an extreme situation since it involves counterhelicity merging and complete annihilation of helicity. Nevertheless, it seems likely that similar processes would occur in all situations where reconnection takes place. The essential concepts revealed in the Ono et al. experiment are (i) reconnection involves the snapping back of stretched field lines, (ii) plasma attached to these snapping field lines will be accelerated to Alfvenic velocities, (iii) substantial ion sheared flow develops, and (iv) the shear is dissipated by ion viscosity resulting in significant ion heating. Classical ion viscosity depends on ion-ion collisions and so should be negligible in a collisionless plasma. It is not clear whether the viscous damping observed by Ono et al. was classical or anomalous.

Other possible mechanisms which could explain the anomalously hot ions include stochastic ion heating and ion Landau damping. Stochastic ion heating due to turbulent drift waves has been observed by McChesney, Stern, and Bellan[148] in Encore, a small research tokamak at Caltech, and occurs when the displacement of ions due to polarization drift becomes comparable to the perpendicular wavelength of the electrostatic wave causing the drift. Ion Landau damping results when the parallel phase velocity of an electrostatic wave is comparable to or somewhat higher than the ion thermal velocity; this damping is a well-known feature of ion acoustic waves[149]. Bellan[88] has shown that ion Landau damping should be important for kinetic Alfvén waves since these waves have parallel dynamics similar to ion acoustic waves. Bellan has proposed that reconnection is mediated by the excitation of a shear Alfvén wave and so if this wave is a kinetic Alfvén wave there could be substantial ion heating via ion Landau damping. The shear Alfvén wave constitutes the dynamical untwisting of field lines during reconnection and so is effectively a self-consistent dynamical description of the slingshot concept discussed by Ono et al.

Finally, Jarboe et al.[67] noted that when confinement was improved in a spheromak (e.g., by using a better flux conserver design and/or by gettering), the ratio of ion to electron temperature would be reduced compared to situations with poor confinement. This behavior occurs because better confinement means less dynamo activity is required to sustain the configuration, less dynamo activity means less reconnection, and less reconnection means less direct ion heating. Thus, a high T_i/T_e ratio is a sign of poor confinement and associated strong dynamo activity striving to replenish the losses.

CHAPTER 14
Some Important Practical Issues

While it is relatively easy to form a basic spheromak, it takes considerable effort to form a spheromak with high temperature, long life, and good confinement properties. This effort entails minimizing non-ideal phenomena that can easily overwhelm the desired spheromak attributes if left unattended. In this chapter we examine several practical issues which been of critical importance for the spheromak experiments of the last two decades. Undoubtedly new issues will arise in future experiments as parameter regimes change.

14.1 Breakdown and Paschen curves

It has been tacitly assumed that one can create a plasma at will, but most experimentalists have suffered the embarrassment of not achieving breakdown in a new experimental configuration because of an ill-chosen combination of gas pressure P, electrode voltage V, and electrode spacing d. Paschen made an empirical study of this issue over a century ago and showed that the voltage V at which breakdown occurs is a function of the product Pd. Most notably, Paschen showed that for each gas

1. there exists some optimum value of Pd at which the breakdown voltage is at a minimum
2. there exists a minimum value of Pd below which breakdown is impossible at any voltage.

During the early 20th century Townsend and co-workers conducted extensive studies of breakdown and developed an analytic formula which reproduced Paschen's empirical formula (commonly called Paschen's law) from first principles. Townsend's formula depends on collision cross sections and effective ionization potentials. Determining these parameters precisely is often difficult so that in practice, the coefficients in Paschen's law are usually determined experimentally.

We now provide a highly simplified derivation of Paschen's law. It must be understood that the values of the physical parameters are nominal; the actual values will differ somewhat because of geometric, statistical, or other considerations.

Suppose N_0 electrons are initially emitted from a cathode, perhaps because of photo-ionization or some other weak process. These electrons, called primary electrons, are accelerated towards the corresponding anode by the applied electric field

$E = V/d$ and may ionize neutral atoms before striking the anode. Ionization creates more electron-ion pairs. The newly formed ions are accelerated to the cathode and bombard the cathode where they cause secondary emission of electrons. The cycle of events repeats itself, with the secondary electrons produced at the cathode constituting the starting group for each successive cycle. If the number of secondary electrons exceeds N_0, then the regenerative process will avalanche; i.e., at each repetition more electrons will be produced at the cathode than in the preceding cycle. The avalanche occurs quickly and leads to the formation of a fully ionized plasma, i.e., an arc.

The process can be analyzed semi-quantitatively by defining α as the number of electron-ion pairs produced per cm by each primary electron; the parameter α is called the first Townsend coefficient. The number of electrons produced in a distance dx by N electrons will be

$$dN = N\alpha dx \tag{14.1}$$

or on integrating from $x = 0$ to $x = d$

$$N = N_0 e^{\alpha d} \tag{14.2}$$

where N_0 is the number of electrons initially emitted by the cathode. The number of new electrons resulting from ionizing collisions will therefore be

$$N_e^{new} = N_0 \left(e^{\alpha d} - 1\right). \tag{14.3}$$

The number of ions resulting from ionizing collisions will be the same as the number of new electrons, i.e.,

$$N_i = N_0 \left(e^{\alpha d} - 1\right). \tag{14.4}$$

Defining γ_i as the number of secondary electrons created at the cathode because of bombardment, the total number of secondary electrons created at the cathode will be $\gamma_i N_i$ and so avalanching will occur if

$$\gamma_i \left(e^{\alpha d} - 1\right) > 1. \tag{14.5}$$

A primary electron can only ionize when its kinetic energy exceeds the ionization potential V_i of the atom. In fact, V_i is just the threshold for ionization, and not all impacts by electrons with energy eV_i result in ionization; in order for there to be a high probability of ionization, the primary electron energy must be eV_i^{eff} where V_i^{eff} is some small multiple of V_i. For most atoms V_i^{eff} is in the range 50-100 volts.

We define l as the distance the primary electron must move to attain V_i^{eff} so

$$l = V_i^{eff}/E = V_i^{eff} d/V \tag{14.6}$$

where V is the applied voltage across the electrodes. Let λ be the mean free path for an electron to be collisionally scattered and lose its directed kinetic energy. The probability that an electron will be able to travel between x and $x + dx$ without being scattered is $p(x) = \beta \exp(-x/\lambda)$ where β is a normalization factor. Since

$$\int_0^\infty p(x)dx = 1 \tag{14.7}$$

the normalization factor must be $\beta = \lambda^{-1}$ so $p(x) = \lambda^{-1} \exp(-x/\lambda)$. The probability of an electron traveling a distance greater than l is

$$p_{>l} = \int_l^\infty p(x)dx = \exp(-l/\lambda). \tag{14.8}$$

This can also be interpreted as the chance that for a given collision, the electron will have traveled long enough before the collision to have gained sufficient energy to ionize when colliding. The number of collisions per cm is λ^{-1}. Thus α, the number of ionizing collisions per cm, is the total number of collisions per cm multiplied by the probability that the collision is ionizing, i.e.,

$$\alpha = \frac{1}{\lambda} \exp(-l/\lambda). \tag{14.9}$$

The electron mean free path is related to the collision cross-section σ and gas density n by

$$\lambda = 1/\sigma n \tag{14.10}$$

in which case

$$\alpha = \sigma n \exp(-\sigma n d V_i^{eff}/V). \tag{14.11}$$

Equation (14.5), the avalanche condition, can therefore be expressed as

$$\sigma n d \exp(-\sigma n d V_i^{eff}/V) > \ln\left(1 + \frac{1}{\gamma_i}\right). \tag{14.12}$$

Solving for the breakdown voltage V gives

$$V > \frac{V_i^{eff} \sigma n d}{\ln(\sigma n d) + \ln\left(1/\ln\left(1 + \gamma_i^{-1}\right)\right)}. \tag{14.13}$$

This shows that the breakdown voltage is a function of nd and has a minimum when

$$\sigma nd = 2.73 \ln\left(1 + \frac{1}{\gamma_i}\right); \quad (14.14)$$

the value of the minimum breakdown voltage is

$$V_{\min} = 2.73 V_i^{eff} \ln\left(1 + \frac{1}{\gamma_i}\right). \quad (14.15)$$

The secondary ionization coefficient γ_i depends on the gas, the ion energy, the electrode material, and the surface conditions. For typical conditions $\gamma_i \sim 1 - 5 \times 10^{-2}$ so that

$$\left(\frac{V}{V_i^{eff}}\right)_{\min} \sim 8 - 12. \quad (14.16)$$

Since the neutral gas is typically at room temperature, the density is just proportional to pressure so that Eq.(14.13) can be expressed[150] as

$$V > \frac{BPd}{\ln(Pd) + C} \quad (14.17)$$

where the pressure is measured in Torr (mm Hg) and the distance is in cm. Consideration of the denominator of this expression, shows that the breakdown voltage becomes infinite when $Pd = \exp(-C)$ and so for $Pd < \exp(-C)$ breakdown becomes impossible.

Typical values of the parameters B and C are tabulated below[150] for several gases of interest using $\gamma = 10^{-2}$. Figure 14.1 plots the Paschen breakdown curve for hydrogen gas (H_2). The minimum voltage at which breakdown can occur is approximately 300 volts and this occurs when $Pd \sim 2$ Torr-cm. When $Pd < 0.85$ Torr-cm breakdown in hydrogen is impossible at any voltage.

Gas	B	C
H_2	139	0.16
Air	365	1.17
He	34	-0.43
A	180	0.96

Breakdown does not occur instantaneously because the avalanching process takes a finite time. The applied voltage collapses after a delay which might be as short as a microsecond or as long as a fraction of a millisecond. This delay is mainly due to the finite transit time for the ions created by electron impact to accelerate to the cathode

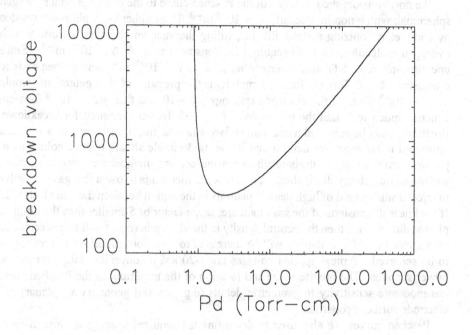

Fig.14.1 Paschen curve for hydrogen gas showing breakdown voltage v. Pd in Torr-cm. Spheromak pressures are normally to the left of the minimum.

from the location where they were created. If an electrode system has complex geometry with varying interelectrode distance, breakdown initiates at the location where the interelectrode spacing corresponds to Paschen minimum. Because magnetic forces act to increase the circuit inductance, magnetic forces typically accelerate the initial arc away from the location of initial breakdown to a location where the interelectrode distance is larger.

The shape of the Paschen breakdown curve can be understood qualitatively as follows: In the region far to the left of the minimum, the neutral gas is so diffuse that most primary electrons reach the anode without colliding with any neutrals; thus there is little ionization and breakdown is difficult. In contrast, for the region far to the right of the minimum, the neutral gas density is so high that primary electrons collide so often that they never attain the energy eV_i^{eff} needed for ionization; again there is little ionization and breakdown is difficult.

We now consider the relevance of the Paschen curve to the operation of a hydrogen spheromak with a nominal density $n = 10^{20}$ m^{-3}. If the spheromak plasma is created by completely ionizing a static fill, i.e., filling the vacuum chamber uniformly with hydrogen molecules, then the required fill density is $n_{H_2} = 5 \times 10^{19}$ m^{-3}. Since one atmosphere at STP has a density $n_{atm} = 2.7 \times 10^{25}$ m^{-3} and corresponds to a pressure of 760 Torr (1 Torr = 1 mm Hg), the pressure of this neutral fill would be $p \sim 10^{-3}$ Torr. If the electrode spacings are $\sim$10 cm, then $pd \sim 10^{-2}$ Torr-cm which is much less than the minimum $Pd \sim 0.85$ Torr-cm required for breakdown. Breakdown can be achieved using a much larger neutral fill, but this produces a plasma immersed in an excessive neutral density, an undesirable situation since collisions of plasma particles with neutrals result in plasma cooling, magnetic energy dissipation, and magnetic helicity dissipation. A much better method is to use a fast gas puff valve to inject a small cloud of high density neutrals in the region between the gun electrodes. If the linear dimensions of the gas cloud are, say, a factor of 5 smaller than the ultimate plasma dimensions, then the neutral density in the electrode region will be concentrated by a factor of $\sim 10^2$ so that pd will be increased to ~ 1 Torr-cm allowing breakdown to be achieved. Typical applied voltages are 1-20 kV, a convenient range for pulsed power supplies. The precise value and location of the minimum in the Paschen curve has moderate sensitivity to geometric details (e.g. coaxial geometry v. planar) and electrode surface properties.

Paschen curves are also used to determine the required spacing for insulation of high voltage conductors at atmospheric pressure. Because atmospheric pressure is high, most atmospheric configurations will be well to the right of the Paschen minimum so that better insulation is achieved by increasing spacing between conductors or by pressurizing the space between conductors. This is exactly the opposite of the behavior inside the vacuum chamber where operation is usually in the vicinity of the Paschen minimum. One must remember that in the low pressure situations typical of spheromaks, prevention of breakdown is achieved by reducing the distance between conductors.

14.2 Gas puff valves

The previous section showed that breakdown optimization requires concentrating the neutral gas in the region between the electrodes. This is obviously impossible in steady state, but by using a fast gas puff valve, a localized gas cloud can be created transiently.

At room temperature the thermal velocity of hydrogen is $v_{H_2} \approx 10^3$ m/s. It is desired to concentrate the high pressure gas into a region of the order of the interelectrode

14.2 Gas puff valves

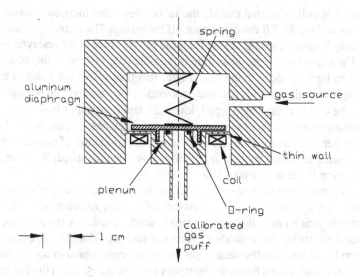

Fig.14.2 Fast gas puff valve. Current pulse in coil repels aluminum diaphragm which moves upward a small distance allowing plenum gas to flow into vacuum chamber.

dimension d which for typical spheromaks is ~ 10 cm. Thus, the valve opening and closing time should be much smaller than d/v_{H_2}, so that the valve should be open for < 100 μs. The valve must also be able to transfer the gas inventory required to make the initial spheromak. This inventory is in the range $10^{19} - 10^{20}$ molecules for a nominal spheromak with density $n \sim 10^{20} - 10^{21}$ m^{-3} and volume $V \sim 0.1 - 1$ m^3.

One cubic centimeter of gas at STP has 2.6×10^{19} particles so that the gas valve must transfer approximately 1 cm^3 of gas at STP in approximately 100 μs. Multiple valves may be used to provide a more even distribution of gas and, of course, the numbers cited here are nominal and will vary depending on the actual spheromak size and operating density.

Because this high throughput and fast response substantially exceed the performance of commercially available gas valves, spheromak researchers have had to develop [151] valves with the required performance. Typically [104] the valve operation involves a small but extremely rapid displacement of an aluminum diaphragm covering a small plenum filled with the desired gas inventory. Figure 14.2 sketches a typical fast gas puff valve. The diaphragm is seated on an o-ring seal, but when lifted, allows gas to pass from the plenum through a small tube into the spheromak volume. The diaphragm is lifted transiently by a fast-pulsed coil arranged to induce currents in the diaphragm; this coil is located outside the valve body. The induced currents repel the coil currents and briefly lift the diaphragm off the o-ring. The restoring force for the diaphragm may be from a spring or else from the gas pressure in the reservoir outside the plenum. The valve body is made as thin as possible (typically < 1mm) so that the coil windings can be very close to the diaphragm; this maximizes the induced currents in the diaphragm and also ensures that eddy currents in the coil body do not shield the coil currents. The coil is pulsed by a small, low inductance capacitor bank connected by a coaxial transmission line so as to minimize stray inductance. Furthermore, the coil has a small number of turns to minimize inductance; minimization of circuit inductance is required in order to obtain the fastest possible valve opening speed. With modest effort ~ 25 μs opening times can be achieved.

The valve speed is measured using a fast ion gauge designed to operate at high pressures. This gauge is very simple and consists of a small filament (cathode) biased to emit electrons which are collected by a small nearby anode. A third electrode, negatively biased relative to the cathode collects ion current. The ion current results from the electron beam ionizing the neutral gas and so is proportional to the neutral density. By keeping the dimensions between electrodes small (e.g., $\leq$ 1 cm) the ion gauge has a fast response time. The ion current is amplified by an operational amplifier in transconductance mode (i.e., current in, voltage out). The valve throughput is obtained by firing it a large number of times into an isolated, pumped-down vacuum chamber. The gas inventory injected per valve firing can be deduced from the pressure rise in the vacuum chamber and the volume of the vacuum chamber. The size of the injected gas cloud can be determined by mounting the fast ion gauge on a moveable probe and measuring the time dependence of the gas profile

14.3 Wall desorption and contamination

At first sight one would expect that the particle inventory of the spheromak plasma comes from the gas introduced into the chamber by the experimenter. In fact, for most experiments the plasma density is not proportional to the injected or pre-filled gas density. Instead, plasma density is typically proportional to the plasma current and the main effect of changing the amount of injected gas is to change the threshold for breakdown. This lack of dependence of plasma density on gas fill occurs because there is typically a very large number of gas atoms (or molecules) loosely bound to the wall surface and consequently easily released by plasma bombardment. The total number of particles sticking to the wall usually exceeds the plasma inventory so that the plasma density is determined by the rate at which plasma particles bombard the wall and detach the loosely bound wall particles. This behavior is called recycling and has been analyzed quite thoroughly in the context of tokamaks[152, 153, 154].

14.3 Wall desorption and contamination

Impurities can also be loosely bound to the wall so that plasma bombardment of the wall can also cause impurities to enter the plasma. Impurity recycling is characterized by the situation where best plasma parameters (e.g., temperature, confinement) are obtained only after a sequence of plasma shots with apparently identical initial conditions (e.g., gas, power supply voltage, coil currents).

Metal wall surfaces hold approximately one gas atom per wall atom and there can also be absorption of gas by metal atoms below the surface layer. Since the linear spacing between the wall atoms is about 3Å, the surface density of wall atoms is $\sim 10^{19}$ m^{-2}. Because there can be more than one monolayer and also absorption below the surface, the actual wall storage can be several times this value. Thus, one square meter of wall can hold $10^{19} - 10^{20}$ gas atoms. If one considers a typical spheromak plasma with density $n = 10^{20}$m^{-3}, cylindrical flux conserver radius $a = 0.5$ meters and length $h = 0.75$ meters (so that the volume is 0.6 m^3 and the flux conserver surface area is 4 m^2) it is seen that the wall has an order of magnitude more gas inventory than the spheromak plasma. The loosely bound gas on the wall is easily released when bombarded by escaping plasma. Unless extreme measures are taken to remove absorbed gas from the wall beforehand, the spheromak plasma will be fueled mainly by atoms desorbed from the wall. If there are no contaminants, this simply means that the spheromak plasma draws from an essentially unlimited reservoir of wall particles. A steady state is established when the flux of particles recycled from the wall balances the flux of spheromak particles escaping to the wall.

Unfortunately, the sponge-like wall can also become loaded with impurities in which case recycling will inject impurities into the spheromak plasma. The most common impurities injected in this manner are oxygen and carbon. Oxygen comes from water vapor absorbed by the wall surface before pumpdown. Carbon dioxide is a significant component of air and is also absorbed from atmospheric exposure before pumpdown. Carbon can be also result from contamination by vacuum pump oil, a serious problem for many early spheromaks. Typically a high vacuum pump (diffusion or turbomolecular) would be backed by an oil-filled rotary vane mechanical pump. These pumps use ultra-low vapor pressure oils, but ultra-low vapor pressure merely means that any oil that backstreams into the working chamber condenses onto the walls. Thus, a vacuum system could easily have an impressively low vacuum base pressure and yet have walls coated with many monolayers of carbon-rich pump oil. Wall contamination can be reduced substantially by firing a large number of plasma shots, each removing some wall contaminants that are then pumped out of the system. If a device using oil-based pumps is not operated for several days, impurities from pump oil backstreaming redeposit on the wall and the cleaning process must be repeated to remove this accumulation.

To see how long atmospheric contaminants would take to cover the wall, consider a typical base pressure of 10^{-7} Torr (resulting from the balance between the influx of air from small leaks and the speed of the vacuum pump). At room temperature this pressure corresponds to an air density $n_{air} \sim 3 \times 10^{15}$ m^{-3}. For indoor humidity levels, the water content of indoor air is about 1% so the water vapor density will be $n_{H_2O} \sim 3 \times 10^{13}$ m^{-3}. The mean thermal velocity of a water molecule is $v_{T,H_2O} \sim 4 \times 10^2$ m/s so the flux of water to the walls will be $\Gamma_{H_2O} \sim 10^{16}$ m^{-2}s^{-1}. It will consequently take a few hours to form a surface coating of 10^{20} m^{-2} atoms. The wall would have much more absorbed water vapor after a machine opening. Similar arguments apply to carbon dioxide.

Several methods have been used to remove impurities from the wall. Discharge cleaning by low density rf-produced plasma can scour impurities from the wall. Alternatively, firing many ($10^1 - 10^2$) cleaning spheromak shots before taking serious data has been found to be effective. Baking the wall at temperatures of 100-300 C for several hours in conjunction with cleansing shots is even more effective. Oil contamination can be avoided altogether by using an oil-free pumping system such as a cryopump or a turbopump backed by a molecular drag/diaphragm pump.

Another approach is to use titanium sublimation pumping, a process where clean titanium is evaporated on the wall interior prior to a plasma shot. The freshly deposited thin titanium film absorbs all impinging gas atoms and acts as an extremely high speed vacuum pump. Because the titanium is freshly deposited, it does not have the embedded oxygen, carbon, or hydrogen typical of a conventional wall. Any plasma atoms striking the titanium are absorbed and not recycled so that the spheromak density is determined by the injected gas and not by wall recycling. The highest temperature spheromak plasmas were achieved in the CTX experiment using titanium gettering.

14.4 Impurity line radiation

Non-atomic radiation processes (i.e., synchrotron radiation and bremsstrahlung) are negligible for the densities and temperatures of existing and planned spheromaks. A spheromak fusion reactor would confine a plasma consisting of deuterium and tritium; these hydrogen isotopes have no bound electrons and so do not emit line radiation. Laboratory spheromak devices typically use hydrogen and so, again the majority species does not emit line radiation. In these cases line radiation comes from ionization of neutrals and from impurities. Impurity radiation can easily be a dominant power loss mechanism and so impurity reduction can profoundly affect spheromak power balance.

Steady-state plasmas with the densities and temperatures typical of spheromaks will attain a coronal equilibrium after sufficient time; this means that the rate of ion/electron pair production resulting from electron impact ionization of neutrals balances the pair loss rate from radiative recombination. Radiative recombination, the inverse of photon-induced ionization, is the process whereby an ion captures a fast electron into a low-lying energy level and emits the excess energy as a photon. Carolan and Piotrowicz [155] calculated the time τ required to attain coronal equilibrium and presented the results of these calculations in the form of $n_e\tau$ plotted versus electron temperature. These plots indicate that times of the order of milliseconds are required for carbon to attain coronal equilibrium at 10 eV and $n_e = 10^{20}$ m^{-3}.

Coronal equilibria depend strongly on the electron temperature and, in particular, the hotter the electrons, the more highly ionized the impurity ions become. Each additional level of ionization requires substantially more energy than the previous level because the additional electron being removed is bound by a less shielded ion. Cold plasmas will contain mainly neutral and singly ionized impurities. As the plasma is heated, the singly ionized impurities become doubly ionized and the neutrals become ionized. With further heating the atoms become doubly ionized, then triply ionized and so on. An approximate electron temperature can be inferred from the degree of ionization of impurities and also from ratios of line intensities.

Passing through each successive degree of ionization is akin to a thermodynamic phase change — when the plasma has not completed the phase change, adding additional heat speeds up the phase change process, but does not significantly increase the temperature. The electron temperature only starts increasing when the phase change is completed.

The power emitted by ionization state j due to radiative recombination is

$$P_j = \nu_j n_j E_j \quad (14.18)$$

where ν_j is the radiative recombination rate, n_j is the density of the state, and E_j is the energy of the emitted photon. The radiative recombination rate is $\nu_j = n_e < v_e \sigma_j(v_e) >$ where n_e is the electron density, v_e is the electron velocity, and σ_j is the cross-section for radiative recombination. Thus for coronal equilibrium

$$P_j = n_j n_e <v_e \sigma_j(v_e)> E_j \quad (14.19)$$

so that $P_j \sim n_e n_j$. Summation over all the ionization states shows[155, 156] that

$$P = \alpha(T_e) n_e n_I \quad (14.20)$$

where n_I is the density of the impurity (i.e., is the sum of the densities of all ionization states for the impurity) and the coefficient $\alpha(T_e)$ depends on the species. Tables and graphs of $\alpha(T_e)$ for 47 different elements with $2 < Z < 92$ have been calculated by Post et al. [156] for the temperature range 20 eV-100 keV. Carolan and Piotrowicz[155] provided graphs in the range 1 eV-1000 eV for impurities of relevance to fusion research and also the time to reach coronal equilibrium.

For both carbon and oxygen $\alpha \sim 10^{-31}$ Watts m^3. For carbon α peaks at about 9 eV while for oxygen α peaks at about 20 eV [155].

As an example, consider a plasma with density $n = 3 \times 10^{20}$ m^{-3} and 10% carbon; this would radiate $P = 10^9$ Watts m^{-3}. If the plasma characteristic dimensions are 0.5 m, corresponding to a volume of 0.1 m^3, the radiated power would be 100 megawatts. Thus, if the power available to heat the electrons is less than 100 megawatts, the electron temperature cannot rise above 9 eV and all input power goes into carbon radiation. If the input power substantially exceeds 100 megawatts, then the carbon cannot radiate away all the input power and the electron temperature will increase. If there were also 10% oxygen impurity, the temperature would increase to about 20eV but then would not be able to rise higher if the power was inadequate to overcome oxygen radiation. Thus, the electron temperature does not rise smoothly with input power, but instead has plateaus at levels corresponding to the peaks of impurity radiation rates. Passing over these peaks requires enough power to overwhelm the demands of the impurity radiation and is sometimes referred to as "burning through" the radiation barrier of a particular impurity.

If either the plasma duration or the ion confinement time is shorter than the coronal equilibration time, the plasma does not attain coronal equilibrium and ionization levels lower than the levels assumed in coronal equilibrium will dominate. In this case, radiative power may be higher than predicted by coronal equilibrium because the plasma has not yet burned through the lower radiation barriers. For a given impurity ratio, radiated power scales as the square of the electron density. This puts a premium on avoiding unnecessarily high plasma densities. An important example of the severity of radiation losses is the experience of the MS spheromak [107] at the University of Maryland. Operation at low density (i.e., densities lower than $6 - 8 \times 10^{20}$ m^{-3}) of this large and carefully designed $z - \theta$ device turned out not to be possible and carbon/oxygen impurity radiation clamped the electron temperature to just 15 eV even though the internal plasma currents were substantial (as high as 650 kA).

The measurement of impurity radiation can be accomplished several ways. The CTX group used bolometers to measure the integrated radiation over all wavelengths. These could be absolutely calibrated for radiation power flux. Calibration was verified by demonstrating that the power radiated by an Argon spheromak (expected to be completely radiation dominated) equalled the input power.

Because of complications with calibration it is difficult to determine absolute impurity concentrations from intensities of optical line emissions. However, these concentrations can be estimated by injecting controlled amounts of additional impurities (impurity doping) and then scaling from the observed increase in radiated power[157].

Also, measurement[158] of the deleterious effect of injected impurities gives an indication of the effect of intrinsic impurities released from wall surfaces.

Because impurities come from the wall and electrode surfaces, small devices are much more susceptible to impurity radiation than large devices because small devices have a greater surface to volume ratio. However, if the same input power is used for a given small and a given large device, the power density in the large device is smaller which negates the advantage of size. Experiments on the CTX device at Los Alamos did not show beneficial scaling with size when input power was kept constant[139].

14.5 Refractory electrode materials

Spheromaks involve high current densities so care must be taken to avoid melting and sputtering. This suggests that the electrodes and walls should be constructed from refractory metals (see for example Brown et al.[159]). However, refractory metals are hard to fabricate and are not good electrical conductors. Since the spheromak needs to be bounded by a flux-conserving wall, it would be desirable for the plasma facing components to have maximum electrical conductivity, i.e., be made out of copper.

A compromise between these opposing requirements has been achieved two different ways. The first and more widely used method is to construct gun electrodes and flux conservers from copper, and then spray-coat plasma-facing surfaces with a thin layer of tungsten. The second method is to line the plasma-facing surfaces with thin tantalum metal. These liners can be fabricated to fit just inside the copper flux conserver, or else copper can be wrapped around a tantalum form. At high plasma temperatures, these refractory materials might become a hindrance because they have large Z and so can emit copious line radiation when highly ionized. Radiation from highly stripped high Z impurities has been a problem for tokamaks which now use low Z materials (e.g., carbon tiles, boron oxides) for plasma-facing surfaces.

14.6 Skin effect and the wall as a flux conserver

Spheromak equilibria and stability depend on wall image currents. In principle, the flux-conserving wall could be eliminated by having the equilibrium field provided by an external coil system and the stabilizing fields provided by an active feedback system; this might be appropriate in future devices having large, long-duration plasmas. At the time of writing, it has proved convenient to use passive walls made of thick copper. The decay time of wall image currents can be estimated by modeling the wall as a very long cylinder carrying an azimuthal current I, i.e., a solenoid with one turn. If r is the wall radius, δ is the wall thickness, and l is the length of the solenoid, then the internal magnetic field will be $B = \mu_0 I/l$. The flux will be $\Phi = B\pi r^2 = \pi \mu_0 I r^2/l$ so that the inductance is $L = \pi\mu_0 r^2/l$. This current will be impeded by the resistance R of the conductor where $R = 2\pi\eta r/\delta l$. The solenoid can be considered as a resistor in series with an inductor and this system has a decay time

$$\tau_{decay} = L/R = \mu_0 r\delta/2\eta. \tag{14.21}$$

Thus, the wall decay time is proportional to both the wall thickness and the wall radius. For a copper wall with radius $r = 0.5$ m, and thickness $\delta = 1$ mm, the decay time will be $\tau_{decay} = 18$ ms. The decay time for finite length walls will be shorter, because the inductance per length will be less than for a long solenoid whereas the resistance per length will be the same.

The skin penetration into a planar conductor for a pulse with rise time τ is

$$\delta_{skin} \sim \sqrt{\frac{\eta\tau}{\mu_0}}, \tag{14.22}$$

a relation that has important implications for pulsed power systems with fast rise times. For example, if the rise time of a capacitor bank power supply is 10 μs then $\delta_{skin} \sim 0.4$ mm for copper and $\delta_{skin} \sim 2$ mm for stainless steel. Since the current flows within the skin layer, large currents will flow on the surfaces of the metals comprising the transmission line and the spheromak gun while negligible current will flow beneath the surface.

14.7 Inductance budget

Because the spheromak has low β, most of its internal energy is magnetic. From the point of view of an external power supply, the spheromak can be considered as an inductor with stored energy $W = LI^2/2$. If the external circuit inductance is much larger than the characteristic spheromak inductance, most of the stored energy of the entire system will lie in the external circuit and not the spheromak. The spheromak inductance scales as $L \sim \mu_0 R$ where R is the characteristic dimension of the spheromak. Since R is typically 0.1-1 m, the nominal inductance of the spheromak load will be 0.1-1 μH; therefore the external circuit inductance ought to have an inductance in this range or smaller. In certain cases it might be desirable to have a large external inductance in which case stringent inductance reduction would be unnecessary. Minimizing inductance generally means minimizing the space between conductors and having many parallel conduction paths. Minimizing space between conductors is precisely the opposite of the requirement of maintaining adequate separation between conductors to avoid voltage breakdown and arcing. This indicates that the forward and return conductors in the external circuit must be separated by a thin, high breakdown voltage insulating material (e.g., a few layers of Mylar or Kapton sheet) so that the conductors can be as close together as possible and still stand off high voltages.

14.8 Mechanical forces

Care must be taken in the mechanical design of spheromak power supplies, transmission lines, and electrode structures because substantial mechanical forces can act on these structures, especially in fault conditions. These forces are well balanced in coaxial systems but not in non-coaxial systems. As an extreme example, a 1 megamp current flowing on a 10 cm wide strip transmission line produces internal magnetic fields in the transmission line of $B \sim \mu_0 I/L \sim 12$ T so that the magnetic pressure tending to push the transmission line strips apart would be $B^2/2\mu_0 = 5 \times 10^7$ Pascals or about 500 atmospheres. For a short pulse, this is not a problem, but for a long pulse it would be necessary to provide strong mechanical restraints.

14.9 Noise radiation from pulsed power supplies

Spheromak formation often involves ramping up large currents quickly. Typically a capacitor bank power supply provides these currents and the bank is switched by a high power switching device such as an ignitron or triggered spark gap. Large electromagnetic noise signals can be radiated from unshielded segments of power supply cables when the power supply voltage is first applied across the electrode gap. This intense switching noise occurs just before and during breakdown of the initial plasma and can easily interfere with diagnostics and control systems.

Switching noise can be substantially reduced by terminating all high power transmission lines with non-inductive high power resistors matching the cable impedance. These termination resistors have negligible effect on the plasma operation, because the cable impedance is typically two or three orders of magnitude larger than the plasma impedance (i.e., ohms compared to milliohms).

Termination reduces noise radiation because without termination there is an open circuit load on the transmission line before plasma breakdown. When the switch is closed the capacitor bank voltage propagates down the transmission line to the electrode in a time l/c_l where l is the length of the transmission line and c_l is the propagation velocity in the transmission line (typically 2/3 the speed of light for the usual dielectric insulators). The plasma takes at least a microsecond to break down and so before breakdown, the gun electrodes appear as an open circuit load, i.e., a mismatched impedance. The propagating voltage pulse reflects from this mismatch and heads back to the power supply. However, the power supply is also not matched to the line impedance and so the pulse is reflected again. The net result is that the initial voltage pulse sloshes back and forth on the transmission line during breakdown. The sloshing frequency is $f = c_l/2L$ and any unshielded section of line will radiate this signal. For a typical transmission line length of 5 m and a propagation velocity $c_l = 2 \times 10^8$ m/s, the radiation will be at $f \sim 20$ MHz and will be extremely strong because the sloshing current will be of the order of the capacitor bank voltage divided by the line impedance. Termination of the line at the load end with the line characteristic impedance absorbs these reflections, but has no adverse effect on plasma breakdown. The resistors must be able to dissipate the full capacitor bank energy in fault situations (i.e., when the plasma does not break down) and so high energy, non-inductive resistors are required.

14.10 Ground loops

The spheromak involves large pulsed magnetic fields. Any circuit that links stray magnetic fields $\mathbf{B}_{stray}(t)$ will enclose a time-dependent magnetic flux $\Phi_{stray}(t) = \int d\mathbf{s} \cdot \mathbf{B}_{stray}(t)$ and so have an induced EMF $\oint \mathbf{E} \cdot d\mathbf{l} = -d\Phi_{stray}/dt$. The concept of electrical ground becomes meaningless in this situation because different points on a conductor forming the loop will be at different voltages. Serious problems with ground loops occur when multiple instruments are connected to each other because typically there are at least two ground connections for each instrument, namely the power cord and the signal cable shield.

Ground loops cause troublesome spurious signals and can even damage or destroy instruments. In spheromak situations it is not unusual for ground loops to raise the level of nominal grounds to several hundred volts or even kilovolts.

There are three general ways to avoid ground loops. The first is to have only one ground so that a loop is not formed. This method is not advisable for safety reasons because it requires disconnecting instrument power supply grounds to break the ground loop. A second and often effective method to break the ground loop is to insert a transformer into the signal cable so there is no direct connection between the grounds of instruments. The third and preferred method is to communicate all signals using optical fibers. This is easily done for trigger pulses but is more difficult and expensive for analog signals which must be faithfully reproduced. Optical fibers are particularly recommended for trigger pulses because optical fibers provide complete reverse isolation, i.e., a spurious power pulse cannot travel backwards through a trigger cable and destroy the triggering system.

CHAPTER 15
Basic Diagnostics for Spheromaks

Spheromaks are diagnosed in much the same way as tokamaks, stellarators, and reversed field pinches. This chapter surveys basic spheromak diagnostics and gives a detailed discussion of some of the more essential ones. Because spheromaks are relatively simple to build, a full complement of diagnostics can easily cost more than the spheromak itself.

Diagnostics measure gun voltage, gun current, internal magnetic fields, density, electron and ion temperatures, optical radiation, impurity concentrations, plasma motion, and x-ray emission. The ideal diagnostic would have low cost and provide unambiguous time- and space-resolved data requiring no deconvolution. Real diagnostics rarely approach this ideal and there is usually a substantial compromise between cost and performance.

15.1 Magnetic field and electric current measurement

Because spheromaks have a very dynamic magnetic topology, much effort has been directed towards studying the evolution of this topology. Magnetic fields are most commonly measured by small diagnostic coils that are either inserted into the plasma or else located at the plasma boundary. Figure 15.1 shows a typical magnetic probe. The probe consists of N turns of small diameter wire, each turn of radius a and linking the magnetic field B to be measured so that the net linked flux is $\Phi = NB\pi a^2$. Measurement of Φ can therefore be used to determine B.

Electric current is typically measured by a Rogowski coil which is an N turn coil wound in the poloidal direction on a toroidal coil form having major radius R and minor radius a (here toroidal and poloidal refer to the coil form, not to the plasma). Figure 15.2 shows the geometry of a Rogowski coil; note that the return connection is strung toroidally back through the poloidal windings so as to avoid having a net toroidal turn. The current I to be measured passes through the hole in the torus and produces a toroidal field $B_\phi = \mu_0 I / 2\pi R$ at the center of each poloidal turn. The flux linked by the N turn coil is therefore $\Phi = \mu_0 N I a^2 / 2R$ and so measurement of Φ can be used to determine I.

The flux linked by these diagnostic coils produces a voltage at the coil terminals $V = -d\Phi/dt$; integration of V allows one to determine Φ and hence the magnetic field linked by a magnetic probe or the current linked by a Rogowski coil. Martin et al.[160] described a combined Rogowski-magnetic probe which measured both J_z and B_z along the coil axis and so allowed determination of $\lambda = \mu_0 J_z / B_z$.

For both magnetic probes and Rogowski coils, the required integration can be accomplished: (i) by a passive integrator, (ii) by an active integrator (operational amplifier), or (iii) numerically. Passive integration is preferable when feasible, because this method is simplest and is least susceptible to noise, has essentially unlimited dynamic range, and works well for fast pulses. Active integrators may be better for long pulses or weak signals. Numerical integration is susceptible to noise overload and dynamic range limitations, but is convenient because no hardware is required and usually the signal is already being computer processed.

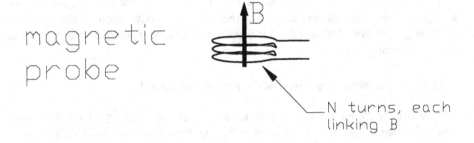

Fig.15.1 A magnetic probe is characterized by its radius a and number of turns N. The total linked flux is $\Phi = NB\pi a^2$. Usually, the probe is made as small as possible so as not to disturb the plasma.

The theory of Rogowski coils and magnetic probes is similar to the theory of electric transformers because the magnetic flux being measured results from currents flowing in an external circuit. If the magnetic probe or Rogowski coil is connected to a passive integrator such as in Fig. 15.3 then the circuit equation is

$$\frac{dB_{ext}}{dt} NA + L\frac{dI}{dt} + IR + V_{cap} = 0 \tag{15.1}$$

where B_{ext} is the external magnetic field linking the Rogowski coil/magnetic probe, N is the number of turns in this coil, A is the area of each turn, L is the coil self-inductance, R is the integrator resistor, and V_{cap} is the voltage on the capacitor C. The coil can be thought of as a transformer winding where the first term in Eq.(15.1) is the time derivative of mutual flux and the second term is the time derivative of self flux.

15.1 Magnetic field and electric current measurement

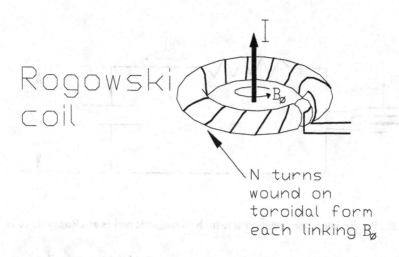

Fig. 15.2 A Rogowski coil is characterized by its major radius R, minor radius a, and number of turns. Each turn links the flux $B_\phi \pi a^2$ produced by the magnetic field $B_\phi = \mu_0 I/2\pi R$ so the total linked flux is $\Phi = N\mu_0 I a^2/2R$.

The voltage on the capacitor is

$$V_{cap} = \frac{\int^t I dt}{C}. \tag{15.2}$$

If the inductive voltage drop in the diagnostic coil is so small that

$$L\frac{dI}{dt} \ll IR \tag{15.3}$$

and if V_{cap} is so small that

$$V_{cap} \ll IR \tag{15.4}$$

then Eq.(15.1) simplifies to

$$I = -\frac{dB_{ext}}{dt}\frac{NA}{R} \tag{15.5}$$

so using Eq.(15.2)

$$B_{ext} = -\frac{RC}{NA}V_{cap}. \tag{15.6}$$

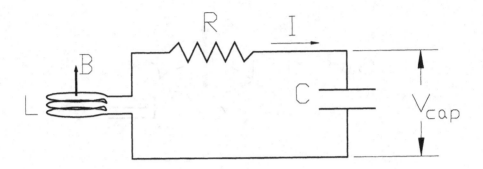

Fig.15.3 Passive integrator circuit suitable for both magnetic probes and Rogowski coils.

Thus, measurement of V_{cap} directly gives the magnetic field linking the coil.

For a signal with time dependence $\sim \exp(i\omega t)$ the conditions Eq.15.3 and 15.4 can be expressed as

$$\omega \ll \frac{R}{L}, \text{ to avoid inductive voltage drop} \tag{15.7}$$

and

$$\frac{1}{\omega C} \ll R \text{ to avoid voltage drop across the capacitor.} \tag{15.8}$$

These may be combined to give

$$\frac{1}{RC} \ll \omega \ll \frac{R}{L} \tag{15.9}$$

as the bandwidth of the integrator/probe system. Making C large improves the low frequency response, but lowers the signal level. Increasing the number of turns increases the signal level but increases L thereby worsening the high frequency response.

The integrator output goes to a signal processing instrument, normally a transient digitizer that converts the integrated signal into digital data suitable for computer processing. The digitizer's preamplifier will have a characteristic input impedance R_{inst} which acts as a further constraint on the system because if R_{inst} is too small, the capacitor charge will be drained away by R_{inst} thereby corrupting the measurement. To avoid significant charge depletion by R_{inst} during the measurement time, the capacitor

15.2 Equilibrium reconstruction using measurements at the wall

must be large enough that

$$\frac{1}{R_{inst}C} \ll \omega \qquad (15.10)$$

for the lowest frequency of interest. Comparing Eqs.(15.9) and (15.10) shows that the optimum value for the integrator resistor is $R = R_{inst}$, since at this value both inequalities are satisfied for the same range of frequencies. Equation (15.10) also determines C and so the only free parameter is the coil inductance L which increases with both N and A. Thus N and A should be made as large as possible consistent with (i) satisfying the high frequency response constraint $\omega \ll R/L$ and (ii) the geometrical constraint that the probe is small enough so as not to perturb the plasma.

Because the amplitude of the integrated signal is much smaller than the amplitude of the raw input signal and because the raw signal is inherently spiky, the integrator is vulnerable to spurious capacitive coupling of the input signal to the output circuit. These stray capacitively coupled currents have a magnitude $I_{stray} = C_{stray}dV/dt$ where C_{stray} is the stray capacitance to the integrator input from any source which has a time-dependent voltage. To minimize capacitive coupling and coupling from noise signals, it is essential to shield the coil and the cables going from the coil to the integrator so that all capacitively coupled currents drain to ground via the shield and are not collected by the integrator capacitor. Figure 15.4 shows an appropriate shielding geometry.

It is best to locate the integrator at the measuring instrument rather than at the coil so that the transmission line between the coil and the measuring instrument will be carrying the large, un-integrated signal and not the small integrated signal. Thus, any noise leaking onto the transmission line via stray coupling will be small compared to the signal.

However, if all of the integration resistance R is located at the instrument end, then the transmission line will be mismatched at both ends and the spiky raw signal will reflect back and forth between the two unmatched ends of the transmission line. These reflections will appear as spurious oscillations superimposed on the desired signal. To avoid such reflections, the resistance R should be subdivided into two physical resistors, $R = R_1 + R_2$, where R_1 corresponds to the characteristic transmission line impedance Z_0 and is located at the coil end of the transmission line. Resistor R_1 absorbs pulses reflected from the instrument end of the transmission line.

Magnetic coils have a low frequency cutoff because they rely on integration of a pulse. If DC magnetic fields must be measured, then commercially available Hall probes can be used.

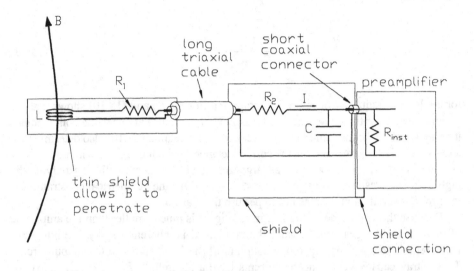

Fig.15.4 A shielded passive integrator circuit.

15.2 Equilibrium reconstruction using measurements at the wall

The toroidal magnetic field vanishes at the wall but the poloidal magnetic field is finite at the wall. Since $\mathbf{B}_{pol} = (2\pi)^{-1}\nabla\psi \times \nabla\phi$ measurement of $\mathbf{B}_{pol}$ at the wall (or equivalently the toroidal current in the wall) as a function of poloidal position gives $\nabla\psi$ at the wall as a function of poloidal position. This measurement can then be used to reconstruct the internal flux profile and also determine the flux on the magnetic axis using the technique [132] described in Sec. 12.4.

15.3 Voltage measurements

According to electrostatic theory, all points on a conductor are at the same potential. It is traditional in circuit analysis to define one equipotential in a circuit as the "ground potential" to which all other potentials are referenced. However, because of the large transient magnetic fluxes associated with spheromak operation, voltages are produced not only by electric potential drops but also by induction. If time-dependent magnetic fluxes are linked by a circuit being measured, then contrary to electrostatic theory, the voltage at two different points on a conductor will differ. Furthermore, the observed difference will depend on the geometry of the measuring circuit. In particular, if the measuring leads of a voltage probe link a magnetic flux Φ, then the voltage measurement will include the EMF $\oint \mathbf{E} \cdot d\mathbf{l} = -d\Phi/dt$ and this induced EMF will depend on the three dimensional contour formed by the leads. To avoid these geometry-dependent inductive voltages, the probe cable should be coaxial so that no flux is linked by the cable and the voltage measurement should always be between adjacent points.

When large currents are flowing in ground circuits, the various points of the nominal ground circuit will be at substantially different voltages, so that measurements relative to ground are ill-defined. In this case it is best to make differential voltage measurements rather than ground-referenced measurements.

15.4 Density measurement

15.4.1 Langmuir probe (triple probe)

Langmuir probes are small conductors with well-defined surface area that are inserted into the plasma for the purpose of collecting electrons and ions in situ. While this technique provides a simple, low cost method for measuring the plasma density n and the electron temperature T_e in low temperature plasmas, the interpretation of Langmuir probe data can be quite ambiguous. In particular, densities inferred from Langmuir probe measurements typically have an imprecision of the order of a factor of 2-5 and electron temperature measurements have an imprecision of 20-50%. Langmuir probe data is conventionally obtained by scanning an applied bias voltage and then measuring the current I flowing from the plasma to the probe. The probe I-V characteristic is then used to deduce n and T_e.

Scanning the Langmuir probe bias voltage is usually impractical in spheromak experiments because the spheromak time scale is too short to allow a reasonable scan time. Also, if the bias voltage is ground-referenced, the actual applied voltage at the probe tip will be susceptible to spurious voltages induced by unknown, transient magnetic fluxes linking the probe wiring. Thus, it is desirable to avoid both ground referencing and voltage scanning when using Langmuir probes to diagnose spheromaks.

The triple probe technique developed by Chen and Sekiguchi[161] eliminates voltage scanning or ground referencing and so is attractive for spheromak measurements, but is less accurate than the traditional scanning method[162].

In order to understand the operation of a triple probe, we first consider the theory for a single probe. The electron current collected by a single probe is

$$I_e = nq_e v_{Te} A \exp(-q_e(V - V_s)/\kappa T_e) \tag{15.11}$$

250 Basic Diagnostics for Spheromaks

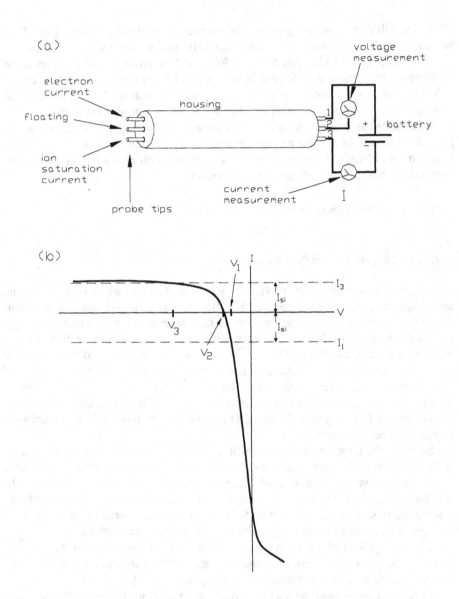

Fig.15.5 (a) Schematic of triple probe. The current meter measures the current coming from probe 3 (ion saturation current) and going to probe 2 (electron current). The voltmeter measures the voltage drop between probe 1 and probe 2 which is floating (no current). (b) Langmuir probe characteristic showing the voltages and currents of the three probes.

15.4 Density measurement

where V_s is the plasma space potential and A is the area of the probe. The corresponding ion current is $I_{si} = nq_i c_s A$ where $c_s = \sqrt{\kappa T_e/m_i}$ is the ion sound velocity. The total current collected by a single probe is therefore

$$I = I_{si} - I_{se} \exp(-q_e(V - V_s)/\kappa T_e) \tag{15.12}$$

where $I_{se} = |nq_e v_{Te} A|$ is the electron saturation current. More sophisticated models take into account particle orbit effects and the dependence of Debye sheath size on bias voltage; these models show that there is a numerical factor of the order of unity multiplying I_{si} and also that I_{si} has a weak dependence on bias voltage. These subtleties will be ignored in the simplified analysis presented here.

As shown in Fig.15.5 a triple probe consists of three adjacent identical Langmuir probes. The probe tips are located close enough to each other so that they sample the same plasma, but not so close that they interfere with each other; this latter stipulation requires the probes to be separated by a few Debye lengths so that their respective sheaths do not interact. The probes are labeled 1, 2 and 3, I_j denotes the current collected from probe j, and V_j denotes the voltage on probe j. A high impedance voltmeter measures $V_1 - V_2$ and a battery is connected between probes 1 (positive battery terminal) and 3 (negative battery terminal). The battery voltage is set to be much larger than both the expected electron temperature and the expected plasma potential so that $I_{se} \exp(-q_e(V_{bat} - V_s)/\kappa T_e) \to 0$. This means that probe 3 is biased in the ion-saturation current regime, i.e., using Eq.(15.12) $I_3 = I_{si}$.

The current entering the positive battery terminal must equal the current leaving the negative battery terminal; thus $I_1 = -I_3$. Because probe 2 is floating, $I_2 = 0$. We define primed currents as the current relative to the ion saturation current; i.e., $I' = I - I_{si}$. Using this definition $I'_3 = 0$, $I'_2 = -I_{si}$, $I'_1 = -I_3 - I_{si} = -2I_{si}$ and Eq.(15.12) can be written as

$$I' = -I_{se} \exp(-q_e(V - V_s)/\kappa T_e). \tag{15.13}$$

Thus

$$\frac{I'_1}{I'_2} = 2 = \exp(e(V_1 - V_2)/\kappa T_e) \tag{15.14}$$

so that the electron temperature in eV is

$$T_e = \frac{V_1 - V_2}{\ln 2} \tag{15.15}$$

where the voltage difference $V_1 - V_2$ is measured by the high impedance voltmeter. The density can be determined from $n = I_{si}/q_i c_s A = I_3/\left[eA(\kappa T_e/m_i)^{1/2}\right]$. Thus, electron density and temperature are measured directly without scanning and without a ground reference.

15.4.2 Interferometry

The line-averaged density can be measured with good accuracy from the phase shift of an electromagnetic wave traversing the plasma provided the wave frequency ω is much larger than the plasma frequency ω_{pe}. Such a wave has dispersion relation $\omega^2 = \omega_{pe}^2 + k^2 c^2$ so that the phase shift ϕ_p incurred on traversing a length L of plasma is

$$\phi_p = \int_0^L k\, dx \approx \frac{\omega}{c} \int_0^L dx \left(1 - \frac{\omega_{pe}^2(x)}{2\omega^2}\right) \tag{15.16}$$

if $\omega^2 \gg \omega_{pe}^2$. The density of short-pulse laboratory spheromaks is well suited to the use of interferometers based on He-Ne lasers. A quadrature design originally described by Buchenauer and Jacobson [163] has proved particularly useful because it provides good noise immunity and does not have phase ambiguity. The quadrature configuration involves measuring both sine and cosine components of the wave so that the phase shift can be determined using the arctangent function. Determination of the individual polarities of the sine and cosine functions resolves phase ambiguities of the arctangent function. An essential feature of the Buchenauer-Jacobson design is that the same optical path is used for both the sine and cosine components so that mechanical vibration does not cause a spurious phase difference between the sine and cosine components.

The interferometer layout is sketched in Fig.(15.16). The output of a linearly polarized He-Ne laser is oriented so that its **E** field is out of the plane of the page (y direction). The laser beam is divided into a scene beam and a reference beam by the non-polarizing beam splitter BS1. The scene beam reflects from mirror M1, enters the plasma through polarization-preserving window W1, exits through polarization-preserving window W2, passes through non-polarizing beam splitter BS2 and enters the Wollaston prism. The Wollaston prism is rotated so that its axis is at a 45^0 angle relative to the y direction. Thus, if $\hat{x}'$ and $\hat{y}'$ are unit vectors in the Wollaston prism coordinate system, the scene beam polarization is $(\hat{x}' + \hat{y}')/\sqrt{2}$.

The reference beam reflects from mirror M2, becomes circularly polarized by the quarter wave plate Q1, reflects from beam splitter BS2 and enters the Wollaston prism co-aligned with the scene beam.

In the Wollaston prism coordinate system the scene beam electric field is

$$\mathbf{E}_s = E_0 \frac{(\hat{x}' + \hat{y}')}{\sqrt{2}} \cos(k_0 L_s - \Delta\phi_p - \omega t) \tag{15.17}$$

15.4 Density measurement

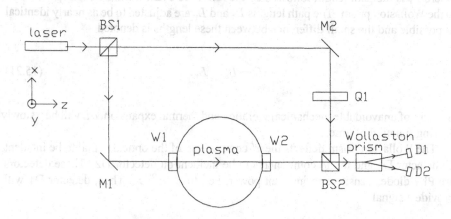

Fig.15.6 Quadrature density interferometer. The laser beam is polarized so that its **E** field is out of the plane of the page (y direction). Quarter wave plate Q1 changes the linearly polarized reference beam into a circularly polarized beam. The Wollaston prism axis is rotated about the z axis by 45^0 so its x', y' axes are 45^0 relative to the x, y axes.

where

$$\Delta\phi_p = \frac{1}{2\omega c}\int_0^L \omega_{pe}^2(x)dx \qquad (15.18)$$

is the phase shift due to presence of plasma, L is the path length through the plasma, and L_s is the path length of the scene beam from beamsplitter BS1 to the Wollaston prism. In practical units the phase shift resulting from passage through the plasma is

$$\Delta\phi_p = 2.8 \times 10^{-15}\lambda_0 \int_0^L n(x)dx \qquad (15.19)$$

where λ_0 is the laser wavelength. For example, if a He-Ne laser ($\lambda_0 = 632$ nm) is used, the phase shift resulting from traversing 0.1 m of plasma with density $n = 10^{20}$ m^{-3} will be $\Delta\phi_p = 1.8 \times 10^{-2}$ radians.

The circularly polarized reference beam has an electric field in the Wollaston coordinate system given by

$$\mathbf{E}_r = E_0\left[\hat{x}'\cos(k_0L_r - \omega t) + \hat{y}'\sin(k_0L_r - \omega t)\right] \qquad (15.20)$$

where L_r is the path length followed by the reference beam from the beamsplitter BS1 to the Wollaston prism. The path lengths L_s and L_r are adjusted to be as nearly identical as possible and the small difference between these lengths is denoted

$$\delta L = L_s - L_r. \tag{15.21}$$

Because of unavoidable mechanical vibration and thermal expansion, δL will be a slowly varying function of time.

The Wollaston prism deflects the x' components of the optical signal to be incident on detector D1 and the y' components to be incident on detector D2. These detectors are PIN diodes sensitive to incident power, i.e., to $<E^2>$. Thus, detector D1 will provide a signal

$$\begin{aligned} S_1 &= \alpha < (E_{rx'} + E_{sx'})^2 > \\ &= \frac{\alpha E_0^2}{2} \cos\left(\Delta\phi_p - k_0 \delta L\right) \end{aligned} \tag{15.22}$$

and detector D2 will provide a signal

$$\begin{aligned} S_2 &= \alpha < (E_{ry'} + E_{sy'})^2 > \\ &= \frac{\alpha E_0^2}{2} \sin\left(\Delta\phi_p - k_0 \delta L\right). \end{aligned} \tag{15.23}$$

Here α is the PIN diode responsivity. Thus, the measured plasma-induced phase shift will be

$$\Delta\phi_p = \arctan(S_2/S_1) + k_0 \delta L. \tag{15.24a}$$

The value of $k_0 \delta L$ is determined from measurement of S_1 and S_2 at the instant before breakdown.

Measurement of $\Delta\phi_p$ provides the line-averaged density. In principle, the density profile could be determined using multiple beams and tomographic analysis. This has not yet been done for spheromak plasmas but has been successfully used on other magnetically confined plasmas.

15.5 Ion temperature measurement

15.5.1 Impurity Doppler shift

In fusion applications the majority species is normally hydrogen (or a hydrogen isotope). Since ionized hydrogen does not have bound electrons, spectroscopy cannot be used. However, impurity ions do have bound electrons and emit a rich spectrum of atomic lines. The temperature of impurity ions can be determined from Doppler broadening providing this is the dominant broadening mechanism. If the impurity ions are in thermal equilibrium with the majority species then the majority species temperature can be inferred from the impurity Doppler broadening. However, in order for the majority species to be in thermal equilibrium with impurities, the experiment duration and particle confinement time must be longer than the majority-impurity energy equilibration time so that majority ions and impurity ions have opportunity to equilibrate before being lost.

The ion-ion collision time is

$$\tau_{ii} = 2 \times 10^{13} \frac{\mu^{1/2} T_i^{3/2}}{Z^4 n \ln \Lambda} \qquad (15.25)$$

where T_i is in eV and μ is the ion mass expressed in units of proton mass. For a hydrogen plasma with parameters $n = 10^{20}$ m^{-3}, $Z = 1$ this gives nominal equilibration times of τ_{ii} =0.6 μs for $T_i = 10$ eV, τ_{ii} =14 μs for $T_i = 100$ eV, and τ_{ii} =370 μs for $T_i = 1000$ eV, with corresponding longer times for heavier ions. In the case of collisions between ions of different mass, the reduced mass should be used for μ. For energy equilibration, these times should be multiplied by the ratio of the heavier to the lighter ion, just as electron-ion equilibration is a factor m_i/m_e slower than electron-electron equilibration.

The Doppler broadened ion temperatures measured in spheromaks are typically in the 100-1000 eV range. It is not obvious that the impurities are in equilibrium with the majority ions because the ion-ion equilibration time at these temperatures is comparable to the experiment duration and the particle confinement time.

Assuming for now that impurity Doppler broadening does characterize the temperature of the majority species, we now calculate the magnitude of this broadening. If ω_0 is the intrinsic frequency of an optical emission line, then the optical radiation emitted by an ion moving with velocity v will be Doppler shifted to $\omega = \omega_0 + k_0 v$ where $k_0 = \omega_0/c$. Defining $\Delta \omega = \omega - \omega_0$ and using $\omega = 2\pi c/\lambda$ shows that $\Delta \omega = -2\pi c \Delta \lambda / \lambda_0^2$

so the velocity can be expressed as

$$v = \frac{\Delta\omega}{k_0} = -\frac{\Delta\lambda}{\lambda_0} c. \tag{15.26}$$

If the impurity ions are thermalized, they will have a velocity distribution

$$f \sim \exp(-m_i v^2/2\kappa T_i) \tag{15.27}$$

and so the observed spectral line will have a wavelength dependence

$$I(\lambda) \sim \exp(-m_i c^2 (\Delta\lambda)^2 / 2\kappa T_i \lambda_0^2). \tag{15.28}$$

The impurity temperature can be determined from the full width at half maximum (FWHM) of the spectral profile and in units of eV is

$$T_i = 1.7 \times 10^8 \mu \left(\frac{\Delta\lambda_{FWHM}}{\lambda_0}\right)^2. \tag{15.29}$$

Conversely, given an impurity ion temperature, the FWHM will be

$$\Delta\lambda_{FWHM} = 7.7 \times 10^{-5} \lambda_0 \sqrt{\frac{T_i}{\mu}}. \tag{15.30}$$

As an example, consider a carbon III emission line at 465 nm and suppose this impurity has a temperature of 25 eV. Since $\mu = 12$, the spectral line FWHM would be 0.05 nm; Gibson et al.[164] report C III Doppler measurements in this parameter range.

At high densities, Stark broadening can become larger than Doppler broadening. Kepple and Griem [165] have calculated Stark broadening for various low temperature plasmas and characterize the broadening as

$$\Delta\lambda = 2.5 \times 10^{-23} \alpha_{1/2}^n N_e^{2/3} \tag{15.31}$$

where $\alpha_{1/2}^n$ is coefficient which depends on the ion and also on the emission line; Welch et al.[166] used Stark broadening to measure tokamak edge densities comparable to typical spheromak densities. For H_α and H_β emission lines the Stark coefficients are of the order of 0.015 and 0.08 respectively for temperatures of a few eV. For example, in a plasma with density $n = 10^{21}$ m^{-3} the Stark broadening of the H_β line would be of the order of $\Delta\lambda \sim 0.2$ nm which is the same as the Doppler broadening for $T_i \sim 30$ eV.

15.5.2 Neutral particle analysis

Mayo et al.[167] used the time of flight distribution of neutral particles ejected from a spheromak to infer the temperature of the majority species. Unlike Doppler measurements where there is uncertainty about majority-impurity equilibration, neutral particle analysis provides unambiguous information on the temperature of the majority species; the disadvantage is complexity of the instrument and poor time resolution. This measurement has proved important because it confirmed the large ion temperatures inferred from Doppler broadening of impurity emission lines.

Mayo et al.'s neutral particle analysis (NPA) measurement took advantage of the relatively high neutral concentration of 1-10% in the core of the CTX spheromak. Hot ions charge-exchange with these neutrals, creating hot neutrals which escape isotropically from the plasma and so have exactly the same velocity distribution as the hot ions from which they were born. A narrow beam of escaping neutrals is formed by a collimator and then chopped by a high speed rotating mechanical shutter. This beam drifts down a long tube (Mayo et al. used a 2 m length) and then collides with a copper-beryllium plate, producing secondary electrons. The secondary electrons are amplified by an electron multiplier and recorded as a function of time. Since fast neutrals arrive earlier than slow neutrals, the time dependence of the electron multiplier signal can be deconvolved to provide the neutral particle energy distribution and hence the energy distribution of the hot ions in the core plasma.

15.6 Electron temperature measurement

15.6.1 Langmuir probes

See discussion of triple probes under density measurement heading.

15.6.2 Thomson scattering

Thomson scattering of light from a high power laser beam provides an unambiguous measurement of electron temperature. This diagnostic is technically challenging because extremely weak signals must be detected in a noisy environment.

The method, originally developed for tokamaks, consists of measuring the very weak electromagnetic radiation of electrons quivering in the oscillating electric field of a high intensity laser beam. The radiated power from the quivering electrons can be considered as a partial scattering of the incident laser beam. The electron velocity distribution is determined from a small Doppler shift in the scattered radiation. The scattering cross-section per electron is $\sigma_T = 8\pi r_e^2/3$ where $r_e = 2.8 \times 10^{-15}$ m is the classical electron radius. Because this cross-section is extremely small, the scattered power is many orders of magnitude lower than the illuminating laser beam. Hence great care is necessary to reduce spurious scattering from wall and window surfaces to a level below the Thomson scattering signal. Another consequence of the extremely low level of the Thomson scattering signal is that a very high power laser must be used in order for the scattered signal to exceed intrinsic detector noise.

The Doppler shift associated with Thomson scattering results from a time-retardation effect inherent in the radiation field of an accelerated charge. According to classical electrodynamics[168] the radiation electric field produced by a non-relativistic accelerating electron is

$$\mathbf{E}(\mathbf{r},t) = \frac{q_e}{4\pi\varepsilon_0 c^2} \left[\frac{\hat{n}(t') \times (\hat{n}(t') \times \ddot{\mathbf{r}}_e(t'))}{|\mathbf{r} - \mathbf{r}_e(t')|} \right]_{t'=t-|\mathbf{r}-\mathbf{r}_e(t')|/c} \quad (15.32)$$

where $\mathbf{r}$ is the observer's location, $\mathbf{r}_e(t)$ is the electron trajectory, $\hat{n}$ is the instantaneous unit vector in the direction from the electron to the observer, i.e.,

$$\hat{n}(t') = (\mathbf{r} - \mathbf{r}_e(t'))/|\mathbf{r} - \mathbf{r}_e(t')|, \quad (15.33)$$

and t' is the retarded time. The equation of motion of an electron in the laser beam is

$$\ddot{\mathbf{r}}_e = \frac{q_e \mathbf{E}_l}{m_e} \exp(i\mathbf{k}_l \cdot \mathbf{r}_e(t) - i\omega_l t) \quad (15.34)$$

where $\mathbf{E}_l$ is the laser electric field and ω_l is the laser frequency. If the displacement due to the laser acceleration is small compared to the laser wavelength, then the electron trajectory in the right hand side of Eq.(15.34) can be approximated as $\mathbf{r}_e(t) \approx \mathbf{r}_e(0) + \mathbf{v}_0 t$ where $\mathbf{v}_0$ is the electron initial (unperturbed) velocity, giving

$$\ddot{\mathbf{r}}_e = \frac{q_e \mathbf{E}_l}{m_e} \exp\left[i\mathbf{k}_l \cdot \mathbf{r}_e(0) - i(\omega_l - \mathbf{k}_l \cdot \mathbf{v}_0)t\right]. \quad (15.35)$$

The Doppler shift $\mathbf{k}_l \cdot \mathbf{v}_0$ is assumed small compared to ω_l. Substitution into Eq.(15.32) gives

$$\mathbf{E}(\mathbf{r},t) = \frac{q_e^2}{4\pi\varepsilon_0 m_e c^2} \frac{\hat{n} \times (\hat{n} \times \mathbf{E}_l)}{r} \exp\left[i\mathbf{k}_l \cdot \mathbf{r}_e(0) - i(\omega_l - \mathbf{k}_l \cdot \mathbf{v}_0)t'\right] \quad (15.36)$$

15.6 Electron temperature measurement

where $r \simeq |\mathbf{r} - \mathbf{r}_e(t')|$ has been assumed for the denominator. The retarded time may be approximated as

$$t' \simeq t - \frac{1}{c}(r - \hat{n} \cdot \mathbf{r}_e(t'))$$
$$= t - \frac{r}{c} + \frac{\hat{n}}{c} \cdot (\mathbf{r}_e(0) + \mathbf{v}_0 t') \qquad (15.37)$$

so that

$$t' = \frac{t - \frac{r}{c} + \frac{\hat{n}}{c} \cdot \mathbf{r}_e(0)}{1 - \frac{\hat{n}}{c} \cdot \mathbf{v}_0}$$
$$\approx \left(t - \frac{r}{c}\right)\left(1 + \frac{\hat{n}}{c} \cdot \mathbf{v}_0\right) + const. \qquad (15.38)$$

Omitting constant terms and dropping terms that are second order in v_0/c, the phase in Eq.(15.36) can be written as

$$(\omega_l - \mathbf{k}_l \cdot \mathbf{v}_0) t' = (\omega_l - \mathbf{k}_l \cdot \mathbf{v}_0) t \left(1 + \frac{\hat{n}}{c} \cdot \mathbf{v}_0\right) - \frac{r}{c}(\omega_l - \mathbf{k}_l \cdot \mathbf{v}_0)\left(1 + \frac{\hat{n}}{c} \cdot \mathbf{v}_0\right)$$
$$= \left(\omega_l - \mathbf{k}_l \cdot \mathbf{v}_0 + \frac{\omega_l}{c}\hat{n} \cdot \mathbf{v}_0\right) t - \left(\omega_l - \mathbf{k}_l \cdot \mathbf{v}_0 + \frac{\omega_l}{c}\hat{n} \cdot \mathbf{v}_0\right) r/c. \qquad (15.39)$$

Defining the frequency and wavenumber of the scattered signal as

$$\mathbf{k}_s = \frac{(\omega_l - \mathbf{k}_l \cdot \mathbf{v}_0 + \omega_l \hat{n} \cdot \mathbf{v}_0/c)}{c}\hat{n}$$
$$\omega_s = (\omega_l - \mathbf{k}_l \cdot \mathbf{v}_0 + \omega_l \hat{n} \cdot \mathbf{v}_0/c) \qquad (15.40)$$

we see that the scattered wave has a phase $\omega_s t - \mathbf{k}_s \cdot \mathbf{r}$.

The change in wavenumber is

$$\Delta k = k_s - k_l$$
$$= \frac{(\omega_l - \mathbf{k}_l \cdot \mathbf{v}_0 + \omega_l \hat{n} \cdot \mathbf{v}_0/c)}{c} - \frac{\omega_l}{c}$$
$$= \left(\frac{\omega_l}{c}\hat{n} - \mathbf{k}_l\right) \cdot \mathbf{v}_0$$
$$= 2\omega_l \frac{v_0}{c} \sin(\theta/2) \qquad (15.41)$$

where θ is the angle between the incident wavevector $\mathbf{k}_l$ and the viewing direction $\hat{n}$, and $\mathbf{v}_0$ is in the direction of $\mathbf{k}_l - \mathbf{k}_s$. The corresponding wavelength shift will be $\Delta\lambda = -2\pi k^{-2} \Delta k$ or

$$\frac{\Delta\lambda}{\lambda} = -2\frac{v_0}{c}\sin\frac{\theta}{2}. \tag{15.42}$$

The radiated intensity will map the electron velocity distribution of electrons moving in the direction $\mathbf{k}_l - \mathbf{k}_s$,

$$f \sim \exp(-m_e v_0^2 / 2\kappa T_e) \tag{15.43}$$

into a wavelength-dependent intensity

$$I_{scat}(\lambda) \sim \exp\left[-m_e c^2 (\Delta\lambda)^2 / 8 \lambda^2 \kappa T_e \sin^2\frac{\theta}{2}\right]. \tag{15.44}$$

The electron temperature will be

$$T_e = 1.1 \times 10^4 \left(\frac{\Delta\lambda_{FWHM}}{\lambda \sin(\theta/2)}\right)^2 \quad \text{(eV)} \tag{15.45}$$

where $\Delta\lambda_{FWHM}$ is the full width at half maximum. Typically a scattering angle $\theta = 90^0$ is used so as to minimize the stray laser light entering the detector. Near perfect beam dumps are required to prevent even the slightest reflection of incident laser light back into the plasma because any such reflection would scatter from the walls and reach the detector.

15.7 Impurity radiation measurements

15.7.1 Spectroscopic identification of ionization states

The degree of ionization of impurity ions depends on the electron temperature [155] and so identifying impurity lines can provide an estimate of electron temperature. The more highly stripped impurities will be located in the hotter regions of the plasma so that an approximate T_e profile can be constructed from the distribution of impurity lines. Measurements of the time dependence of line ratios (e.g. OVI/OV ratio) can give an easy and important indication of the time evolution of the electron temperature.

15.7.2 Bolometry

Bolometers measure the power flux of incident electromagnetic radiation without regard to wavelength and can provide an absolute measurement of radiated power, especially line radiation from impurities and from ionization. Bolometers consist of a thin layer of broadband-absorbing, plasma-facing material heated by the incident radiation; the resulting temperature rise can be used to determine the incident broadband power flux.

As an example of a bolometer, Fernandez et al.[139] described a design in which a 5μm thick gold film acted as the broad-band energy absorber. This film was deposited on a 1.5 μm thick mylar film which provided mechanical support. A 50 Ω resistor was formed on the backside of the mylar film by depositing a 0.2 μm thick gold zigzag. The device was oriented so that the 5 μm gold layer faced the plasma. An approximately 1 m long collimator provided a well-defined path for the electromagnetic radiation. The collimator also prevented plasma particles from striking the bolometer since most particles would strike the collimator walls and be absorbed.

In operation, incident broadband radiation heated the gold layer, producing a temperature rise of the order of 1^0 C. The heat quickly diffused through the thin mylar to the resistor so that the resistor rapidly attained thermal equilibrium with the absorber. The equilibration time was approximately 25 μs as determined by calibration with a short laser pulse applied to the heat absorbing layer. The slight change in resistance due to change in temperature was measured and used to infer the power flux of incident broadband electromagnetic radiation.

CHAPTER 16
Applications of Spheromaks

16.1 The spheromak as a fusion reactor

Nuclear fusion requires heating a plasma to a temperature where colliding ions have enough kinetic energy to overcome their mutual electrostatic repulsion. This kinetic energy requirement is expressed quantitatively as a temperature-dependent fusion reaction cross-section. The reaction which has a useful cross-section at lowest ion temperature is the deuterium-tritium reaction; this produces an alpha particle and a neutron:

$$D + T \rightarrow He^4 + n + 17.6 \text{ MeV}.$$

The cross-section rises rapidly with ion temperature and becomes reactor-relevant [169] in the range $T_i \sim$ 5-10 keV. The 17.6 MeV output energy consists of 3.5 MeV in alpha particle kinetic energy and 14.1 MeV in neutron kinetic energy. Since the alpha particle is charged, it becomes a high energy plasma component and, if well confined, its energy eventually becomes collisionally transferred to the bulk plasma, thereby heating the plasma and sustaining the fusion burn. On the other hand, the neutron, being neutral, immediately escapes the magnetic confinement system and so cannot heat the plasma; instead the neutron energy can be captured outside the plasma and used to provide the reactor output energy. Typical fusion reactor designs employ a lithium blanket to capture the neutron energy and turn this energy into heat. The heat is then used to drive an electric generator via a steam cycle; the lithium also produces tritium so as to have a closed tritium cycle.

Merely heating a plasma until $T_i \sim$ 5-10 keV is not sufficient for a workable fusion reactor, because if the plasma cools off immediately after being heated, not enough fusion output energy will be collected to pay back for the energy invested in heating the plasma. Thus, a critical parameter for fusion is the energy confinement time τ_E. The net fusion energy output per $D - T$ pair will therefore be $n < \sigma v > \tau_E E_{out}$ where $n < \sigma v >$ is the reaction rate and E_{out} is the energy liberated in the fusion reaction. As indicated above, this output energy must exceed the thermal energy $\sim 4 \times 3\kappa T/2$ invested in heating the $D - T$ pair to 5-10 keV. The factor of 4 comes from there being 2 ions and 2 electrons for each $D - T$ pair. The condition for net fusion output thus has the form $n\tau_E > 6\kappa T/ < \sigma v > E_{out}$. This criterion is called the Lawson condition and quantitatively is

$$n\tau_E > 3 \times 10^{20} \text{ m}^{-3}\text{s}. \tag{16.1}$$

Tokamak development[169] at the end of the 20th century has reached the point where the temperatures and $n\tau_E$ required for fusion are clearly achievable and a prototype tokamak fusion reactor could be built. However, such a reactor would be so large, complex, and costly as to be non-competitive with other power sources. If controlled fusion is to become an economically viable energy source, it is necessary to develop fusion devices which would be smaller, simpler, and less costly than implied by extrapolation of existing tokamak concepts.

Because the spheromak configuration is created by plasma self-organization and because the spheromak vacuum vessel is simply connected (unlike tokamaks which require a doubly connected vacuum vessel), a spheromak-based fusion reactor offers the intrinsic advantage of simplicity. Furthermore, since spheromak magnetic fields are mainly provided by internal plasma currents, the requirements for external coils are modest. This intrinsic simplicity of the spheromak vacuum vessel and coil system suggest that inexpensive compact systems could be built.

Unfortunately these enticing advantages are completely offset at the present time by two serious shortcomings: (i) the spheromak confinement time τ_E is many orders of magnitude shorter than corresponding tokamak confinement times, and (ii) even with large power inputs, no spheromak has yet been heated to reactor-relevant temperatures. If spheromaks are to become a viable alternative to tokamaks, these shortcomings will have to be overcome.

Tokamak performance is often characterized by a figure of merit $n\tau_E T$ that takes into account both the Lawson criterion and the temperature dependence of the reaction rate cross-section. The figure of merit [169] required for a reactor is $n\tau_E T > 3 \times 10^{24}$ m^{-3}s eV and, to a large extent, tokamak history can be viewed as a steady march towards this goal. The first tokamaks (circa-1950 Soviet devices) had $n\tau_E T \sim 10^{17}$ m^{-3}s eV and so were seven orders of magnitude lower than the reactor requirement. By the late 1990's, tokamak devices achieved figures of merit very close to the reactor requirement. For example, in 1998 the JT-60U tokamak[170] obtained $n = 4 \times 10^{19}$ m^{-3}, $\tau_E = 0.97$ s, and $T_i = 16$ keV giving $n\tau_E T = 6 \times 10^{23}$ m^{-3}s eV; this corresponds to a six order-of-magnitude improvement in figure of merit over four decades. This remarkable improvement is the consequence of a human investment of tens of thousands of man-years (~ 40 years of worldwide effort with ~ 1000 workers each year worldwide) and a several billion dollar financial investment.

In contrast, the best spheromak performance[66] at the time of writing is $n = 5 \times 10^{19}$m^{-3}, $T \sim 10^2$ eV, and $\tau_E \sim 0.2$ ms giving $n\tau_E T \sim 10^{18}$ m^{-3}s eV; this is comparable to tokamak performance of the early 1960's[171] and almost six orders of magnitude lower than state-of-the-art tokamaks. To make a fair comparison with tokamaks it should be taken into consideration that spheromaks have received only about one decade of intensive research effort, totaling hundreds of man-years, and a financial investment of a few tens of millions of dollars. Present-day spheromak performance might be much better had there simply been more investment of time and effort.

Of course, there is no assurance that more investment would have significantly improved performance, and in fact, it is essential that there be no intrinsic flaw preventing spheromaks from functioning as viable fusion reactors. The critical question is whether τ_E and T can be substantially increased in spheromaks by continued investment in spheromak research. Since spheromak confinement is not understood well enough to be calculated from first principles, this question can only be answered by building and testing actual devices.

If a spheromak plasma achieved fusion-relevant parameters, the most important engineering issues would be wall loading and β limits. Because the maximum heat flux that conventional wall materials can bear is 10-20 MW/m^2 it seems there is no advantage in making a reactor so compact as to exceed this limit. However, Moir[172] noted that spheromak fusion reactors would be excellent candidates for liquid first walls, i.e., schemes where the plasma-facing wall is molten lithium or a lithium compound. The liquid wall provides tritium breeding capability, is continuously replaced, and can be subject to heat loads as high as 900 MW m^{-2}. The high heat load capability would permit extremely compact reactors with high power densities.

High β means more fusion power per investment in hardware and it seems that a β of at least several percent is desirable for a spheromak reactor to be of economic interest. Spheromak experiments have achieved β's of several percent and if this performance persists into higher τ_E and T regimes, β should not be a serious limitation. However, because achieved β's are higher than predicted by classical MHD models, it is possible that this anomalously good behavior might disappear as confinement and temperature increase.

A critical and almost paradoxical issue for spheromaks is that helicity injection (i.e., dynamo action) requires at least the temporary destruction of the axisymmetric flux surfaces required for good confinement. If helicity can be injected with only a minimal and transient degradation of confinement, then spheromaks will be an attractive fusion confinement concept. However, if helicity injection and associated Taylor relaxation invariably cause a serious degradation of confinement, then spheromaks will not be a successful fusion confinement system. This issue can only be resolved by further experiments and analysis.

There have been a few limited preliminary engineering studies of what a spheromak-based fusion reactor might look like, but because the spheromak is so many orders of magnitude away from the fusion regime these studies must be considered as specula-

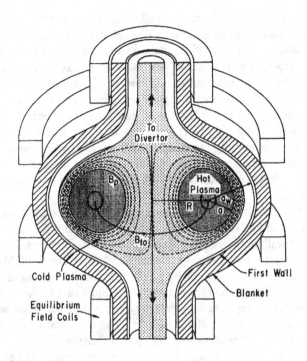

Fig.16.1 Conceptual spheromak-based fusion reactor design considered by Katsurai and Yamada [58].

tive. In 1982 Katsurai and Yamada[58] made the first conceptual study of spheromak fusion reactors; Fig. 16.1 shows the layout of a reactor design resulting from their analysis. The nominal engineering parameters for this DT spheromak fusion reactor were 3 meters major radius, 1 GW fusion output power, 4 MW wall loading, and a 4 T wall magnetic field. These engineering parameters were based on assumed physics parameters of $\beta = 4.5\%$, 15 keV plasma temperature, $n = 2 \times 10^{20}$ m^{-3} plasma density, and energy confinement time $\tau_E = 1.4$ s. The stored magnetic energy would be 1.6 GJ and the toroidal and poloidal plasma currents would be 29 MA and 110 MA respectively. These are attractive numbers for a fusion reactor; the key issue is whether it is indeed possible for a spheromak to achieve the assumed plasma temperature and energy confinement time.

Hagenson and Krakowski[61] made an engineering optimization of spheromak reactor design which resulted in parameters similar to those of Katsurai and Yamada. A particularly interesting parameter is the mass of the fusion power core which Hagenson and Krakowski estimated would be only 500 tons. This is to be compared to the 20,000 ton mass of ITER, the international tokamak reactor prototype which, at the time of writing, has been designed, but may not be built because of high cost.

Unlike a tokamak, a spheromak's toroidal field is entirely due to plasma currents. These currents are poloidal and several times larger than the toroidal current. Consequently a spheromak with the same toroidal current as a given tokamak will have a much larger total current. For example, the poloidal current in the Katsurai/Yamada design is 110 MA whereas a tokamak has insignificant internal poloidal currents. The ~ 30 MA toroidal current of the Katsurai/Yamada design is comparable to the ITER toroidal current. Thus, the total Ohmic heating power $\eta(J_{pol}^2 + J_{tor}^2)$ of a spheromak is about an order of magnitude larger than the Ohmic power ηJ_{tor}^2 of a tokamak reactor with the same toroidal current.

Because of this much larger Ohmic power input, Fowler et al. [53] and Mayo[54] proposed that a spheromak could be heated to fusion temperatures using Ohmic heating alone. In contrast, tokamaks inevitably require non-Ohmic heating to attain fusion-relevant temperatures, because at these high temperatures tokamak Ohmic heating power (i.e., ηJ_{tor}^2) becomes inadequate.

Fowler et al.'s model takes into account the radial profiles of the temperature, magnetic field, current density, and transport. For simplicity we neglect these geometrical considerations and consider instead a zero-dimensional analysis which outlines the main issues. The essential ingredients are the temperature dependence of Spitzer resistivity [173]

$$\eta = \frac{5 \times 10^{-5} \ln \Lambda}{T_e^{3/2}} \text{ Ohm-m} \tag{16.2}$$

and power balance in the reactor core.

The current density is determined from the Taylor state $\mu_0 \mathbf{J} = \lambda \mathbf{B}$. The choice of h/a is a trade-off between tilt stability and maximizing the ratio of poloidal to toroidal flux (recall that confinement depends on poloidal flux). Equation (9.24) shows that for a given toroidal flux, the poloidal flux scales as $1/\lambda$ suggesting that small λ gives more confinement per toroidal field. Since $\lambda = (x_{11}/a)\sqrt{1 + \pi^2 a^2/x_{11}^2 h^2}$, for a given a the poloidal flux and hence confinement increases with h/a. However, for tilt stability the maximum allowed prolateness is $h \approx 1.6a$. Using this ratio in Eq.(9.11) gives $\lambda \approx 4/a$. Thus, the current density of a tilt-stable, maximum poloidal flux spheromak will be

$$J \sim \frac{4B}{a\mu_0} = 3\frac{B}{a} \text{ Megamp/meter}^2 \tag{16.3}$$

and so the Ohmic heating input power will be

$$P_{OH} = \eta J^2 = \frac{5 \times 10^8 B^2 \ln \Lambda}{T_e^{3/2} a^2} \text{ W m}^{-3}. \qquad (16.4)$$

In equilibrium this input power must balance the power lost from the plasma,

$$P_{loss} = \frac{\frac{3}{2} n \times 1.6 \times 10^{-19} (T_e + T_i)}{\tau_E} \text{ W m}^{-3} \qquad (16.5)$$

where τ_E is the energy confinement time and temperatures are in eV. We assume that $T_e = T_i = T$, balance the Ohmic input power with the power loss, assume $\ln \Lambda \approx 10$ and solve for the required confinement time as a function of temperature, obtaining

$$\tau_E = 10^{-28} \frac{T_e^{5/2} n a^2}{B^2} \text{ s}. \qquad (16.6)$$

In their discussion, Fowler et al. defined the ignition temperature as the temperature where plasma heating from alpha particles exceeds the power lost from the plasma by bremsstrahlung radiation; this criterion gave an ignition temperature $T = 4.2$ keV. Using this ignition temperature, and assuming $B = 10$ T, $a = 2$ m, and $n = 2 \times 10^{20}$ m^{-3}, Eq.(16.6) shows that a core energy confinement time of $\tau_E \approx 1$ s is required. Fowler et al. argued that such a confinement time is plausible for these plasma parameters if one assumes a conservative confinement scaling law such as gyro-Bohm [see Eq.(13.21)].

The spatial profile of the magnitude of the magnetic field given by Eq.(9.12a-9.12c) is

$$\frac{B(r,z)}{B_0} = \sqrt{\frac{J_1^2(x_{11}r/a) \left[(\pi a/x_{11}h)^2 \sin^2(\pi z/h) + (\lambda a/x_{11})^2 \cos^2(\pi z/h) \right]}{+ J_0^2(x_{11}r/a) \cos^2(\pi z/h)}}. \qquad (16.7)$$

where the z axis origin is now defined to lie in the midplane.

Figure (16.2) plots $B(r,z)/B_0$ assuming nominal values $h = 1.5a$ and $\lambda = 4/a$ while Fig.(16.3) plots the variation on the midplane. The maximum field (and therefore maximum current density) occurs at $r = 0$, $z = 0$ and the maximum wall field is $B/B_0 = 0.4$ which occurs on the outer wall at $r = a, z = 0$.

Figures (16.2) and (16.3) show that the maximum Ohmic heating occurs at the geometric origin $r = 0$, $z = 0$. However, since the flux surface passing through this point is also the wall-facing flux surface, the particles which receive the most heating input are on the outermost flux surface where they will easily lose their energy to the wall. On the other hand, the magnetic field (and current density) on the magnetic axis will be

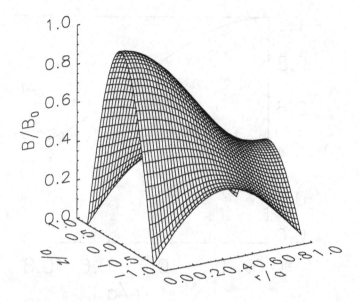

Fig.16.2 Plot of $B(r,z)/B_0$ for $h/a = 1.5$ and $\lambda = 4/a$.

$B/B_0 = 0.5$ so that the Ohmic power input where the best confinement occurs (magnetic axis) is only one quarter of the heating at the location where confinement is at its worst (outermost flux surface).

16.2 Accelerated spheromaks

Because the spheromak configuration is a minimum magnetic energy state, it is very robust and persists even when the entire configuration is physically translated (provided the required wall boundary conditions are maintained). Translation of spheromaks is not difficult because the spheromak, being simply connected, is not physically linked to any external structure.

Hammer, Hartman et al.[31] investigated spheromak translation and acceleration using the Ring Accelerator Experiment (RACE) shown in Fig.(16.4). In this device a

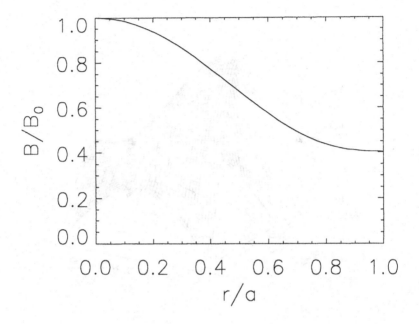

Fig.16.3 B/B_0 v. r/a on the midplane of a cylindrical spheromak. The magnetic axis is at $r/a = 0.63$; the magnetic energy density at the magnetic axis is only one quarter of the magnetic energy density on the geometric origin, $r = 0$, $z = 0$.

spheromak was created by a coaxial plasma gun and then transferred to a long coaxial rail gun which accelerated the spheromak to high velocity along the direction of the geometric axis (z axis). The coaxial plasma gun center electrode supported the long cantilevered rail gun center electrode, and the spheromak functioned as the moving armature of the coaxial rail gun. A separate, isolated pulsed power supply drove current along the center conductor, through the spheromak, and back through the outer conductor. The length of the system was 6 m, with a spheromak length of about 0.5 m. The magnetic $\mathbf{J} \times \mathbf{B}$ force of the rail gun (or equivalently the magnetic pressure behind the spheromak) accelerated the spheromak to high velocities.

Figure (16.5) illustrates how the dynamics of a spheromak accelerator can be modeled as an electro-mechanical system where the coaxial rail gun acts as a variable in-

16.2 Accelerated spheromaks

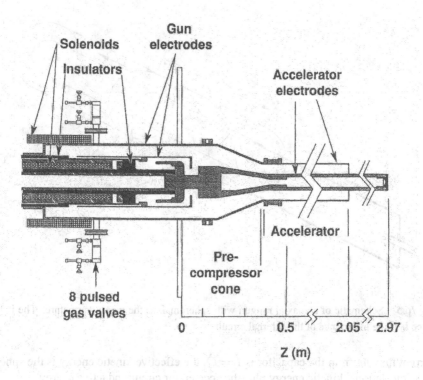

Fig.16.4 Layout of the Ring Accelerator Experiment (RACE) from Hammer et al. [31].

ductor with inductance

$$l = l'z. \tag{16.8}$$

Here $l' = (2\pi)^{-1}\mu_0 \ln(b/a)$ is the inductance per length of a coaxial transmission line consisting of inner conductor having radius a and outer conductor having radius b, and z is the spheromak axial position.

The governing equations are found by constructing a Hamiltonian-Lagrangian model of the system. The potential energy of this system is

$$U = \frac{Q^2}{2C} \tag{16.9}$$

where Q is the charge stored on the power supply capacitor C. Thus the generalized coordinates are the spheromak axial position z and the capacitor charge Q. Since the

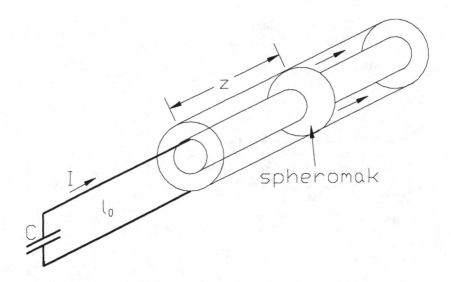

Fig. 16.5 Schematic of a coaxial railgun with spheromak as the moving armature. The inductance l_0 is the inductance of the external circuit.

current flowing from the capacitor is $I = \dot{Q}$, the effective kinetic energy is the spheromak translational kinetic energy plus the accelerator circuit inductive energy

$$T = \frac{1}{2}m\dot{z}^2 + \frac{1}{2}(l_0 + l'z)\dot{Q}^2; \quad (16.10)$$

here l_0 is the fixed inductance in the capacitor bank and transmission line. The Lagrangian of this system is

$$L = \frac{1}{2}m\dot{z}^2 + \frac{1}{2}(l_0 + l'z)\dot{Q}^2 - \frac{Q^2}{2C} \quad (16.11)$$

so the canonical momentum associated with z is

$$P_z = \frac{\partial L}{\partial \dot{z}} = m\dot{z} \quad (16.12)$$

and the canonical momentum associated with Q is

$$P_Q = \frac{\partial L}{\partial \dot{Q}} = (l_0 + l'z)\dot{Q}. \quad (16.13)$$

16.2 Accelerated spheromaks

It should be noted that P_Q is the magnetic flux linking the acceleration circuit; i.e., $P_Q = \psi_{acc}$. Thus, if we consider the canonical momentum as a two-dimensional vector $\mathbf{P} = (P_z, P_Q)$, the first component corresponds to the normal mechanical momentum while the second component corresponds to the magnetic flux linked by the accelerator circuit.

The Hamiltonian is

$$H = \sum p_j \dot{q}_j - L$$
$$= \frac{P_z^2}{2m} + \frac{P_Q^2}{2(l_0 + l'z)} + \frac{Q^2}{2C} \qquad (16.14)$$

and Hamilton's equations describe the system dynamical evolution as

$$\dot{\psi}_{acc} = -Q/C$$
$$\dot{Q} = \frac{\psi_{acc}}{l_0 + l'z}$$
$$\dot{P}_z = \frac{\psi_{acc}^2 l'}{2(l_0 + l'z)^2}$$
$$\dot{z} = \frac{P_z}{2m}. \qquad (16.15)$$

Given initial conditions $z = 0$, $Q = Q_0, P_z = 0, \psi_{acc} = 0$, these equations can be integrated to give the spheromak trajectory. Although an analytic solution is not possible, numerical integration is elementary and shows that the most efficient acceleration (i.e., maximum conversion of initial capacitor bank energy into spheromak translational kinetic energy) occurs when the half-cycle time of the capacitor bank ringing into the inductor approximately equals the time for the spheromak to travel the accelerator length. The basis for this optimization can be understood by realizing that if the half cycle-time were much longer than this optimum time, the capacitor would not be fully discharged when the spheromak reached the accelerator end indicating that more energy could have been extracted from the capacitor, whereas if the half-cycle time were shorter, then energy from the inductive circuit would flow back to recharge the capacitor.

Spheromaks with a mass of approximately 20 μg were accelerated on RACE to 140 km/s with efficiencies of the order of 30%. Since 20 μg is the mass of a 0.5 mm×0.5 mm scrap of paper, the RACE device accelerated a macroscopic object (i.e., an object that would be visible if in the form of paper) to 1/2000 the speed of light. The directed kinetic energy was an order of magnitude larger than the internal magnetic energy of the spheromak. Hammer et al. noted that the accelerating magnetic field should be smaller than the spheromak internal fields to avoid destroying the internal equilibrium of the accelerated spheromak. Peterkin[37] characterized this constraint as follows: Information travels across the spheromak at the Alfvén velocity, so if d is the characteristic dimension of the spheromak, then the time for information to be communicated within the spheromak is $t = d/v_A$. In order for the spheromak to remain in equilibrium, the change in velocity in time t due to acceleration must be small compared to the internal Alfvén velocity, i.e., $\ddot{z}d/v_A \ll v_A$. The accelerating force is $F_{acc} = B_{acc}^2 S/2\mu_0$ where S is the surface area on which the force acts. The acceleration is $\ddot{z} = F_{acc}/m = B_{acc}^2/2\mu_0\rho d$ using $m = \rho S d$. Thus the condition that the acceleration be sufficiently small for the spheromak to remain in internal equilibrium corresponds to the accelerating field being small compared to the internal field.

By introducing a slight taper into the accelerator electrodes, the spheromak can be compressed as it is accelerated. Experiments on RACE[174] demonstrated a two-fold radial compression with a corresponding field amplification of 2-3. Spheromaks with up to 2 Tesla internal fields were obtained this way.

16.3 Tokamak Fuel injection

A fusion reactor needs to be refueled after its fuel has burned. Ideally, the refueling system would inject fuel into the hot plasma core where most of the thermonuclear burning occurs. Since a tokamak reactor is likely to have a minor radius of 2-3 m, injected fuel would have to penetrate through 2-3 m of hot plasma to reach the central core. Neutral gas injected at the tokamak edge is ionized immediately and would have to diffuse to the core. Because the tokamak is purposely designed to have good magnetic confinement, diffusion of fuel to the core would be very slow and probably inadequate. Thus, some means faster than diffusion is desirable. The injection schemes that have received most attention are neutral beam injection and pellet injection, but there is concern about whether fuel injected using these methods could reach the reactor core.

High velocity injection of accelerated spheromaks has been proposed as an alternate fueling scheme. The injected spheromaks need not have good confinement properties because the time of flight for injection would be extremely short, of the order of microseconds. The main criterion for spheromak injection into a tokamak is that the spheromak kinetic energy density exceed the tokamak magnetic field energy density, i.e., $\rho_{sph}U_{sph}^2/2 > B_{tok}^2/2\mu_0$. This is because the spheromak acts on the injection time scale as a flux-conserving diamagnetic body which displaces tokamak toroidal field and so must have enough translational energy to do so. The injection condition is thus similar to Archimedes' buoyancy principle.

In order to maximize fuel transfer and minimize disruption to the tokamak it is best to have the highest possible spheromak density. This gives a small dense spheromak that will briefly displace a small volume of tokamak field before merging with the tokamak.

16.4 Helicity injection current drive in tokamaks

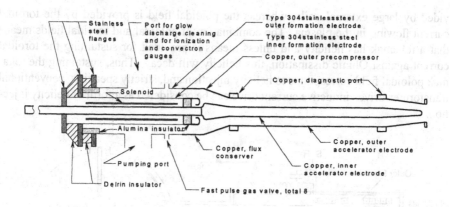

Fig.16.6 Spheromak accelerator used by Raman et al. [35] for refueling studies on the Tokamak de Varennes.

The RACE-like spheromak accelerator at Tokamak de Varennes successfully demonstrated [35] spheromak injection into a tokamak; a sketch of this accelerator is shown in Fig. 16.6. Spheromaks with mass in the range 10-60 μg were accelerated to velocities of 120-220 km/s and penetrated the 0.7-1 T field of the tokamak. The tokamak density increased by 30-90% after injection and the electron temperature decreased from 600 eV to 190 eV because of the introduction of the cold spheromak plasma. A similar device built at Caltech was used on the TEXT tokamak, but because of tokamak toroidal field trapped in the outer accelerator electrode, this device was not able to penetrate more than a 0.4 T field which was below the operating regime of the TEXT tokamak. Investigation of this behavior by Yee and Bellan[175] showed that the penetration condition depends strongly on the vacuum toroidal field profile in the vicinity of the injector muzzle.

McLean, Hwang et al. at UC Davis have developed a high repetition rate spheromak accelerator[176] designed for refueling applications where the breakdown of the spheromak plasma itself acts as the electronic switch. This device has attained repetition rates of 0.2 Hz.

16.4 Helicity injection current drive in tokamaks

Tokamak particle confinement is a direct consequence of the poloidal magnetic field and its axisymmetry. The much larger toroidal field prevents kink instabilities but does not provide confinement. The tokamak toroidal field is essentially a vacuum field provided by large external coils, whereas the poloidal field is provided by the toroidal current flowing in the plasma. The combination of poloidal and toroidal fields means that a tokamak has helicity and unless means is provided for sustaining the toroidal current against Ohmic dissipation, this helicity will decay. Thus, sustaining the tokamak poloidal field corresponds to helicity injection and, strictly speaking, conventional transformer-driven inductive current drive can be considered as a form of helicity injection.

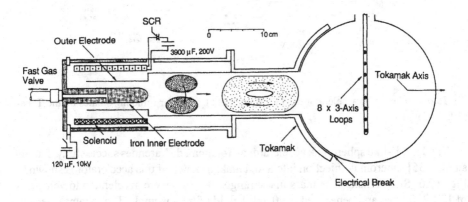

Fig.16.7 Setup for injecting spheromaks into the Encore small research tokamak used by Brown and Bellan [40]. Magnetic probe measurements indicate that spheromak tilts after leaving the gun region (as shown).

Helicity could be injected in other ways as well. For example, when a spheromak containing substantial helicity merges with a tokamak plasma, the plasma resulting from the merging would have the algebraic sum of the original helicities. Because helicity is a scalar quantity, the spatial orientation of the injected spheromak does not matter and, in fact, the helicity conservation principle shows that the geometric details will sort themselves out in such a way that the tokamak toroidal current would increase if the injected spheromak has the same sign of helicity as the tokamak. This helicity injection concept was demonstrated by Brown and Bellan [40] on the Caltech Encore research tokamak. In this experiment (cf. Fig. 16.7) a small spheromak produced by a coaxial gun was injected into a modest pre-existing tokamak plasma. The tokamak toroidal current, measured by a Rogowski coil located far from the injection region, increased when the spheromak helicity was of the same sign as the tokamak helicity and decreased (or did not increase) when the signs were opposite. The magnitude of the increment of tokamak helicity corresponded to the injected helicity after accounting for losses.

For maximum helicity injection efficiency, the spheromak gun should have the largest possible physical dimensions, i.e., dimensions similar to the tokamak dimensions. This is because helicity scales as B^2L^4 whereas energy scales as B^2L^3 so the most helicity per energy is obtained by having the largest possible characteristic length L. This optimization to the largest possible spheromak linear dimension for helicity injection is in contrast to the optimization for fueling where the most compact spheromak is desirable.

Placing a rod down the geometric axis of a spheromak tends to lower q_{wall} and therefore increases the shear. If a current is made to flow down this rod, then q_{wall} can be actively modified. Bruhns et al.[64] added a current carrying rod to a small spheromak at the University of Heidelberg. The ratio of h/a in this spheromak was so large that the spheromak was tilt-unstable without the rod. By passing a very large current through the axial rod, the overall q profile was raised and made to be tokamak-like, i.e., increase from the magnetic axis. This configuration was found to be stable and since it was tokamak-like and had a unity major to minor axis ratio, it was called an Ultra Low Aspect Ratio Tokamak (ULART). Browning et al.[177] similarly placed a rod down the geometric axis of the SPHEX spheromak, and found that the effect of the rod was to improve the coupling efficiency between the coaxial plasma gun and the spheromak plasma.

This combining of coaxial plasma gun helicity injection with tokamak q profiles has been extensively developed in the Helicity Injected Tokamak (HIT) at the University of Washington. This device is an ultra low aspect ratio tokamak in which the toroidal current is provided entirely by helicity injection from a magnetized coaxial plasma gun integral to the tokamak[42]. As shown in Fig.16.8 the coaxial gun has its axis along the tokamak axis and injects helicity directly into the tokamak. The coaxial gun links a poloidal conductor on the geometric axis; current flow in this conductor produces the tokamak toroidal magnetic field. HIT has a major radius $R = 0.3$ m, minor radius $a = 0.2$ m, and toroidal magnetic field $B = 0.46$ T. The device is diagnosed with a set of magnetic probes which measure the edge poloidal field. A computer reconstruction fits the measured poloidal field profile to a Grad-Shafranov solution which then gives internal flux contours from which the toroidal current can be determined. This fitting indicates a toroidal current on closed field lines of $\sim$200 kA, while Thomson scattering measurements indicate electron temperatures of $\sim$100 eV.

16.5 Colliding spheromaks to investigate magnetic reconnection

Magnetic reconnection is localized and yet changes the global magnetic topology;

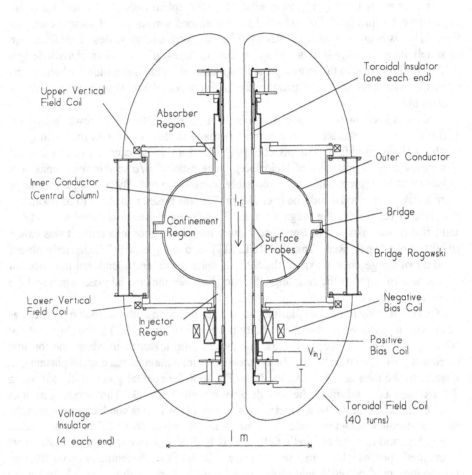

Fig.16.8 Sketch of the Helicity Injection Tokamak (HIT) from Nelson et al. [42]. The central column (z axis) contains axially directed currents which produce the tokamak toroidal field. This column goes through the center of the helicity injector at the bottom.

16.5 Colliding spheromaks to investigate magnetic reconnection

thus, even though reconnection is typically a microscopic process it is a critical behavior of magnetized plasmas. There has been substantial theoretical effort directed towards understanding magnetic reconnection, but relatively few experimental investigations. This is because it is difficult to conduct a "clean" experiment where the various aspects of reconnection are well-defined and easily measured. The difficulty arises because reconnection is a boundary layer phenomenon involving a balance between two different types of dynamics having vastly different characteristic length scales. Diagnosis of multiple-scale phenomena requires measurements having extremely high spatial resolution; such high resolution is usually unattainable in plasma experiments.

Reconnection theories typically involve simplifying geometric assumptions in order to make analysis tractable; these symmetries can be difficult to arrange in an experiment. Furthermore, actual reconnection might be turbulent and therefore inherently not reproducible, making high resolution diagnosis essentially impossible.

Yamada et al.[46, 178] and Ono et al.[47] circumvented many of these problems in a series of experiments on the TS-3 device at the University of Tokyo. As shown in Fig.16.9 these experiments involved merging two axisymmetric, reproducible coaxial spheromaks. The merging process necessarily involved magnetic reconnection and so these experiments provided well-defined reproducible information on magnetic reconnection in a symmetric, yet realistic geometry. The reconnection rate had a strong dependence on the relative helicities of the merging spheromaks. In particular, counter-helicity merging (i.e., merging two spheromaks with opposite helicities) was found to proceed about three times as fast as co-helicity merging and the reconnection rate was proportional to the impact velocity of the colliding spheromaks. As discussed in Sect.13.5, these experiments also showed that the magnetic energy dissipated during magnetic reconnection is converted directly into ion heating.

At Princeton, Yamada et al. [48] made high resolution measurements of reconnection using the Magnetic Reconnection Experiment (MRX), a device specifically designed for the investigation of magnetic reconnection. This device creates two coaxial spheromak-like plasmas using the flux core method originally developed for the S-1 Spheromak. The two spheromak-like plasmas were merged in a highly controlled fashion, leading to well-defined, reproducible reconnection layers that could be examined in detail using magnetic probes. Besides colliding co- and counter-helicity spheromaks, the MRX device was also used to create non-spheromak plasmas with null helicity (no toroidal field) and merge these with each other. These measurements indicated that merging for co-, counter-, and null-helicity are very different. However, in all cases, merging occurs spontaneously because the two initial plasmas have parallel toroidal currents (i.e., same sense of poloidal field) and therefore attract each other.

Co-helicity merging means that the poloidal currents also have the same sense (i.e., same sense of toroidal field) so that merging maintains finite toroidal field. This type of merging retains the low β character of the spheromak. In contrast, null helicity merging involves annihilation of the entire magnetic field in the region between the two plasmas because equal and opposite fields are being superimposed. The local annihilation of the entire magnetic field results in a high β region, large Larmor radii in the vicinity of the field null, and a well-defined current sheet associated with the sharp magnetic field gradient.

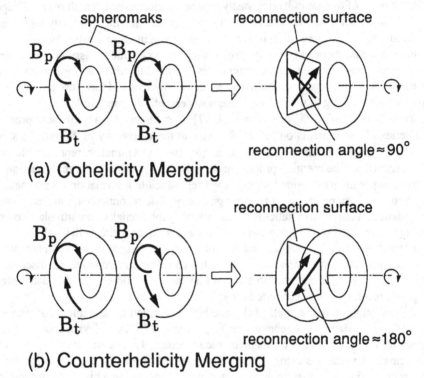

Fig.16.9 Merging spheromaks investigated by Ono et al.[47] and Yamada et al.[178].

16.6 Proposed additional spheromak applications

16.6.1 Pulsed high power X-ray radiation sources

Dietz et al. at Kirtland Air Force Base proposed using accelerated spheromaks as pulsed x-ray sources[179]. In this scheme a spheromak would be created with as much mass as possible by using heavy ions (e.g., argon or krypton), high density, and large volume. The cold, heavy spheromak (typically 1-10 mg mass) is accelerated to high velocity using a RACE-like accelerator and then made to impact a wall. Most of the translational kinetic energy is in ion motion and the velocity just before impact is U =500-2000 km/s. The translational kinetic energy per ion $m_i U^2/2$ can be characterized as equivalent to the energy $3\kappa T_i/2$ of thermal ions with temperature

$$(T_i)_{equiv} = 3.5 \times 10^{-9} \mu U^2 \text{ eV} \qquad (16.16)$$

where μ is the atomic mass of the ion. For example, the equivalent ion temperature of a 10 mg Argon spheromak traveling at 700 km/s would be $(T_i)_{equiv} \sim 70$ keV and the total translational kinetic energy would be 2 MJ.

When this spheromak collides with the wall, its directed velocity becomes zero (stagnation). A shock wave reflects back from the impact region and then compressionally heats the plasma. Some of the ion kinetic energy is converted into electron thermal energy and the hot electrons excite various atomic transitions causing emission of x-ray photons. The pulse duration τ_{pulse} is of the order of l/U where l is the spheromak length. Thus, for $l \sim 5$ cm and $U \sim 700$ km/s the pulse length will be $\tau_{pulse} < 100$ ns. Dietz et al.[179] presented numerical calculations showing that approximately half of the initial translational energy could be converted into x-radiation and that this would be continuum radiation predominantly in the range of a few keV. A stagnating spheromak is predicted to produce a <100 ns pulse of keV x-radiation with total emitted energy in the MJ range. The requirements for success are substantially beyond what has been achieved so far in laboratory experiments; in particular heavy spheromaks have not been accelerated to high enough velocities and have not been compressed to the required densities.

16.6.2 Opening switches

This concept, shown in Fig.16.10, uses an accelerated spheromak as a power multiplier [38]. Here a load is connected across a short gap in the center electrode of a coaxial rail gun. The spheromak is accelerated by the accelerator circuit magnetic field and the energy stored in this field is comparable to the spheromak energy. This energy is just the inductive energy $W_B = \psi^2/2(l_0 + l'z)$ where $\psi = \int_0^t V dt$ is the total flux in the acceleration region and V is the acceleration bank voltage. When the spheromak passes over the gap, the current $I = \psi/(l_0 + l'z)$ commutates to flow through the load. Thus, the system acts as a rapidly switched high current source. The switching time is the spheromak transit time across the gap so, for a nominal 10 cm gap and a nominal 1000 km/s spheromak velocity, the switching time will be 100 ns. If the inner/outer electrode configuration extends past the gap, then the spheromak will continue on its way, maintaining the switched circuit. Thus, not only is there a fast switching time but also the load circuit remains intact for a considerable time after switching. For example, if the electrodes extend another 1 meter past the gap, then current is forced through the load for 1 μs. In this way energies of the order of 1 MJ could be switched to low impedance loads in times of the order of 100 ns.

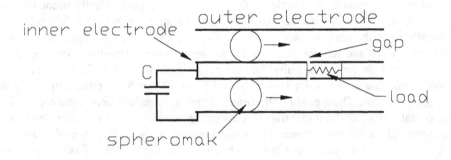

Fig. 16.10 Coaxial spheromak accelerator with load across gap in inner electrode to form opening switch. Accelerator circuit current is forced to flow through load after sheromak crosses gap.

CHAPTER 17
Solar and Space Phenomena Related to Spheromaks

Certain solar and space plasmas have low β and so, like spheromaks, have behavior dominated by magnetic forces. However, unlike spheromaks, these solar and space plasmas are either unbounded or else bounded on only one side. The difference in boundary conditions changes the geometry and topology, but much of the phenomenology is quite similar to that of spheromaks. In particular, these solar and space phenomena involve magnetic helicity conservation, magnetic reconnection, magnetic energy dissipation, and self-organizing relaxation towards minimum energy states. We will discuss solar phenomena first and then work outwards because solar events typically drive subsequent phenomena in interplanetary space and in the magnetosphere. Table 17.1 gives a chronology of selected solar and space papers relevant to helicity and spheromak issues.

Fig.17.1 Photograph of a large solar prominence (lower right) from NASA Skylab Mission, 1973.

Solar corona

Eclipse observations of luminescent structures protruding from the solar corona have been documented since the 13th century; the historical record has been reviewed by Tandberg-Hanssen [194]. Scientific study of these structures, generically called prominences, began in the mid 19th century with the advent of photography. Probably the first book containing a scientific discussion of prominences was by Secchi in 1875 [180]. Since then a great variety of morphologies and sizes have been categorized using various taxonomies[194]. Figure 17.1 shows an example of a large prominence. Prominences may be twisted or untwisted (the example in Fig. 17.1 is quite twisted). While the filamentary shape of prominences and the frequent appearance of twist suggested that prominences are plasma-filled magnetic flux tubes, it was not until the 1960's that high resolution optical emission measurements by Zirin and Severny [181] of Zeeman splitting confirmed that prominences were indeed magnetic.

Simultaneous with the development of better observational techniques, astrophysicists developed models describing low β plasma equilibria. In 1950 Lundquist[2] gave the Bessel function solution to the $\mathbf{J} = \lambda \mathbf{B}$ force-free equation and in 1954 Lüst and Schlüter [57] discussed the properties of force-free magnetic fields. As noted in Chapter 4, Woltjer argued in 1958 that an astrophysical plasma relaxes to the lowest possible magnetic energy state consistent with the constraint of helicity conservation. In a 1961 monograph, Ferraro and Plumpton [195] showed that force-free equilibria are of particular relevance to space plasmas and considered these equilibria in various geometries.

Using measured values of solar prominence density, temperature, and magnetic field Nakagawa et al.[182] concluded in 1971 that prominences have low β and should be in a force-free equilibrium. Heyvaerts and Priest[30] applied the Taylor relaxation hypothesis in 1984 to solar coronal structures but used different boundary conditions from spheromaks to take into account the semi-infinite geometry of the solar corona.

Over the years, a number of solar observers noted[183, 196, 197] that structures with left-handed twist predominate in the north solar hemisphere whereas structures with right-handed twist predominate in the southern hemisphere; this is called hemispherical segregation of chirality (handedness). Vrsnak et al.[187] investigated the dependence of prominence stability on the magnetic pitch angle defined in Fig.17.2(a) for a large number of prominences and found that prominences with magnetic pitch angle much less than 45^0 were typically stable, whereas those with pitch angles of the order of 45^0 tended to erupt. A chart summarizing Vrsnak et al.'s correlation between pitch angle and stability is shown in Fig.17.2(b).

Interplanetary phenomena

Burlaga[186] analyzed spacecraft measurements of a specific type of interplanetary magnetic disturbance that lasted several hours and had a characteristic rotation of the magnetic field vector. Burlaga called this phenomenon a magnetic cloud and showed

Year	Description
1860	Secchi[180] photographs prominences during eclipse
1950	Lundquist[2] proposes force-free equilibrium $\mathbf{J} \times \mathbf{B} = 0$ and Bessel function solution
1954	Lüst and Schlüter [57] discuss force-free magnetic fields
1958	Woltjer[4] shows that $\mathbf{J} = \lambda \mathbf{B}$ is minimum energy state
1961	Zirin and Severny[181] measure prominence magnetic fields using the Zeeman effect
1971	Nakagawa[182] shows that prominences are low β, discusses $\mathbf{J} = \lambda \mathbf{B}$ solutions
1978	Gigolashvili[183] notes hemispheric helicity segregation on sun
1979	Russell and Elphic[184] observe flux ropes in Venusian ionosphere
1984	Berger and Field[76] introduce relative helicity, discuss helicity
1984	Heyvaerts and Priest[30] apply Taylor's hypothesis to the solar corona
1985	Konigl and Choudri[185] apply Taylor's hypothesis to astrophysical jets
1988	Burlaga[186] identifies magnetic clouds, shows they satisfy Bessel function model
1989	Wright and Berger[85] interpret magnetic reconnection of solar wind flux tube using cross-over helicity
1991	Vrsnak et al.[187] note correlation between prominence twist and instability
1991	Detman et al.[188] propose spheromak-like structure for magnetic cloud
1994	Moldwin and Hughes[189] observe helical fields in magnetotail plasmoids
1994	Rust[50] shows helicity in magnetic clouds corresponds to helicity shed by erupting prominences
1995	Rust[190] proposes solar helicity is created by subsurface dynamo and then bubbles to surface
1995	Feynman and Martin[191] observe merging of prominences prior to eruption
1996	Rust and Kumar [51] associate S-shape with prominence instability, helicity
1997	Lites and Low [192] propose spheromak-like model for solar prominences
1997	Pevtsov et al.[193] correlate S-shape with λ

Table 17.1   Selected chronology of spheromak-relevant solar and space publications

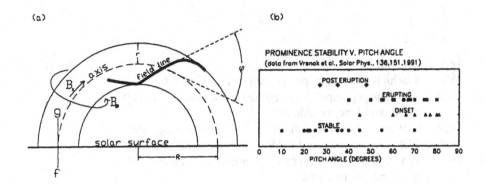

Fig.17.2 (a) Simplified schematic of the field geometry of a prominence showing the magnetic pitch angle φ; (b) stability v. φ for 49 prominences from Vrsnak et al. [187].

that the observed rotation of the magnetic field direction could be explained as a convection past the spacecraft of a helical magnetic field prescribed by Eq.(1.2). In general, the helix axis would be oriented at some arbitrary angle; this angle was determined by deconvolving the measured data. The agreement with Eq.(1.2) strongly suggests that magnetic clouds are essentially Taylor states convecting past the spacecraft. Magnetic clouds have dimensions of the order of 0.25 AU (1 AU= distance from Earth to Sun), and can disturb the Earth's magnetosphere if they collide with it. At the present time it is not known whether magnetic clouds remain attached to the sun as in Fig. 17.3(a) or detach as in Fig.17.3(b). Figure 17.3 is not to scale and in fact, the magnetic cloud is much larger than the magnetosphere.

By correlating magnetic clouds with solar prominence eruptions that took place approximately 80-100 hours earlier, Rust [50] showed that magnetic clouds are most likely spawned by prominence eruptions. In addition, Rust demonstrated a correlation between the handedness of a magnetic cloud and the solar hemisphere from which it was spawned. Because of the hemispheric segregation of chirality, this correlation suggests that magnetic clouds contain helicity shed by an erupting prominence.

Prior to the late 1990's, models of solar prominences generally assumed that prominence twist results from differential rotation of the solar surface twisting the footpoints of the prominence. In 1995 Rust and Kumar [190] argued that the magnitude of solar surface differential rotation is inadequate to explain observed prominence twist and proposed instead that a twisted prominence results from the merging of small twisted flux tubes buoyantly rising from beneath the solar surface. These emerging twisted flux tubes are presumed to be created by some subsurface dynamo. Because of the ten-

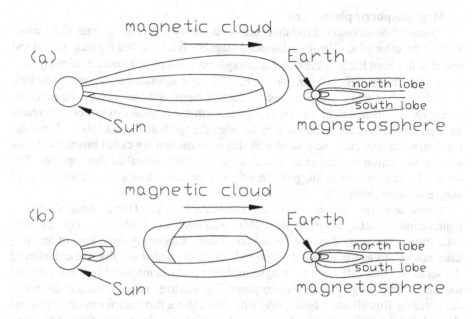

Fig.17.3 The global geometry of magnetic clouds is not yet known; the two possibilities are shown here: (a) the magnetic cloud is just a distended prominence which remains attached to the sun; (b) the magnetic cloud is detached from the sun (like the closed field lines of a spheromak).

dency for hemispheric segregation of chirality, it is further presumed that the structures in each hemisphere are driven by a hemispherical dynamo of the appropriate chirality. According to this model, helicity accumulates as the emerging small twisted flux tubes coalesce to form a prominence. At a certain threshold, the prominence becomes too twisted to remain in equilibrium and erupts, shedding a magnetic cloud containing the excess helicity. The principle of helicity conservation characterizes each stage of this multi-step process.

Rust and Kumar [51] examined soft x-ray photos of the solar corona from the Yohkoh spacecraft and noted that structures on the coronal surface sometimes had a shape like the letter S. Typically, reversed S−shapes occurred in the northern hemisphere and forward S−shapes in the southern hemisphere. These S-shapes appeared to be a precursor for eruption. Figure 17.4 shows a particularly good example of an S−shape visible in the northern hemisphere; this example is oriented opposite to the more typical orientation. Pevtsov et al.[193] studied S−shapes and showed that S shape curvature increases with λ and that high curvature is a precursor for eruption.

Magnetospheric phenomena

Spacecraft observations show that Earth's magnetosphere is blown back like a windsock by the solar wind. The long distended magnetic field is called the magnetotail and stretches out from Earth in the direction away from the sun for a distance of many hundreds of Earth radii (R_E). As shown in Fig. 17.3 the magnetotail incorporates two long flux tubes called lobes which connect to Earth's North and South magnetic poles respectively. At distances of 40-200 R_E from Earth the magnetotail typically contains an x-point where reconnection occurs between the north and south lobes. The magnetospheric disturbances associated with this reconnection are called magnetic storms and include tailwards ejection of helical magnetic structures called flux ropes or plasmoids. These observations suggest that magnetic helicity is also an important aspect of magnetospheric dynamics.

Magnetospheric physicists have known since the 1960's that the collision of a south-pointing interplanetary magnetic field (IMF) with Earth's magnetosphere drives a magnetic storm. The storm does not occur immediately, suggesting some kind of intermediate energy storage is involved. Dungey[198] first pointed out that the collision of the south-pointing IMF with the magnetosphere results in magnetic reconnection and transfer of IMF energy to the magnetosphere. The localized nature of reconnection was discussed by Russell and Elphic [199] who called this a flux transfer event (FTE) and showed that the FTE would lead to tailwards convection of magnetic flux. According to the FTE model (which is supported by spacecraft observations), flux accumulates in the magnetotail until an instability occurs causing the excess flux to return back to its starting position.

These magnetospheric phenomena have recently been considered from a magnetic helicity point of view. The magnetospheric situation does not have as close a correspondence to spheromaks as do solar prominences, because magnetospheric plasmas (except for the tail lobe) typically have high β making kinetic effects important. Nevertheless, helicity still provides useful insights regarding topological evolution. In 1989 Wright and Berger[85] applied helicity concepts to the IMF-magnetosphere reconnection process and showed that if an impinging IMF magnetic flux tube (e.g., from a magnetic cloud) is viewed from the sun, the IMF flux tube typically crosses over the adjacent Earthward magnetospheric flux tube. Thus, prior to reconnection there is cross-over magnetic helicity as discussed in Sec.3.10; after reconnection the cross-over helicity is converted into half-twist helicity. The sense of the half twist depends on whether the south-pointing IMF had east-west or west-east orientation. The half-twists emit torsional Alfvén waves which propagate along the flux tube equilibrating the twist along the length of the flux tube. Furthermore, there is a magnetic force $B^2 \hat{B} \cdot \nabla \hat{B}$ oriented so as to straighten out the hairpin bend of the reconnected flux tube. A pictorial example of half twist was provided in 1982 by Cowley[200], but helicity issues were not

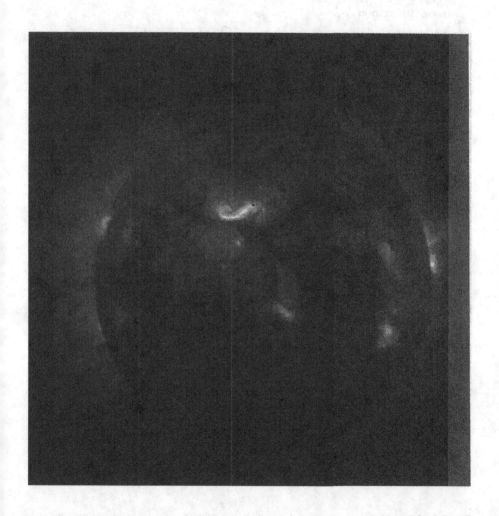

Fig.17.4 S-shape on solar coronal surface imaged by Yohkoh spacecraft soft x-ray camera (photo courtesy of Yohkoh team).

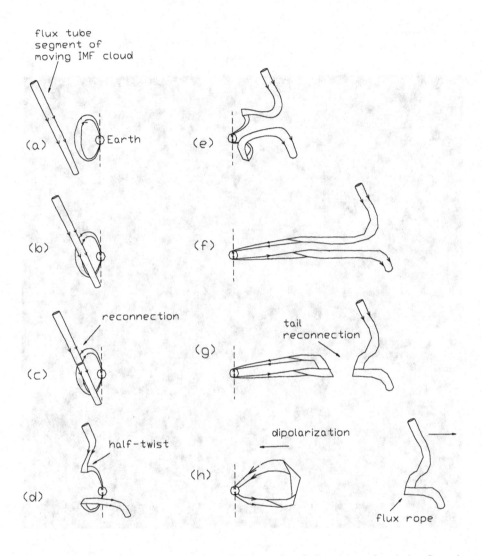

Fig. 17.5 Magnetic storm sequence from a helicity point of view: (a) south-pointing flux tube segment from a huge magnetic cloud approaches Earth's magnetosphere; (b) overlap with dipolar flux tube connecting auroral regions; (c) reconnection (as discussed by Wright and Berger [85]) forms half-twists; (d) cloud flux tube continues to convect to right, pulling along auroral region field lines; (e) more convection to right; (f) stretching of field lines contributes to magnetotail as cloud moves to far right; (g) reconnection of stretched magnetotail; (h) dipolarization of reconnected tail field lines and outward propagating flux rope in IMF.

considered.

Figure 17.5 shows the combination of Wright and Berger's half-twist picture, the FTE field-line straightening, and the effect of convection of the IMF (magnetic cloud) flux tube; this figure shows how the magnetosphere would likely respond to the passage of the magnetic cloud shown in Fig.17.3. Because the magnetic cloud has linear dimensions two orders of magnitude larger than Earth's magnetosphere, the impinging portion of the cloud can be approximated as a large straight flux tube aligned at some angle with respect to Earth and moving in the anti-sunward direction. This moving straight flux tube is often called the interplanetary magnetic field (IMF) acting on the magnetosphere. Figure 17.5(a) shows that as the IMF flux tube moves past the Earth it first reconnects with some of the Earth's dipolar field lines on the sunward side. On moving tailwards it then pulls the reconnected flux tubes tailwards (Fig. 17.5(d),(e)), stretching them (Fig. 17.5(f)), until they reconnect (Fig. 17.5(g)) and snap back (Fig. 17.5(g)) to re-form dipolar field lines (Fig. 17.5(h)). This process constitutes the magnetic storm and involves slow storage of energy followed by sudden relaxation. A rough analogy would be water streaming by a ball of bubble gum, pulling a long chunk of gum back into the wake, and stretching this chunk until it breaks and snaps back. Unlike two dimensional reconnection models, helicity is an important consideration and constrains the topological evolution. Twisted magnetic fields naturally result from reconnection topology and do not depend on any specific physical model (e.g., high β or low β). Observations exist supporting this point of view: for example in 1992 Macwan[201] reported an analysis of ISEE 3 spacecraft data indicating that the magnetotail has substantial twist while in 1993 Farrugia et al.[202] analyzed storm data where the south-pointing magnetic field was identified as being from a magnetic cloud.

17.1 Sun-Earth connection viewed as helicity flux/relaxation

In the author's opinion much of Sun-Earth plasma dynamics can be explained qualitatively using two simple helicity-based concepts closely related to spheromak physics:

1. Magnetic helicity tends to be conserved for all phenomena evolving faster than the resistive diffusion time.
2. A magnetic configuration with characteristic dimension L becomes unstable when it becomes excessively twisted. In particular, instability occurs when $\mu_0 I/\psi > L^{-1}$ where I is the current along a typical flux tube ψ. Since I/ψ is proportional to twist, this corresponds to a twist threshold.

These two concepts imply that helicity can accumulate in a configuration until the configuration abruptly becomes unstable and erupts, shedding its excess helicity. Eruption is analogous to the formation of a detached spheromak plasma that occurs when

λ exceeds the appropriate criterion. Slow accumulation of some critical quantity followed by sudden instability was first postulated by Gold and Hoyle[203] in an early pre-helicity model for prominence eruption.

Using the two concepts listed above, magnetically dominated Sun-Earth dynamics can be postulated as consisting of the following sequence:

1. A subsurface hemispherical solar dynamo creates helicity which bubbles to the surface and coalesces to form prominences [190].
2. When solar prominences accumulate excessive helicity, they erupt, shedding excess helicity in the form of magnetic clouds [190] which propagate into interplanetary space.
3. If a magnetic cloud happens to impact the Earth's magnetosphere with a southward pointing magnetic field, it transfers magnetic helicity to the Earth's magnetosphere [85] as in Fig.17.5.
4. The transferred helicity accumulates in the magnetotail.
5. When the magnetotail becomes excessively twisted, it becomes unstable, and reconnects.
6. The tailward part of the reconnected magnetotail is connected to the IMF and is twisted (plasmoid flux rope) while the Earthward part is connected to the Earth's poles and contracts (dipolarization).

Magnetic energy and flux are not conserved when helicity is accumulated nor when helicity is shed during a reconnection event. Dissipation of magnetic energy causes significant plasma heating, and sudden annihilation of flux causes large electric fields. These large electric fields generate energetic particles that in turn can emit high energy electromagnetic radiation.

To first approximation, these theories assume that magnetic helicity is perfectly conserved. However, the reality is that while magnetic helicity is much better conserved than magnetic energy for microscopic dissipative processes, magnetic helicity is in fact eventually dissipated. Chapter 12 showed that when this small but finite dissipation is taken into account, λ decays on moving away from the helicity source. Furthermore, there is a relationship between the gradient of λ, the local energy dissipation, and the helicity flux [cf. Eq.(12.56)]. The dynamo electric field associated with the helicity flux heats the plasma Ohmically so that magnetic energy is ultimately turned into thermal energy; this interpretation is similar in spirit but slightly more complicated than Kumar and Rust's analysis of magnetic cloud decay [204] where it was assumed that helicity is absolutely conserved and that only magnetic energy decay heats the plasma. Allowing λ gradients and a stationary state of net outward helicity flow in the solar system is analogous to a very leaky thermodynamic system [cf. Fig. 8.1(d)] where all the heat leaving the source goes away, never to return. The scale length for helicity decay would be quite large, so assuming locally uniform λ would be reasonable for small scale phenomena (much like the assumption of local thermodynamic equilibrium is reasonable for thermodynamic systems having gentle temperature gradients).

17.2 A spheromak-like laboratory model of solar prominences

Bellan and Hansen[52] constructed a spheromak-based magnetized plasma gun experiment designed to produce laboratory scale versions of solar prominences. This experiment is based on the hypothesis that solar prominences evolve through a sequence of Taylor states, accumulating helicity until becoming unstable. Experimental details will be deferred until Sec.17.7; the theoretical basis will be presented here and is similar to previous analyses made by Nakagawa[182], Heyvaerts and Priest[30], and Browning[205] but includes additional geometrical features and interpretations.

The physical assumptions are identical to those used for spheromaks, i.e., the plasma is assumed to be low β and to evolve through a sequence of Taylor states each characterized by

$$\nabla \times \mathbf{B} = \lambda \mathbf{B}. \quad (17.1)$$

The essential difference is in the boundary conditions. Spheromaks are surrounded on all sides by a flux-conserving wall, whereas prominences effectively have only one wall, the solar surface, and otherwise extend into infinite half space. This difference in boundary conditions can be accommodated by characterizing the prominence as a relaxed state with magnetic field lines intercepting one wall only (the solar surface); all other walls are at infinity. In order to have bounded energy it is necessary for all fields to vanish at infinity and, in addition, to fall off sufficiently quickly on going to infinity.

As shown in Fig. 17.6 solar prominences arch from a region of positive magnetic field to a region of negative magnetic field. A cylindrical coordinate system is defined such that the z coordinate corresponds to the altitude above the solar surface. The ground plane loci where B_z vanishes are called neutral lines; two representative neutral line geometries, straight and circular, are shown in Fig. 17.6.

The solution of Eq.(17.1) for the geometry of Fig. 17.6 can be written as

$$\mathbf{B} = \lambda \nabla \chi \times \nabla z + \nabla \times (\nabla \chi \times \nabla z) \quad (17.2)$$

where

$$\nabla^2 \chi + \lambda^2 \chi = 0. \quad (17.3)$$

Unlike the spheromak analysis in Sec.11.1, axisymmetry is not invoked and so in cylin-

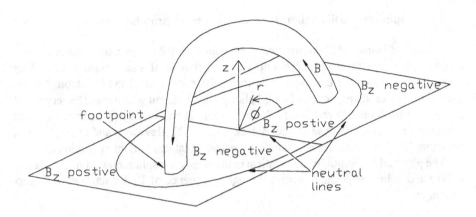

Fig.17.6 Sketch showing ground plane ($z = 0$ plane) and two types of neutral line separating regions of positive and negative B_z in the ground plane. A cylindrical coordinate system is defined with z axis corresponding to altitude above the ground plane. The z axis passes through the midpoint of the bisection of the line between the two footpoints of the prominence.

drical geometry Eq.(17.3) becomes

$$\frac{1}{r}\frac{\partial}{\partial r}\left(r\frac{\partial \chi}{\partial r}\right) + \frac{1}{r^2}\frac{\partial^2 \chi}{\partial \phi^2} + \frac{\partial^2 \chi}{\partial z^2} + \lambda^2 \chi = 0. \tag{17.4}$$

Thus, spheromaks are distinguished from prominences because spheromaks are bounded at $z = 0, z = h, r = a$, whereas prominences are bounded only at $z = 0$.

The boundary condition at $z \to \infty$ is satisfied by requiring that all quantities vary as $\exp(-kz)$ where k is a positive constant to be determined. With this assumption Eq.(17.4) has solutions of the form

$$\chi(r, \phi, z) = J_m(\kappa r) \cos(m\phi) e^{-kz} \tag{17.5}$$

where

$$\kappa = \sqrt{\lambda^2 + k^2}. \tag{17.6}$$

17.2 A spheromak-like laboratory model of solar prominences

The prominence z fields in Fig. 17.6 are antisymmetric about the straight neutral line and so their Fourier decomposition involves only odd values of the azimuthal mode number m. The simplest non-trivial configuration has $m = 1$ symmetry and so for simplicity we shall assume $m = 1$ in our analysis.

The force-free magnetic field can be expressed as

$$\mathbf{B} = -\lambda \hat{z} \times \nabla \chi - k \nabla \chi + \hat{z} \lambda^2 \chi. \tag{17.7}$$

In the limit $\lambda \to 0$ this becomes the vacuum field $\mathbf{B} = -k \nabla \chi$ which is a potential field going from the hills of χ to the valleys of χ. Because χ depends on z only through $\exp(-kz)$, the χ contours are isomorphic at all z and therefore $\mathbf{B}$ is also isomorphic at all z.

Let $z = 0$ denote the ground plane as in Fig. 17.6 and consider the magnetic field in a constant z plane at some arbitrary altitude z. The vector $-\lambda \hat{z} \times \nabla \chi$ follows a level contour of χ in this plane and so corresponds to a twist in the magnetic field. In fact $-\lambda \hat{z} \times \nabla \chi$ is azimuthal with respect to the z axis passing through a peak or a valley. The z component of the magnetic field is

$$B_z = \kappa^2 J_1(\kappa r) e^{-kz} \cos \phi; \tag{17.8}$$

this shows that the neutral lines are given by the roots of $J_1(\kappa r)$ and by the roots of $\cos \phi$, i.e., the lines $\phi = \pm \pi/2$. This situation of two different kinds of neutral lines is qualitatively different from the $m = 0$ spheromak case where there was only one type of neutral line, namely the roots of $J_0(\kappa r)$.

Field line trajectories are given by the differential equation

$$\frac{d\mathbf{r}}{ds} = \frac{\mathbf{B}}{B} \tag{17.9}$$

where s is the distance along the field line. Equation (17.8) shows that the B_z contours are isomorphic to the χ contours and in particular B_z reverses polarity whenever χ reverses polarity. From Eq.(17.9) it is seen that when B_z is positive, the field line ascends and when B_z is negative the field line descends. Both the r, ϕ components of $\nabla \chi$ and the term $-\lambda \hat{z} \times \nabla \chi$ cause the field line trajectory to veer sideways as it goes up. The accumulation of sideways deflection means that eventually the field line trajectory becomes sufficiently displaced horizontally to cross a neutral line.

When the field line is deflected sideways to cross a neutral line, it begins its descent back towards the ground plane. Thus, arch shaped field lines are intrinsic to the situation prescribed by Eq. (17.5) and the projection on the ground plane of the field line apex is a neutral line. (This picture could also be used to describe the trajectory of field lines emanating from a coaxial helicity gun, the only difference being that $m = 0$ symmetry would apply, and neutral lines would only occur at the roots of the J_0 Bessel functions.)

The two footpoints of the prominence sketched in Fig. 17.6 are effectively magnetic polefaces, one with $+B_z$ and the other with $-B_z$. From Eq.(17.8) it is seen that these polefaces correspond to local maxima and minima of χ; the separation L between the first hill/valley pair of χ corresponds to the separation between prominence footpoints. This first hill/valley pair in χ occurs at the first maximum of $J_1(\kappa r)$, i.e., when $\kappa L = 3.6$ and so

$$k = \sqrt{\left(\frac{3.6}{L}\right)^2 - \lambda^2}. \tag{17.10}$$

This determines the altitudinal decay factor k and furthermore shows that loss of equilibrium occurs when

$$\lambda > \frac{3.6}{L}. \tag{17.11}$$

Let σ be a surface on the $z = 0$ ground plane defined to include the local maximum of χ at the location $\phi = 0$, $\kappa r = 3.6$ and extending to the first neutral line. Thus, σ is the surface associated with the first and dominant hill of χ. The meaning of λ can be established by integrating Eq.(4.18) over S

$$\int_\sigma d\mathbf{s} \cdot \nabla \times \mathbf{B} = \lambda \int_\sigma d\mathbf{s} \cdot \mathbf{B} \tag{17.12}$$

which gives

$$\mu_0 I = \lambda \psi \tag{17.13}$$

where I is the total current leaving σ and ψ is the total magnetic flux leaving the surface σ. Thus

$$\lambda = \frac{\mu_0 I}{\psi} \tag{17.14}$$

and so the criterion for instability becomes

$$\frac{\mu_0 I}{\psi} > \frac{3.6}{L} \tag{17.15}$$

which is essentially the same as the spheromak formation condition, except that $m = 1$ symmetry has been used instead of $m = 0$ symmetry. Equations (17.14) and (17.15) predict the following properties for prominences:

1. If λ is small the prominences will not be twisted and the magnetic field will be similar to a vacuum field.
2. As λ increases, the twist increases, and when the instability threshold is crossed, loss of equilibrium occurs resulting in an erupting prominence.

As was shown in Chapter 12 the Taylor state assumption of spatially uniform λ is an oversimplification. In reality λ is uniform along a field line but has a gradient across field lines. Magnetic instabilities feed from gradients in λ and have the tendency of making λ more uniform, and so make the plasma approach a Taylor state. However, because of dissipative losses, the plasma typically does not actually reach the Taylor state.

17.3 S-shapes

A qualitative analysis of Eqs. (17.1) and (17.8) provides an explanation for the S-shaped loops often seen on the sun (see example in Fig.17.4). These S shapes have recently been studied by Pevtsov, Canfield and McClymont[193] and appear to be a precursor for prominence eruption. To understand the relation between S-shapes and force-free fields, suppose λ is positive and consider a field line rising out of a positive B_z footpoint. Equation (17.1) shows that

$$\hat{z} \cdot \nabla \times \mathbf{B} = \lambda B_z; \qquad (17.16)$$

thus, when B_z is positive, the z component of the curl of $\mathbf{B}$ must also be positive. This means that the field line has a counterclockwise twist as projected on the surface of the sun. The field line ascends because B_z is positive, but veers sideways in a counterclockwise sense until it reaches its apex where $B_z = 0$. If the field is self-similar with altitude (e.g., exponential z dependence), then the apex corresponds to a neutral line in the ground plane. After the apex, B_z becomes negative in which case the field line starts its descent back towards the solar surface. Equation (17.16) shows that $\hat{z} \cdot \nabla \times \mathbf{B}$ is now negative and so the field line projection on the surface of the sun will have a clockwise twist. The entire field line therefore starts with a counterclockwise projection and then ends with a clockwise projection, i.e., the projection has the shape of the letter S. If λ is negative, then a reversed S projection results.

Since the sign of λ corresponds to the sign of the helicity, regions of positive helicity will have forward S projections whereas regions of negative helicity will have backwards S projections. The existence of S-shapes on the sun supports the force-free model and the assumption of altitude self-similarity.

17.4 Flux tube bifurcation and breakup

The $m = 1$ behavior discussed differs qualitatively from the $m = 0$ solution (relevant to coaxial spheromak guns), because the $m = 1$ solution has two different kinds of neutral lines (i.e., zeroes of $J_1(\kappa r)$ and zeroes of $\cos\phi$) whereas the $m = 0$ solution has only one kind of neutral line (zero of $J_0(\kappa r)$). This means that a prominence-type field line can cross different kinds of neutral lines whereas the spheromak-type field line can only cross one type.

Numerical integration of field lines resulting from the $m = 1$ solution shows that a non-trivial bifurcation of field line trajectories develops as λ is increased[52]. To see this, suppose that an equilibrium described by Eq.(4.6) begins with small λ and then has λ slowly increase with time so that the system evolves through a sequence of Taylor states. Figure 17.7 shows the sequence of states obtained when

$$\chi(r, \phi, z) = J_1(\kappa r) \cos(\phi) \exp(-\sqrt{\kappa^2 - \lambda^2} z) \qquad (17.17)$$

is used in Eq.(17.7). If $\lambda = 0$, then Eq.(17.7) shows that χ is a potential function for field lines and so if λ is small, χ can still be considered as approximately a potential function for field lines. Thus, field lines flow from large χ to small χ.

Consider a flux tube which at small λ goes from the main χ hill to the main χ valley; the apex is at the neutral line $\phi = \pm\pi/2$; in other words, this flux tube goes from the positive magnetic pole to the negative magnet pole. Because λ is small, the flux tube is nearly a potential flux tube and has little twist. However, as λ increases the flux tube becomes more twisted and its magnetic axis (also a field line) veers sideways. For sufficiently large λ, the field lines are deflected sideways so much that they cross the neutral line associated with the first zero of $J_1(\kappa r)$ (and not the neutral line $\phi = \pm\pi/2$). In this case, the field line descends towards a ground plane landing point associated with a sign reversal of $J_1(\kappa r)$. This landing point is far from the original landing point (which was associated with the zero of $\cos\phi$); thus there is a discontinuous change in field line trajectory when λ increases continuously. This bifurcation of field line trajectory does not happen simultaneously to all the field lines comprising the initial flux tube but rather only to those field lines which happen to intercept the zero of $J_1(\kappa r)$ before having a chance to intercept the zero of $\cos\phi$. Because only some field lines comprising the original flux tube bifurcate, the flux tube fractures into two smaller flux tubes that have widely separated end footpoints. Numerical calculations show that the field lines become braided when λ becomes a large fraction of its maximum allowed value in which case flux surfaces cease to exist altogether. This destruction of flux surfaces does not occur in the $m = 0$ case because it has only one type of neutral line so field lines cannot discontinuously change their trajectory. The existence of flux surfaces for

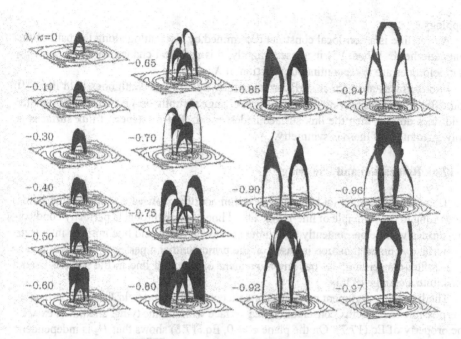

Fig.17.7 Field line trajectories from Ref. [52] for $\chi(r,\phi,z) = J_1(\kappa r)\cos(\phi)\exp\sqrt{\kappa^2 - \lambda^2}\,z)$. A pair of flux tubes is plotted, one starting from footpoint with positive B_z and integrating forward, the other starting from a footpoint with negative B_z and integrating backwards. As λ/κ increases the field lines swerve, cross different neutral lines (Bessel zero rather than zero of cosine) and bulge outwards. Projection in ground plane has S shape.

$m = 0$ symmetry but not always for $m = 1$ symmetry is consistent with the notion that axisymmetry implies the existence of flux surfaces.

17.5 Comparison of magnetic field, field lines, flux tubes

The change of magnetic topology as λ is increased highlights substantive differences in the topological behavior of the magnetic field, field lines, and flux tubes. Equations (17.2) and (17.5) show that the magnetic field is a continuous function of λ. An observer at a fixed point in space would see a continuous change in the direction and magnitude of the local magnetic field vector as λ is increased and from this local measurement would not be able to deduce any properties of the overall magnetic topology.

A field line is a non-local construct determined by integrating along the path of the magnetic field. When λ is increased slightly, this path can change abruptly, showing that field lines are not a continuous function of λ.

Flux surfaces are tubular surfaces formed by a family of contiguous field lines. If some field lines constituting the flux surface change abruptly as λ is increased, but other field lines do not, then the flux surface is destroyed. The existence of flux surfaces is only guaranteed if there is symmetry.

17.6 Relaxation and line tying

There is a contradiction between relaxation and the oft-used concept of line tying. According to the principle of line-tying, a field line passing through a perfectly conducting surface will be permanently tied to the point where the field line initially intersects the surface. This supposition is based on the principle that a perfectly conducting surface is flux-conserving, so that any movement of the field line in the surface would constitute a change of flux.

The line tying argument confuses the concepts of flux and field lines. In particular, it is possible to have flux conservation in a surface without line tying, and this is exactly the property of Eq.(17.7). On the plane $z = 0$, Eq.(17.8) shows that B_z is independent of λ because the only quantity depending on λ is $k = \sqrt{\kappa^2 - \lambda^2}$. Variation of λ does not change B_z in the ground plane and therefore does not change the flux in the ground plane. Hence, the ground plane is a flux conserver. On the other hand, increasing λ causes discontinuous jumps in the field line trajectory.

17.7 Prominence simulation experiment

Bellan and Hansen[52] have used spheromak technology to create magnetized plasmas designed to simulate solar prominences. The plasma gun has the shape of a horseshoe magnet with a gap so that a voltage difference can be applied across the pole faces. This geometry is to be contrasted with the coaxial geometry used for conventional spheromak guns. To lowest order, the prominence gun can be considered as having $m = 1$ symmetry. In order to simulate the infinite half space of solar geometry, the horseshoe gun is placed in a vacuum chamber much larger than the gun dimensions. The wall on which the gun is mounted represents the solar surface ($z = 0$ plane) and all other walls are far away.

The horseshoe gun produces arch-shaped plasmas similar to solar prominences. Bifurcation of field lines when λ is increased is observed as well as S shapes. Figure 17.8 shows a sketch of the experimental setup and Fig. 17.9 shows a typical simulated prominence. This experiment has shown that the shape of the simulated prominence is sensitive to the potential distribution on the $z = 0$ plane. In particular, the current flows on arched field lines spanning regions on the $z = 0$ plane having different potentials, but does not flow on field lines spanning regions having the same potential. By controlling the potential distribution on the $z = 0$ plane as well as the B_z distribution, various prominence geometries can be formed.

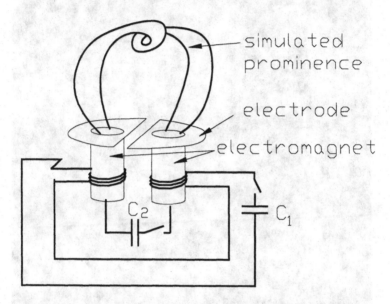

Fig.17.8 Sketch of apparatus used by Bellan and Hansen for simulating prominences: The sequence of operation is as follows: (1) capacitor C_1 is discharged to power electromagnets and establish arched vacuum field between electrodes, (2) a gas puff valve (not shown) injects a gas cloud between electrodes, and (3) capacitor C_2 is discharged to break down gas, form plasma, drive current through plasma, and create twisted simulated prominence.

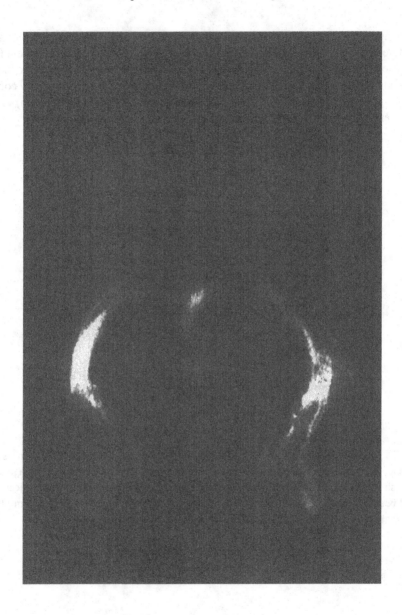

Fig.17.9 Photograph of simulation of a solar prominence from Bellan and Hansen. Image is partially reflected in electrodes at bottom.

REFERENCES

[1] H. Alfven, *Arkiv Mat. Astron. Fysik* **B29** (1943) 2.

[2] S. Lundquist, *Arkiv for Fysik* **B2** (1950) 361.

[3] H. P. Furth, M. A. Levine, and R. W. Waniek, *Rev. Sci. Instrum.* **28** (1957) 949.

[4] L. W. Woltjer, *Proc. Nat. Acad. Sci. (USA)* **44** (1958) 490.

[5] L. W. Woltjer, *Proc. Nat. Acad. Sci. (USA)* **44** (1958) 833.

[6] S. Chandrasekhar and P. C. Kendall, *Astrophys. J.* **126** (1957) 457.

[7] S. Chandrasekhar, *Proc. Nat. Acad. Sci. (USA)* **42** (1956) 1.

[8] H. Alfven, L. Lindberg, and P. Mitlid, *J. Nucl. Energy (Part C)* **1** (1959) 116.

[9] L. Lindberg, E. Witalis, and C. T. Jacobsen, *Nature* **185** (1960) 452.

[10] L. Lindberg and C. T. Jacobsen, *Phys. Fluids Supplement* (1964) S44.

[11] L. Lindberg and C. Jacobsen, *Astrophys. J.* **133** (1961) 1043.

[12] H. A. B. Bodin, *Plasma Phys. Controlled Fusion* **30** (1988) 2021.

[13] W. H. Bostick and D. R. Wells, *Phys. Fluids* **6** (1963) 1325.

[14] D. R. Wells, *J. Plasma Phys.* **3** (1969) 21.

[15] D. R. Wells, *Phys. Fluids* **7** (1964) 826.

[16] E. E. Nolting, P. E. Jindra, and D. R. Wells, *J. Plasma Phys.* **9** (1972) 1.

[17] K. Kawai, Z. A. Pietrzyk, and H. T. Hunter, *Phys. Fluids* **30** (1987) 2561.

[18] J. B. Taylor, *Phys. Rev. Lett.*, **33** (1974) 1139.

[19] M. N. Rosenbluth and M. N. Bussac, *Nucl. Fusion*, **19** (1979) 489.

[20] Y. Nogi, H. Ogura, Y. Osanai, K. Saito, S. Shiina, and H. Yoshimura, *J. Phys. Soc. Japan* **49** (1980) 710.

[21] G. C. Goldenbaum, J. H. Irby, Y. P. Chong, and G. W. Hart, *Phys. Rev. Lett.*, **44** (1980) 393.

[22] W. C. Turner, E. H. A. Granneman, C. W. Hartman, D. S. Prono, J. Taska and A. C.

Smith, Jr., *J. Appl.Phys.* **52** (1981) 175.

[23] W. C. Turner, G. C. Goldenbaum, E. H. A. Granneman, J. H. Hammer, C. W. Hartman, D. S. Prono, and J. Taska, *Phys. Fluids* **26** (1983) 1965.

[24] T. R. Jarboe, I. Henins, H. W. Hoida, R. K. Linford, J. Marshall, D. A. Platts, and A. R. Sherwood, *Phys. Rev. Lett.,* **45** (1980) 1264.

[25] M. Yamada, H. P. Furth, W. Hsu, A. Janos, S. Jardin, M. Okabyashi, J. Sinnis, T. H. Stix, and K. Yamazaki, *Phys. Rev. Lett.* **46** (1981) 188.

[26] K. Watanabe, K. Ikegami, A. Ozaki, N. Satomi, and T. Uyama, *J. Phys. Soc. Japan* **50** (1981) 1823.

[27] H. P. Furth, Compact Tori, *Nucl. Instrum. and Methods* **207** (1983) 93.

[28] T. Sato and T. Hayashi, *Phys. Rev. Letters* **50** (1983) 38.

[29] P. K. Browning, G. Cunningham, S. J. Gee, K. J. Gibson, A. al-Karkhy, D. A. Kitson, R. Martin, and M. G. Rusbridge, *Phys. Rev. Lett.* **68** (1992) 1718.

[30] J. Heyvaerts and E. R. Priest, *Astron. Astrophys.* **137** (1984) 63.

[31] J. H. Hammer, C. W. Hartman, J. L. Eddleman, and H. S. McLean, *Phys. Rev. Letters* **61** (1988) 2843.

[32] L. J. Perkins, S. K. Ho, and J. H. Hammer, *Nucl. Fusion* **28** (1988) 1365.

[33] P. B. Parks, *Phys. Rev. Lett.* **61** (1988) 1364.

[34] P. K. Loewenhardt, M. R. Brown, J. Yee, and P. M. Bellan, *Rev. Sci. Instr.* **66** (1995) 1050.

[35] R. Raman, F. Martin, B. Quirion et al., *Phys. Rev. Letters* **73** (1994) 3101.

[36] J. H. Degnan, R. E. Peterkin, Jr., G. P. Baca et al., *Phys. Fluids B* **5** (1993) 2938.

[37] R. E. Peterkin, Jr., *Phys. Rev. Lett.* **74** (1995) 3165.

[38] R. E. Peterkin, Jr., J. H. Degnan, T. W. Hussey, N. F. Roderick, and P. J. Turchi, *IEEE Trans. Plasma Sci.* **21** (1993) 522.

[39] T. H. Jensen and M. S. Chu, *Phys. Fluids* **27** (1984) 2881.

[40] M. R. Brown and P. M. Bellan, *Phys. Rev. Lett.* **64** (1990) 2144.

[41] T. R. Jarboe, *Fusion Tech.* **15** (1989) 7.

[42] B. A. Nelson, T. R. Jarboe, D. J. Orvis, L. A. McCullough, J. Xie, C. Zhang, and L. Zhou, *Phys. Rev. Letters* **72** (1994) 3666.

[43] M. G. Rusbridge, S. J. Gee, P. K. Browning, G. Cunningham, R. C. Duck, A. al-Karkhy, R. Martin, and J. W. Bradley, *Plasma Phys. Contr. Fusion* **39** (1997) 683.

[44] R. C. Duck, P. K. Browning, G. Cunningham, S. J. Gee, A. al-Karkhy, R. Martin and M. G. Rusbridge, *Plasma Phys. Contr. Fusion* **39** (1997) 715.

[45] T. R. Jarboe, *Plasma Phys. Controlled Fusion* **36** (1994) 945.

[46] M. Yamada, Y. Ono, A. Hayakawa, M. Katsurai, and F. W. Perkins, *Phys. Rev. Lett.* **65** (1990) 721.

[47] Y. Ono, A. Morita, M. Katsurai, and M. Yamada, *Phys. Fluids B* **5** (1993) 3691.

[48] M. Yamada, H. Ji, S. Hsu, R. Kulsrud, N. Bretz, F. Jobes, Y. Ono, and F. Perkins, *Phys. Plasmas* **4** (1997) 1936.

[49] C. G. R. Geddes, T. W. Kornack, and M. R. Brown, *Phys. Plasmas*, **5** (1998) 1027.

[50] D. M. Rust, *Geophys. Res. Lett.* **21** (1994) 241.

[51] D. M. Rust and A. Kumar, *Astrophys. J.* **464** (1996) L199.

[52] P. M. Bellan and J. F. Hansen, *Phys. Plasmas* **5** (1998) 1991.

[53] T. K. Fowler, J. S. Hardwick, and T. R. Jarboe, *Comments Plasma Phys. Controlled Fusion* **16** (1994) 91.

[54] R. M. Mayo, *Nucl. Fusion*, **36** (1996) 1599.

[55] E. B. Hooper, R. H. Bulmer, B. I. Cohen, D. N. Hill, L. D. Pearlstein, K. I. Thomassen, R. D. Wood, A. Turnbull, R. Gatto, T. R. Jarboe, and C. W. Domier, Lawrence Livermore National Laboratory Report UCRL-JC-132034, 1998; paper submitted to the 17th IAEA Fusion Energy Conference, Yokohama, Japan, 1998.

[56] *The Spheromak Path to Fusion Energy*, E. B. Hooper, Editor; C. W. Barnes, P. M. Bellan, M. R. Brown, J. C. Fernandez, T. K. Fowler, D. N. Hill, T. R. Jarboe, L. L. LoDestro, H. S. McLean, M. J. Schaeffer, K. I. Thomassen and M. Yamada, Lawrence Livermore National Laboratory Report UCRL-ID-130429, 1998.

[57] R. Lüst and A. Schlüter, *Z. Astrophysik* **34** (1954) 263.

[58] M. Katsurai and M. Yamada, *Nucl. Fusion* **22** (1982) 1407.

[59] M. Katsurai, M. Arata et al., *Trans. Inst. Electr. Eng. Japan* **103A** (1983) 219.

[60] T. R. Jarboe, I. Henins, A. R. Sherwood, C. W. Barnes, and H. W. Hoida, *Phys. Rev. Lett.* **51** (1983) 39.

[61] R. L. Hagenson and R. A. Krakowski, *Fusion Technology* **8** (1985) 1606.

[62] C. W. Barnes, J. C. Fernandez, I. Henins, H. W. Hoida, T. R. Jarboe, S. O. Knox, *Phys. Fluids* **29** (1986) 3415.

[63] Y. Honda, Y. Kato, N. Satomi, M. Nishikawa, and K. Watanabe, *J. Phys. Soc. Jpn.* **57** (1987) 1273.

[64] H. Bruhns, R. Brendel, G. Raupp, and J. Steiger, *Nucl. Fusion* **27** (1987) 2178.

[65] F. J. Wysocki, J. C. Fernandez, I. Henins, T. R. Jarboe, and G. J. Marklin, *Phys. Rev. Lett.* **61** (1988) 2457.

[66] F. J. Wysocki, J. C. Fernandez, I. Henins, T. R. Jarboe, and G. J. Marklin, *Phys. Rev. Lett.* **65** (1990) 40.

[67] T. R. Jarboe, F. J. Wysocki, J. C. Fernandez, I. Henins, and G. J. Marklin, *Phys. Fluids B* **2** (1990) 1342.

[68] K. Wira and Z. A. Pietrzyk, *Phys. Fluids B* **2** (1990) 561.

[69] A. al-Karky, P. K. Browning, G. Cunningham, S. J. Gee, and M. G. Rusbridge, *Phys. Rev. Lett.*, **70** (1993) 1814.

[70] H. K. Moffatt, *Magnetic Field Generation in Electrically Conducting Fluids* (Cambridge University Press, 1978), p.19.

[71] J. M. Finn, W. M. Manheimer, and E. Ott, *Phys. Fluids* **24** (1981) 1336.

[72] S. Kaneko, A. Takimoto, M.Taguchi, and T. Miyazaki, *J. Phys. Soc. Jpn.* **53** (1984) 201.

[73] J. D. Jackson, *Classical Electrodynamics (Third Edition)*, (John Wiley, New York, 1998), p. 352.

[74] T. G. Cowling, *Monthly Notices Roy. Astron. Soc.* **94** (1934) 39.

[75] A. Boozer, *Phys. Fluids B* **5** (1993) 2271.

[76] M. A. Berger and G. B. Field, *J. Fluid Mech.* **147** (1984) 133.

[77] J. M. Finn and T. M. Antonsen, Jr., *Comments Plasma Phys. Controlled Fusion* **9** (1985) 111.

REFERENCES

[78] M. K. Bevir and J. W. Gray, in *Proceedings of the Reversed Field Pinch Workshop*, edited by H. R. Lewis and R. A. Gerwin (Los Alamos Scientific Laboratory, Los Alamos, NM, 1981), III A-3.

[79] M. Ono, G. J. Greene, D. Darrow, C. Forest, H. Park, and T. H. Stix, *Phys. Rev. Lett.* **59** (1987) 2165.

[80] L. Turner, *IEEE Trans. Plasma Sci.* **PS-14** (1986) 849.

[81] K. Avinash, *Phys. Fluids B* **4** (1992) 3856.

[82] S. R. Oliveira and T. Tajima, *Phys. Rev. E* **52** (1995) 4287.

[83] L. C. Steinhauer and A. Ishida, *Phys. Plasmas* **5** (1998) 2609.

[84] H. Pfister and W. Gekelman, *Am. J. Phys.* **59** (1991) 497.

[85] A. N. Wright and M. A. Berger, *J. Geophys. Res.* **94** (1989) 1295.

[86] E. N. Parker, *Astrophysical J.* **174** (1972) 499.

[87] S. J. Vainshtein and E. N. Parker, *Astrophysical J.*, **304** (1986) 821.

[88] P. M. Bellan, *Phys. Plasmas* **5** (1998) 3081.

[89] D. J. Lees and M. G. Rusbridge, *Proc. 4th Intl. Conf. Ionized Gases* **2** (1959) 954.

[90] J. Mathews and R. L. Walker, *Mathematical Methods of Physics*, (W. A. Benjamin, 1964), p. 313.

[91] A. Bondeson, G. Marklin, Z. G. An, H. H. Chen, Y. C. Lee, and C. S. Liu, *Phys. Fluids* **24** (1981) 1682.

[92] I. B. Bernstein, E. A. Friedman, M. D. Kruskal, and R. M. Kulsrud, *Proc. Roy. Soc.* **A244** (1958) 17.

[93] J. P. Freidberg, *Ideal Magnetohydrodynamics*, (Plenum Press, New York 1987), p. 27.

[94] K. Miyamoto, *Plasma Physics for Nuclear Fusion*, (MIT Press, Cambridge, Mass., 1989), p.243-246.

[95] H. P. Furth, J. Killeen, M. N. Rosenbluth, and B. Coppi, in *Plasma Phys. and Controlled Nucl. Fusion Res. 1964*, (IAEA, Vienna, 1965), Vol. I, p. 103.

[96] J. M. Greene and J. L. Johnson, *Plasma Phys.* **10** (1968) 729.

[97] J. P. Freidberg, *Ideal Magnetohydrodynamics*, (Plenum Press, New York 1987), p.

258.

[98] G. Bateman, *MHD Instabilities*, (MIT Press, Boston, 1978).

[99] J. P. Freidberg, *Ideal Magnetohydrodynamics*, (Plenum Press, New York 1987), p. 298.

[100] B. R. Suydam, *Proc. 2nd United Nations International Conference on the Peaceful Uses of Atomic Energy* **31** (1958) 157.

[101] C. Mercier, *Nucl. Fusion* **1**, 47 (1960).

[102] R. M. Mayo and G. J. Marklin, *Phys. Fluids* **31** (1988) 1812.

[103] C. W. Barnes, private communication.

[104] J. C. Thomas, D. Q. Hwang, R. D. Horton, J. H. Rogers, and R. Raman, *Rev. Sci. Instrum.* **64** (1993) 1410.

[105] J. C. Fernandez, B. L. Wright, G. J. Marklin, D. A. Platts, and T. R. Jarboe, *Phys. Fluids B* **1** (1989) 1254.

[106] M. Yamada, *Nucl. Fusion* **25** (1985) 1327.

[107] C. Chin-Fatt, A.W. DeSilva, G. C. Goldenbaum, R. Hess, C. Coté, A. Filuk, J.-L. Gavreau, and F. K. Hwang, *Phys. Fluids B* **5** (1993) 1816.

[108] M. R. Brown, *J. Plasma Phys.* **57** Pt 1 (1997) 203.

[109] M. Abramovitz and I. A. Stegun, *Handbook of Mathematical Functions*, (National Bureau of Standards, Washington DC, 10th printing, 1972), formula 9.1.30, p.361.

[110] Y. Ono, M. Yamada, A. C. Janos, and F. M. Levinton, *Phys. Fluids B* **3** (1991) 1452.

[111] T. Uyama, Y. Honda, M. Nagata, M. Nishikawa, A. Ozaki, N. Satomi, and K. Watanabe, *Nucl. Fusion* **27** (1987) 799.

[112] B. L. Wright, C. W. Barnes, J. C. Fernandez, I. Henins, H. W. Hoida, T. R. Jarboe, S. O. Knox, G. J. Marklin, R. M. Mayo, and D. A. Platts, in *Plasma Physics and Controlled Fusion Research; Proc. 11th Conf., Kyoto, 1986 (IAEA, Vienna, 1987)* volume II, p. 519.

[113] M. Abramovitz and I. A. Stegun, *Handbook of Mathematical Functions*, (National Bureau of Standards, Washington DC, 10th printing, 1972), formula 10.1.23, p.439.

[114] J. D. Jackson, *Classical Electrodynamics (Third Edition)*, (John Wiley, New York,

1998), p. 101.

[115] A. Janos, G. W. Hart, C. H. Nam, and M. Yamada, *Phys. Fluids* **28** (1985) 3667.

[116] C. Munson, A. Janos, F. Wysocki, and M. Yamada, *Phys. Fluids* **28** (1985) 1525.

[117] F. J. Wysocki, *Phys. Fluids* **30** (1987) 482.

[118] Y. Ono, R. A. Ellis, Jr., A. C. Janos, F. M. Levinton, R. M. Mayo, R. W. Motley, Y. Ueda, and M. Yamada, *Phys. Rev. Lett.* **61** (1988) 2847.

[119] C. W. Barnes, H. W. Hoida, I. Henins, J. C. Fernandez, T. R. Jarboe, and G. J. Marklin, *Phys. Fluids* **28** (1985) 3443.

[120] L. Turner, *Phys. Fluids* **27** (1984) 1677.

[121] M. S. Chu, T. H. Jensen, and P. M. Bellan, General Atomics Report GA-A22913, October 1998, (to be published in *Phys. Plasmas*, May 1999).

[122] A. M. Dixon, P. K. Browning, M. K. Bevir, C. G. Gimblet, and E. R. Priest, *J. Plasma Phys.* **43** (1990) 357.

[123] J. Mathews and R. L. Walker, *Mathematical Methods of Physics*, (W. A. Benjamin, 1964), p. 175.

[124] J. Yee and P. M. Bellan, *Bull. Am. Phys. Soc.* **43** (1998) 1770.

[125] C. W. Barnes, T. R. Jarboe, G. J. Marklin, S. O. Knox, and I. Henins, *Phys. Fluids B* **2** (1990) 1871.

[126] D. A. Kitson and P. K. Browning, *Plasma Phys. Controlled Fusion*, **32** (1990) 1265.

[127] T. Kanki, M. Nagata, T. Uyama, S. Ikuno, and A. Kamitani, *J. Phys. Soc. Japan* **67** (1998) 140.

[128] T. R. Jarboe, C. W. Barnes, D. A. Platts, and B. L. Wright, *Comments Plasma Phys. Controlled Fusion* **9** (1985) 161.

[129] D. J. Campbell, E. Lazzaro, M. F. F. Nave, J. P. Christiansen, J. G. Cordey, F. C. Schuller, and P. R. Thomas, *Nucl. Fusion* **28** (1988) 981.

[130] H. R. Lewis and P. M. Bellan, *J. Math. Phys.* **31** (1990) 2592.

[131] A. Janos, G. W. Hart, and M. Yamada, *Phys. Rev. Lett.* **55** (1985) 2868.

[132] S. O. Knox, C. W. Barnes, G. J. Marklin, T. R. Jarboe, I. Henins, H. W. Hoida, and B. L. Wright, *Phys. Rev. Lett.* **56** (1986) 842.

[133] M. Nagata, T. Kanki, T. Masuda, S. Naito, H.Tatsumi, and T. Uyama, *Phys. Rev. Letters* **71**, 4342 (1993).

[134] A. H. Boozer, *J. Plasma Phys.* **35** (1986) 133.

[135] A. Bhattacharjee and E. Hameiri, *Phys. Rev. Lett.* **57** (1986) 206.

[136] H. R. Strauss, *Phys. Fluids* **28** (1985) 2786.

[137] R. J. Goldston and P. H. Rutherford, *Introduction to Plasma Physics* (Institute of Physics Publishing, Bristol, 1995), p. 199.

[138] F. F. Chen, *Plasma Physics and Controlled Fusion, Second Edition*, Vol 1 (Plenum Press, New York 1984), p. 160.

[139] J. C. Fernandez, C. W. Barnes, T. R. Jarboe, I. Henins, H. W. Hoida, P. L. Klingner, S. O. Knox, G. J. Marklin, and B. L. Wright, *Nucl. Fusion* **28** (1988) 1555.

[140] F. M. Levinton, D. D. Meyerhofer, R. M. Mayo, A. C. Janos, Y. Ono, Y. Ueda, and M. Yamada, *Nucl.Fusion* **30** (1990) 871.

[141] R. M. Mayo, F. M. Levinton, D. D. Meyerhofer, T. K. Chu, S. F. Paul, and M. Yamada, *Nucl. Fusion* **29** (1989) 1493.

[142] R. M. Mayo and L.S. Kirschenbaum, *Phys. Fluids B* **3** (1991) 2096.

[143] R. E. Chrien, J. C. Fernandez, I. Henins, R. M. Mayo, and F. J. Wysocki, *Nucl. Fusion* **31** (1991) 1390.

[144] R. M. Mayo, J. C. Fernandez, I. Henins, L. S. Kirschenbaum, C. P. Munson, and F. J. Wysocki, *Nucl. Fusion* **31** (1991) 2087.

[145] R. M. Mayo, D. J. Hurlburt, and J. C. Fernandez, *Phys. Fluids B* **5** (1993) 4002.

[146] Y. Ono, M. Yamada, T. Akao, T. Tajima, and R. Matsumoto, *Phys. Rev. Lett.* **76** (1996) 3328.

[147] T. W. Kornack, P. K. Sollins, and M. R. Brown, *Phys. Rev. E* **58** (1998) R36.

[148] J. M. McChesney, R. A. Stern, and P. M. Bellan, *Phys. Rev. Lett.* **59** (1987) 1436.

[149] A. Y. Wong, R. W. Motley, and N. D'Angelo, *Phys. Rev.* **133** (1964) A436.

[150] A. Von Engel, *Ionized Gases* (Oxford University Press, Oxford, 1965), p. 195.

[151] I. Henins at Los Alamos pioneered the development of gas puff valves and his contributions to valve design were important to the success of many spheromak experiments.

[152] G. M. McCracken, S. J. Fielding, S. K. Ernets, A. Pospieszczyk, and P. E. Stott, *Nucl. Fusion* **18** (1978) 35.

[153] G. M. McCracken and P. E. Stott, *Nucl. Fusion* **19** (1979) 889.

[154] P. C. Stangeby and G. M. McCracken, *Nucl. Fusion* **30** (1990) 1225.

[155] P. G. Carolan and V. A. Piotrowicz, *Plasma Physics* **25** (1983) 1065.

[156] D. E. Post, R. V. Jensen, C. B. Tarter, W. H. Grasberger, and W. A. Lokke, *Atomic Data and Nuclear Data Tables* **20** (1977) 397.

[157] R. Raman, G. C. Vlases, and T. R. Jarboe, *Nucl. Fusion* **33** (1993)1685.

[158] C. W. Barnes, T. R. Jarboe, I. Henins et al. *Nucl. Fusion* **24** (1984) 267.

[159] M. R. Brown, A. D. Bailey III, and P. M. Bellan, *J. Appl. Phys.* **69** (1991) 6302.

[160] R. Martin, S. J. Gee, P. K. Browning, and M. G. Rusbridge, *Plasma Phys. Control. Fusion* **35** (1993) 269.

[161] S.-L. Chen and T. Sekiguchi, *J. Applied Phys.* **36** (1965) 2363.

[162] M. Weinlich and A. Carlson, *Phys. Plasmas* **4** (1997) 2151.

[163] C. J. Buchenauer and A. R. Jacobson, *Rev. Sci. Instrum.* **48** (1977) 769.

[164] K. J. Gibson, S. J. Gee, and M. G. Rusbridge, *Plasma Phys. Control. Fusion* **37** (1995) 31.

[165] P. Kepple and H. R. Griem, *Phys. Rev.* **173** (317) 1968.

[166] L. Welch, H. R. Griem, J. Terry, B. LaBombard, B. Lipschultz, E. Marmer and G. McCracken, *Phys. Plasmas* **2** 1995) 4246.

[167] R. M. Mayo, J. C. Fernandez, I. Henins, L. S. Kirschenbaum, C. P. Munson, and F. J. Wysocki, *Nucl. Fusion* **31** (1991) 2087.

[168] J. D. Jackson, *Classical Electrodynamics (Third Edition)*, (John Wiley, New York, 1998), p. 664.

[169] J. Wesson, *Tokamaks* (Clarendon Press, Oxford, 1987).

[170] T. Fujita, T. Hatae, T. Oikawa, S. Takeji, H. Shirai, Y. Koide, S. Ishida, S. Ide, Y. Ishii et al., *Nucl. Fusion* **38** (1998) 207.

[171] J. Wesson, *Tokamaks* (Clarendon Press, Oxford, 1987), p. 3.

[172] R. W. Moir, *Nucl. Fusion* **37** (1997) 557.

[173] R. J. Goldston and P. H. Rutherford, *Introduction to Plasma Physics* (Institute of Physics Publishing, Bristol, 1995), p. 177.

[174] J. H. Hammer, J. E. Eddleman, C. W. Hartman, H. S. McLean, and A. W. Molvik, *Phys. Fluids B* **3** (1991) 2236,

[175] J. Yee and P. M. Bellan, *Nucl. Fusion* **38** (1998) 711.

[176] H. S. McLean, D. Q. Hwang, R. D. Horton, R. W. Evans, and S. D. Terry, *Fusion Technology* **33** (1998) 252.

[177] P. K. Browning, G. Cunningham, R. Duck, S. J. Gee, K. J. Gibson, D. A. Kitson, R. Martin, and M. G. Rusbridge, *Phys. Rev. Lett.* **68** (1992) 1722.

[178] M. Yamada, F. W. Perkins, A. K. MacAulay, Y. Ono, and M. Katsurai, *Phys. Fluids B* **3** (1991) 2379.

[179] D. Dietz, T. W. Hussey, N. F. Roderick, M. R. Douglas, and J. H. Degnan, *Phys. Plasmas* **4** (1997) 873.

[180] A. Secchi, *Le Soleil* (Gauthier-Villars, Paris, 1875-77), Vols. 1 and 2.

[181] H. Zirin and A. B. Severny, *Observatory* **81** (1961) 155.

[182] Y. Nakagawa, M. A. Raadu, D. E. Billings, and D. McNamara, *Solar Phys.* **19** (1971) 72.

[183] M. Gigolashvili, *Solar Phys.* **60** (1978) 293.

[184] C. T. Russell and R. C. Elphic, *Nature* **279** (1979) 616.

[185] A. Konigl and A. R. Choudhuri, *Astrophys. J.* **289** (1985) 173.

[186] L. F. Burlaga, *J. Geophys. Res.* **93** (1988) 7217.

[187] B. Vrsnak, V. Ruzdak, and B. Rompolt, *Solar Phys.* **136** (1991) 151.

[188] T. R. Detman, M. Dryer, T. Yeh, S. Han, S. T. Wu, and D. J. McComas, *J. Geophys. Res.* **96** (1991) 9531.

[189] M. B. Moldwin and W. J. Hughes, *J. Geophys. Res.* **99** (1994) 183.

[190] D. M. Rust and A. Kumar, in *Solar Dynamic Phenomena and Solar Wind Consequences* (Proc. 3rd SOHO Workshop, European Space Agency, ESTEC, Noorwijk, 1995).

[191] J. Feynman and S. F. Martin, *J. Geophys. Res.* **100** (1995) 3355.

[192] B. W. Lites and B. C. Low, *Solar Phys.* **174** (1997) 91.

[193] A. A. Pevtsov, R. C. Canfield, and A. N. McClymont, *Astrophys. J.*, **481** (1997) 973.

[194] E. Tandberg-Hanssen, *The Nature of Solar Prominences* (Kluwer, Dordrecht, 1995).

[195] V. C. A. Ferraro and C. Plumpton, *An Introduction to Magneto-Fluid Mechanics* (Oxford Univ. Press, 1961).

[196] S. F. Martin, W. H. Marquette, and R. Bilimoria, in *The Solar Cycle* (ed. K. L. Harvey, ASP Conf. Series), **27** (1992) 53.

[197] A. A. Pevtsov, R. C. Canfield, T. R. Metcalf, *Astrophys. J.* **440** (1994) 109.

[198] J. W. Dungey, *Phys. Rev. Letters* **6** (1961) 47.

[199] C. T. Russell and R. C. Elphic, *Space Sci. Rev.* **22** (1978) 681.

[200] S. W. H. Cowley, *Rev. Geophys. and Space Phys.* **20** (1982) 531.

[201] S. E. Macwan, *J. Geophys. Res.* **97** (1992) 19239.

[202] C. J. Farrugia, M. P. Freeman, L. F. Burlaga, R. P. Lepping, and K. Takahashi, *J. Geophys. Res.* **98** (1993) 7657.

[203] T. Gold and F. Hoyle, *Mon. Not. Roy. Astr. Soc.* **120** (1959) 89.

[204] A. Kumar and D. M. Rust, *J. Geophys. Res.* **101** (1996) 15667.

[205] P. K. Browning, *J. Plasma Phys.* **40** (1988) 263.

APPENDIX A
Vector Identities and Operators

Vector Identities

$$\nabla \cdot (f\mathbf{A}) = \nabla f \cdot \mathbf{A} + f \nabla \cdot \mathbf{A}$$

$$\nabla \times (f\mathbf{A}) = f\nabla \times \mathbf{A} + \nabla f \times \mathbf{A}$$

$$\nabla \cdot (\mathbf{A} \times \mathbf{B}) = \mathbf{B} \cdot \nabla \times \mathbf{A} - \mathbf{A} \cdot \nabla \times \mathbf{B}$$

$$\nabla \times (\mathbf{A} \times \mathbf{B}) = \mathbf{A}\nabla \cdot \mathbf{B} + \mathbf{B} \cdot \nabla \mathbf{A} - \mathbf{B}\nabla \cdot \mathbf{A} - \mathbf{A} \cdot \nabla \mathbf{B}$$

$$\nabla (\mathbf{A} \cdot \mathbf{B}) = \mathbf{A} \times (\nabla \times \mathbf{B}) + \mathbf{A} \cdot \nabla \mathbf{B} + \mathbf{B} \times (\nabla \times \mathbf{A}) + \mathbf{B} \cdot \nabla \mathbf{A}$$

$$\nabla \times \nabla \times \mathbf{A} = \nabla \nabla \cdot \mathbf{A} - \nabla^2 \mathbf{A}$$

$$\nabla \times \nabla f = 0$$

$$\nabla \cdot \nabla \times \mathbf{A} = 0$$

Operators in cylindrical coordinates

$$\nabla f = \hat{r}\frac{\partial f}{\partial r} + \frac{\hat{\phi}}{r}\frac{\partial f}{\partial \phi} + \hat{z}\frac{\partial f}{\partial z}$$

$$\nabla \cdot \mathbf{A} = \frac{1}{r}\frac{\partial (rA_r)}{\partial r} + \frac{1}{r}\frac{\partial A_\phi}{\partial \phi} + \frac{\partial A_z}{\partial z}$$

$$\nabla \times \mathbf{A} = \hat{r}\left(\frac{1}{r}\frac{\partial A_z}{\partial \phi} - \frac{\partial A_\phi}{\partial z}\right) + \hat{\phi}\left(\frac{\partial A_r}{\partial z} - \frac{\partial A_z}{\partial r}\right) + \hat{z}\left(\frac{1}{r}\frac{\partial (rA_\phi)}{\partial r} - \frac{1}{r}\frac{\partial A_r}{\partial \phi}\right)$$

$$\nabla^2 f = \frac{1}{r}\frac{\partial}{\partial r}\left(r\frac{\partial f}{\partial r}\right) + \frac{1}{r^2}\frac{\partial^2 f}{\partial \phi^2} + \frac{\partial^2 f}{\partial z^2}$$

Operators in spherical coordinates

$$\nabla f = \hat{r}\frac{\partial f}{\partial r} + \frac{\hat{\theta}}{r}\frac{\partial f}{\partial \theta} + \frac{\hat{\phi}}{r\sin\theta}\frac{\partial f}{\partial \phi}$$

$$\nabla \cdot \mathbf{A} = \frac{1}{r^2}\frac{\partial (r^2 A_r)}{\partial r} + \frac{1}{r\sin\theta}\frac{\partial (A_\theta \sin\theta)}{\partial \theta} + \frac{1}{r\sin\theta}\frac{\partial A_\phi}{\partial \phi}$$

$$\nabla \times \mathbf{A} = \frac{\hat{r}}{r\sin\theta}\left(\frac{\partial (A_\phi \sin\theta)}{\partial \theta} - \frac{\partial A_\theta}{\partial \phi}\right) + \frac{\hat{\theta}}{r}\left(\frac{1}{\sin\theta}\frac{\partial A_r}{\partial \phi} - \frac{\partial (rA_\phi)}{\partial r}\right)$$
$$+ \frac{\hat{\phi}}{r}\left(\frac{\partial (rA_\theta)}{\partial r} - \frac{\partial A_r}{\partial \theta}\right)$$

$$\nabla^2 f = \frac{1}{r^2}\frac{\partial}{\partial r}\left(r^2\frac{\partial f}{\partial r}\right) + \frac{1}{r^2 \sin\theta}\frac{\partial}{\partial \theta}\left(\sin\theta\frac{\partial f}{\partial \theta}\right) + \frac{1}{r^2 \sin^2\theta}\frac{\partial^2 f}{\partial \phi^2}$$

APPENDIX B
Bessel Orthogonality Relations

Recursion relations

An integral representation for Bessel functions is

$$J_m(z) = \frac{1}{2\pi} \int_0^{2\pi} e^{iz\sin\theta - im\theta} d\theta. \tag{B.1}$$

Thus,

$$\begin{aligned}
J_{m+1}(z) + J_{m-1}(z) &= \frac{1}{\pi} \int_0^{2\pi} e^{iz\sin\theta - im\theta} \cos\theta\, d\theta \\
&= \frac{1}{\pi i z} \int_0^{2\pi} e^{-im\theta} \frac{d}{d\theta} e^{iz\sin\theta}\, d\theta \\
&= -\frac{1}{\pi i z} \int_0^{2\pi} e^{iz\sin\theta} \frac{d}{d\theta} e^{-im\theta}\, d\theta \\
&= \frac{2m}{z} J_m(z).
\end{aligned} \tag{B.2}$$

Similarly,

$$\begin{aligned}
J_{m+1}(z) - J_{m-1}(z) &= \frac{1}{\pi i} \int_0^{2\pi} e^{iz\sin\theta - im\theta} \sin\theta\, d\theta \\
&= -2 \frac{d}{dz} \left(\frac{1}{2\pi} \int_0^{2\pi} e^{iz\sin\theta - im\theta} d\theta \right) \\
&= -2 J_m'(z).
\end{aligned} \tag{B.3}$$

Using $m = 0$, Eq.(B.2) gives $J_1(z) = -J_{-1}(z)$ and so Eq.(B.3) gives

$$J_1(z) = -J_0'(z). \tag{B.4}$$

Adding Eqs.(B.2) and (B.3) gives

$$J_{m+1}(z) = \frac{m}{z} J_m(z) - J_m'(z). \tag{B.5}$$

APPENDIX B

Orthogonality Relations

Let x_{mn} be either the nth root of the Bessel function of order m, i.e., $J_m(x_{mn}) = 0$ or, alternatively, the nth root of its derivative, i.e., $J'_m(x_{mn}) = 0$. We write the equation for J_m twice with different arguments $x_{mn}r/a$ and $x_{mn'}r/a$

$$\frac{1}{r}\frac{d}{dr}r\frac{d}{dr}J_m(x_{mn}r/a) + \left(\frac{x_{mn}^2}{a^2} - \frac{m^2}{r^2}\right)J_m(x_{mn}r/a) = 0 \quad \text{(B.6)}$$

$$\frac{1}{r}\frac{d}{dr}r\frac{d}{dr}J_m(x_{mn'}r/a) + \left(\frac{x_{mn'}^2}{a^2} - \frac{m^2}{r^2}\right)J_m(x_{mn'}r/a) = 0. \quad \text{(B.7)}$$

Equation (B.6) is multiplied by $rJ_m(x_{mn'}r/a)$ and Eq.(B.7) is multiplied by $rJ_m(x_{mn}r/a)$ and the resulting equations are subtracted from each other, giving

$$J_m(x_{mn'}r/a)\frac{d}{dr}r\frac{d}{dr}J_m(x_{mn}r/a) - J_m(x_{mn}r/a)\frac{d}{dr}r\frac{d}{dr}J_m(x_{mn'}r/a)$$
$$+ \frac{r}{a^2}\left(x_{mn}^2 - x_{mn'}^2\right)J_m(x_{mn}r/a)J_m(x_{mn'}r/a) = 0. \quad \text{(B.8)}$$

This is now integrated from 0 to a with the first two terms integrated by parts,

$$\left[rJ_m(x_{mn'}r/a)\frac{d}{dr}J_m(x_{mn}r/a) - J_m(x_{mn}r/a)r\frac{d}{dr}J_m(x_{mn'}r/a)\right]_0^a$$
$$+ \left(\frac{x_{mn}^2 - x_{mn'}^2}{a^2}\right)\int_0^a rJ_m(x_{mn}r/a)J_m(x_{mn'}r/a)dr = 0. \quad \text{(B.9)}$$

Since either $J_m(x_{mn}) = 0$ or $J'_m(x_{mn}) = 0$, the integrated terms vanish at both $r = 0$ and $r = a$ so

$$\left(x_{mn}^2 - x_{mn'}^2\right)\int_0^a rJ_m(x_{mn}r/a)J_m(x_{mn'}r/a)dr = 0. \quad \text{(B.10)}$$

Case where $n' \neq n$

In this case $x_{mn}^2 - x_{mn'}^2 \neq 0$ and so we must have

$$\int_0^a r J_m(x_{mn}r/a) J_m(x_{mn'}r/a) dr = 0. \qquad (B.11)$$

Case where $n' = n$

In this case we define

$$I_n = \int_0^a r J_m^2(x_{mn}r/a) dr = \frac{a^2}{x_{mn}^2} \int_0^{x_{mn}} s J_0^2(s) ds. \qquad (B.12)$$

Integrals of the form $\int s J_m^2(s) ds$ can be evaluated as follows. We write $s = d(s^2/2)/ds$ and integrate by parts obtaining

$$\int_0^s s J_m^2(s) ds = \frac{s^2}{2} J_m^2(s) - \int_0^s s^2 J_m(s) J_m'(s) ds. \qquad (B.13)$$

Bessel's equation can be used to rewrite a factor in the integrand as

$$s^2 J_m(s) = m^2 J_m(s) - s \frac{d}{ds} s J_m'(s) \qquad (B.14)$$

so

$$\begin{aligned}
\int_0^s s J_m^2(s) ds &= \frac{s^2}{2} J_m^2(s) - \int_0^s \left(m^2 J_m(s) - s \frac{d}{ds} s J_m'(s) \right) J_m'(s) ds \\
&= \frac{s^2}{2} J_m^2(s) - \frac{1}{2} \int_0^s \frac{d}{ds} \left(m^2 J_m^2(s) - (s J_m'(s))^2 \right) ds \\
&= \frac{s^2}{2} J_m^2(s) - \frac{1}{2} m^2 J_m^2(s) + \frac{1}{2} (s J_m'(s))^2. \qquad (B.15)
\end{aligned}$$

First orthogonality relation

Here $m = 0$ and the upper limit is at $s = x_{1n} = \gamma_n a$ where γ_n is the nth root of $J_1(s)$. The middle and last terms on the RHS of Eq.(B.15) vanish, leaving

$$\int_0^a r J_0^2(\gamma_n r) dr = \frac{1}{\gamma_n^2} \int_0^{x_{1n}} s J_0^2(s) ds = \frac{a^2}{2} J_0^2(x_{1n}). \qquad (B.16)$$

Combining the results for both cases gives the first orthogonality relation

$$\int_0^a r J_0(\gamma_n r) J_0(\gamma_{n'} r) dr = \frac{a^2}{2} J_0^2(x_{1n}) \delta_{nn'}. \qquad (B.17)$$

Second orthogonality relation

Here $m = 1$ is used and again the upper limit is $s = x_{1n} = \gamma_n a$ where γ_n is the nth root of $J_1(s)$. Using the recursion relations and Eq.(B.15) integrated between 0 and a gives another Bessel recursion relation

$$\begin{aligned}
\int_0^a r J_1^2(\gamma_n r) dr &= \frac{1}{\gamma_n^2} \int_0^{x_{1n}} s J_1^2(s) ds \\
&= \frac{1}{\gamma_n^2} \left[\frac{s^2}{2} J_1^2(s) - \frac{1}{2} J_1^2(s) + \frac{1}{2} (s J_1'(s))^2 \right]_0^{x_{1n}} \\
&= \frac{1}{\gamma_n^2} \frac{1}{2} (x_{1n} J_1'(x_{1n}))^2 .
\end{aligned} \qquad (B.18)$$

However, from Eq.(B.5)

$$J_0(x_{1n}) = -\frac{1}{x_{1n}} J_{-1}(x_{1n}) - J_{-1}'(x_{1n}) \qquad (B.19)$$

and since $J_{-1}(x_{1n}) = 0$ and $J_{-1}(s) = +J_1(s)$ it is seen that Eq.(B.18) becomes

$$\int_0^a r J_1^2(\gamma_n r) dr = \frac{1}{2} [a J_0(x_{1n})]^2 . \qquad (B.20)$$

Combining Eqs.(B.11) and (B.20) gives the second Bessel orthogonality relation

$$\int_0^a r J_1(\gamma_n r) J_1(\gamma_{n'} r) dr = \frac{a^2}{2} J_0^2(x_{1n}) \delta_{nn'} . \qquad (B.21)$$

APPENDIX C
Capacitor Banks

The various pulsed circuits of most contemporary spheromak experiments are powered by a capacitor bank or a pulse forming network (which can be thought of as a chain of capacitor banks). Although these circuits can provide the large pulsed powers required, every circuit component typically has substantial non-ideal characteristics which must be taken into account in order to obtain the desired performance.

A capacitor bank is essentially a damped harmonic oscillator. A capacitor C with charge Q discharges into an inductance L and a resistor R which are in series. Both L and R are made up of the resistance and inductance of the load, the transmission line, the switch, and the capacitor. The circuit equation is

$$L\ddot{Q} + R\dot{Q} + \frac{Q}{C} = 0 \tag{C.1}$$

or equivalently

$$\ddot{Q} + 2\gamma\dot{Q} + \omega_0^2 Q = 0 \tag{C.2}$$

where

$$\gamma = \frac{R}{2L}, \quad \omega_0^2 = \frac{1}{LC}. \tag{C.3}$$

The general solution of Eq.(C.2) is $\exp(i\omega t)$ where $\omega = i\gamma \pm \sqrt{\omega_o^2 - \gamma^2}$. The circuit is called underdamped if $\omega_o^2 > \gamma^2$, critically damped if $\omega_o^2 = \gamma^2$ and overdamped if $\omega_o^2 < \gamma^2$ and the circuit time behavior will be qualitatively different for these three cases.

The capacitor has charge Q_0 and the current from the capacitor is $I = -\dot{Q}$. Since the current is zero at the instant the switch is closed, the boundary conditions at $t = 0$ are $Q = Q_0$ and $\dot{Q} = 0$. The voltage on the capacitor is $V = Q/C$.

For an underdamped circuit, the solution satisfying these boundary conditions will be

$$Q = Q_0 e^{-\gamma t} \left[\frac{\gamma}{\sqrt{\omega_0^2 - \gamma^2}} \sin\left(\sqrt{\omega_0^2 - \gamma^2}\, t\right) + \cos\left(\sqrt{\omega_0^2 - \gamma^2}\, t\right) \right] \tag{C.4}$$

with corresponding current

$$I = -\dot{Q} = \omega_0^2 Q_0 e^{-\gamma t} \frac{\sin\left(\sqrt{\omega_0^2 - \gamma^2}\, t\right)}{\sqrt{\omega_0^2 - \gamma^2}}. \tag{C.5}$$

For a critically damped circuit, the solutions will be

$$Q = Q_0 e^{-\gamma t}(1 + \gamma t) \tag{C.6}$$

and

$$I = -\dot{Q} = \omega_0^2 t Q_0 e^{-\gamma t}. \tag{C.7}$$

For an underdamped circuit, these will be

$$Q = Q_0 e^{-\gamma t}\left[\frac{\gamma}{\sqrt{\omega_0^2 - \gamma^2}} \sinh\left(\sqrt{\omega_0^2 - \gamma^2}\, t\right) + \cosh\left(\sqrt{\omega_0^2 - \gamma^2}\, t\right)\right] \tag{C.8}$$

and

$$I = -\dot{Q} = \omega_0^2 Q_0 e^{-\gamma t} \frac{\sinh\left(\sqrt{\omega_0^2 - \gamma^2}\, t\right)}{\sqrt{\omega_0^2 - \gamma^2}}. \tag{C.9}$$

L, R, and C are best determined from the actual circuit performance because of unavoidable stray inductances and resistances in the circuit. It is not unusual for the transmission line inductance or resistance to dominate circuit behavior because typically the plasma load has very low resistance and inductance. If the transmission line resistance dominates, most of the power supply energy goes into heating up the transmission line, rather than creating and sustaining plasma. If the transmission line inductance dominates, the rise time will be slowed down and the maximum current will be reduced.

The actual values of L, R, and C in an underdamped circuit can be determined from the current and voltage waveforms as follows:

1. Determination of L : At early times when $\gamma t \ll 1$ and $\omega_0 t \ll 1$, the current is nearly zero and so the middle term in Eq.(C.1) is small compared to the other terms. Balancing the first and last terms in Eq.(C.1) gives

$$L = -\frac{V_0}{(dI/dt)_{t=0}} \tag{C.10}$$

where $V_0 = CQ_0$ is the initial capacitor voltage and $(dI/dt)_{t=0} = -\ddot{Q}_{t=0}$ is the initial rate of rise of the current.

2. Determination of R : At the time t_{max} when the current is maximum, the first term in Eq.(C.1) vanishes leaving

$$R = \frac{V(t_{max})}{I(t_{max})} \qquad (C.11)$$

so that the circuit resistance can be determined by measuring the voltage and current at the time when the current is at its maximum.
3. Using the measured values of R and L, the value of γ can be determined.
4. From Eq.(C.5), it is seen that the first zero crossing of the current occurs when $\sqrt{\omega_0^2 - \gamma^2}\, t = \pi$. This will be the first half-period of oscillation and so denoting this time as $t_{1/2}$ it is seen that $\omega_0^2 = \pi^2/t_{1/2}^2 + \gamma^2$. Using the measured $t_{1/2}$ gives ω_0^2 and thus C is determined using $C = 1/\omega_0^2 L$.

APPENDIX D
Selected Formulae

Magnetic curvature, Eq.(2.18), derivation on page 22

$$\kappa = \hat{B} \cdot \nabla \hat{B} = -\hat{B} \times \left(\nabla \times \hat{B}\right) = -\frac{\hat{R}}{R}$$

Magnetic force, Eq. (2.17), derivation on page 22

$$\mathbf{J} \times \mathbf{B} = -\frac{B^2}{\mu_0} \frac{\hat{R}}{R} - \nabla_\perp \left(\frac{B^2}{2\mu_0}\right)$$

Magnetic force, Eq.(2.19), derivation on page 22

$$\mathbf{J} \times \mathbf{B} = \frac{B^2}{\mu_0} \kappa - \nabla_\perp \left(\frac{B^2}{2\mu_0}\right)$$

General form of axisymmetric magnetic field, Eq.(2.22) on page 25

$$\mathbf{B} = \frac{1}{2\pi} \nabla \psi(r,z) \times \nabla \phi + r B_\phi(r,z) \nabla \phi$$

Axisymmetric magnetic field in terms of poloidal, toroidal fluxes, Eq.(3.18) on p. 43

$$\mathbf{B} = \frac{1}{2\pi} \left(\nabla \Phi \times \nabla \theta + \nabla \psi \times \nabla \phi\right)$$

Canonical momentum in terms of poloidal flux, Eq.(2.27) on page 25

$$P_\phi = m r v_\phi + q\psi/2\pi$$

325

Safety factor defined in terms of fluxes, Eq.(2.28) on page 28

$$q(\psi) = \frac{d\Phi}{d\psi}$$

Safety factor in terms of fields, Eq.(2.37) on page 31

$$q = \frac{1}{2\pi} \oint \frac{B_\phi}{rB_r} dr$$

Safety factor on magnetic axis, Eq.(2.47) on page 32

$$q_{axis} = \frac{\left(\sqrt{\psi_{rr}/\psi_{zz}} + \sqrt{\psi_{zz}/\psi_{rr}}\right)}{r_{axis}} \frac{B_\phi^{axis}}{\mu_0 J_\phi^{axis}}$$

Magnetic helicity, Eq.(3.4) on page 40

$$K = \int_V \mathbf{A} \cdot \mathbf{B} \, d^3r$$

Relative helicity, Eq. (3.34) on page 48

$$K_{rel} = \int_V d^3r \, (\mathbf{A} + \mathbf{A}_{vac}) \cdot (\mathbf{B} - \mathbf{B}_{vac})$$

Conservation equation for relative helicity, Eq.(3.48) on page 52

$$\frac{dK_{rel}}{dt} = -2 \int_{V_a} d^3r \mathbf{E} \cdot \mathbf{B} - 2 \int_{S_a} d\mathbf{s} \cdot (\phi \mathbf{B})$$

Helicity accounting rules for twists, links, crossovers, see page 62

Isolated Taylor state equilibrium, Eq.(4.7) on page 75,

$$\mu_0 \mathbf{J} = \lambda \mathbf{B}$$

Selected Formulae

Relationship between energy and helicity of isolated Taylor state, Eq.(4.12) on page 76

$$\lambda = \frac{\int B^2 d^3r}{\int \mathbf{A} \cdot \mathbf{B} d^3r} = 2\mu_0 \frac{W}{K}$$

Relation between toroidal, poloidal magnetic energies for an isolated Taylor state, Eq.(4.14) on page 77

$$W_{tor} = W_{pol}$$

Bessel Function Model (axial uniform, axisymmetric state), Eq.(4.24) on page 78

$$B_r(r,\phi,z) = 0$$
$$B_\phi(r,\phi,z) = \bar{B} J_1(\lambda r)$$
$$B_z(r,\phi,z) = \bar{B} J_0(\lambda r)$$

Taylor State in spherical geometry, Eq.(4.42) on page 83

$$B_r = 2B_0 \frac{a}{r} j_1(\lambda r) \cos\theta$$
$$B_\theta = -B_0 \frac{a}{r} \frac{\partial}{\partial r}(r j_1(\lambda r)) \sin\theta$$
$$B_\phi = \lambda a B_0 j_1(\lambda r) \sin\theta.$$

Poloidal flux in spherical geometry, Eq.(4.41) on page 82

$$\psi(r,\theta) = 2\pi r a B_0 j_1(\lambda r) \sin^2\theta$$

Rate of injection of helicity, Eq.(5.20) on page 92

$$\left(\frac{dK_{rel}}{dt}\right)_{inj} = 2\psi_{open}(\phi_1 - \phi_2)$$

APPENDIX D

Relation between current and open flux ψ_{open} for driven uniform λ Taylor state, Eq.(5.25) on page 93

$$\mu_0 I = \lambda \psi_{open}$$

Impedance of driven Taylor state with uniform resistivity, uniform λ; Eq.(5.27) on page 94

$$Z = \frac{2\mu_0 \eta}{\psi_{open}^2} W$$

Approximate (non-rigorous derivation) β limit, Eq.(6.61) on page 107

$$\left(\frac{q_{axis} - q_{wall}}{(q_{axis} + q_{wall})/2} \right)^2 - 12\beta > 0$$

Radial eigenvalue, Eq.(9.7) on page 130

$$\gamma_1 = \frac{x_{11}}{a}$$

where $x_{11} = 3.83$ is first root of J_1

Isolated cylindrical spheromak field components, Eqs.(9.12a)-(9.12c) on page 131

$$B_r = B_0 \frac{\pi}{\gamma_1 h} J_1(\gamma_1 r) \cos(k_1 (z - h))$$
$$B_\phi = -B_0 \frac{\lambda}{\gamma_1} J_1(\gamma_1 r) \sin(k_1 (z - h))$$
$$B_z = -B_0 J_0(\gamma_1 r) \sin(k_1 (z - h))$$

Eigenvalue of isolated cylindrical spheromak, Eq.(9.13) on page 131

$$\lambda = \pm \sqrt{\frac{x_{11}^2}{a^2} + \frac{\pi^2}{h^2}}.$$

Poloidal flux function of isolated cylindrical spheromak, Eq.(9.18) on page 132

$$\psi = -\frac{2\pi}{\gamma_1} B_0 r J_1(\gamma_1 r) \sin(k_1(z-h))$$

Magnetic axis radius of isolated cylindrical spheromak, Eq.(9.20) on page 132

$$r_{axis} = \frac{x_{01}}{\gamma_1} = \frac{x_{01}}{x_{11}} a = 0.63a$$

Poloidal flux on magnetic axis of isolated cylindrical spheromak, Eq.(9.21) on page 132

$$\psi_{axis} = 2\pi B_0 \frac{x_{01}}{\gamma_1^2} J_1(x_{01})$$

Toroidal flux at wall of isolated cylindrical spheromak, Eq.(9.22) on page 132

$$\Phi_{wall} = 2B_0 \frac{\lambda}{k_1 \gamma_1^2} [1 - J_0(x_{11})]$$

Toroidal current of isolated cylindrical spheromak, Eq.(9.23) on page 133

$$I_\phi = 2B_0 \frac{\lambda^2}{\mu_0 k_1 \gamma_1^2} [1 - J_0(x_{11})]$$

Helicity content of isolated cylindrical spheromak, Eq.(9.30) on page 134

$$K = \pi \lambda a^2 h \frac{B_0^2}{\gamma_1^2} J_0^2(x_{11})$$

Magnetic energy content of isolated cylindrical spheromak, Eq.(9.31) on page 135

$$W = \pi a^2 h \frac{\lambda^2}{\gamma_1^2} \frac{B_0^2}{2\mu_0} J_0^2(x_{11})$$

Safety factor on magnetic axis of isolated cylindrical spheromak, Eq.(9.37) on page 137

$$q_{axis} = \frac{1}{x_{01}}\sqrt{1 + \frac{x_{11}^2 h^2}{\pi^2 a^2}}$$

Safety factor at wall of isolated cylindrical spheromak, Eq.(9.38) on page 137

$$q_{wall} = \frac{1}{4}\sqrt{1 + \frac{x_{11}^2 h^2}{\pi^2 a^2}}$$

λ at electrode of driven cylindrical spheromak, Eq.(11.4) on page 157

$$\lambda = \frac{\mu_0 I_{gun}}{\bar{\psi}_{gun}}$$

Magnetic field components of driven cylindrical spheromak, uniform λ assumption, Eq.(11.17) on page 161

$$B_r(r,z) = -\frac{\bar{\psi}_{gun}}{2\pi}\sum_{n=1}^{\infty} c_n \gamma_n J_1(\gamma_n r)\frac{\partial S_n(z,\lambda)}{\partial z}$$

$$B_\phi(r,z) = \lambda\frac{\bar{\psi}_{gun}}{2\pi}\sum_{n=1}^{\infty} c_n \gamma_n J_1(\gamma_n r) S_n(z,\lambda)$$

$$B_z(r,z) = \frac{\bar{\psi}_{gun}}{2\pi}\sum_{n=1}^{\infty} c_n \gamma_n^2 J_0(\gamma_n r) S_n(z,\lambda)$$

where, if $\gamma_n^2 > \lambda^2$

$$S_n(z,\lambda) = -\frac{\sinh\left(\sqrt{\gamma_n^2 - \lambda^2}(z-h)\right)}{\sinh\left(\sqrt{\gamma_n^2 - \lambda^2}\,h\right)}$$

and, if $\gamma_n^2 < \lambda^2$

$$S_n(z,\lambda) = -\frac{\sin\left(\sqrt{\lambda^2 - \gamma_n^2}(z-h)\right)}{\sin\left(\sqrt{\lambda^2 - \gamma_n^2}h\right)}$$

and from Eq.(11.20) on page 161

$$c_n = \frac{4\pi}{[x_{1n}J_0(x_{1n})]^2} \frac{\int_0^a B_z^{gun}(r) J_0(x_{1n}r/a) r \, dr}{\bar{\psi}_{gun}}.$$

Poloidal flux for driven cylindrical spheromak with uniform λ, Eq.(11.23) on page 162

$$\psi(r,z) = \bar{\psi}_{gun} \sum_{n=1}^{\infty} c_n \gamma_n r J_1(\gamma_n r) S_n(z,\lambda)$$

Impedance of a driven cylindrical spheromak, uniform λ assumption, see page 179

MHD helicity flux, Eq.(12.50) on page 203

$$\mathbf{h} = -\left\langle 2\tilde{\phi}\tilde{\mathbf{B}} + \tilde{\mathbf{A}} \times \frac{\partial \tilde{\mathbf{A}}}{\partial t} \right\rangle$$

Dynamo electric field, Eq.(12.49) on page 203

$$\mathbf{E}_{dyn} = \frac{\mathbf{B}}{2B^2} \nabla \cdot \mathbf{h}$$

Local balance between helicity flux and dissipation, Eq.(12.56) on page 204

$$-\frac{1}{2\mu_0} \mathbf{h} \cdot \nabla \lambda = \eta J^2.$$

INDEX

accelerated spheromaks, 269
   tokamak fuel injection, 274
Alfven waves, 66, 109, 225
   field line topology, 26
axisymmetry
   Cowling's theorem, 35

Bessel function model, 79
Bessel orthogonality relations, 317
beta, 24
   limit, 103
   pressure driven instability, 101
Bohm diffusion, 210
bolometry, 261
bounding surface
   equipotential, 91
   not an equipotential, 92
breakdown, 227

canonical angular momentum, 25, 207
capacitor bank design, 321
Chandrasekhar-Kendall functions, 78
charge exchange, 214
chirality, 284
classical/neoclassical diffusion, 211
coaxial plasma gun, 110
   crow-barred (short-circuited), 91
   overview of, 110
   sequence of operation, 110
coaxial spheromak
   analytic model for driven system, 159
   evolution of topology, 157
conduction, 213
confinement, 207
   particle, 208
confinement time, 209
   magnetic energy, 215
   magnetic helicity, 214

   thermal energy, 215
conflict between confinement and dynamo, 208
conical theta pinch, 72
constraints on dynamo modes, 187
controlled thermonuclear fusion, 263
coronal equilibrium, 237
Cowling's theorem, 35
   relation to spheromak formation, 110
crowbarring (short-circuiting), 176
current drive, 275
current measurement, 243

density measurement
   interferometry, 252
   Langmuir probe, 249
deviation from Taylor state, 196
   observations of, 197
discharge cleaning, 236
doubly connected volume, 49
drift wave turbulence, 212
driven spheromak, 155
dynamo
   Cowling's theorem, 36
   impossibility of axisymmetric, 36
   relaxation mechanisms, 185
dynamos
   along geometric axis, 188
   along magnetic axis, 189
   effect on flux function, 200
   fluctuation growth rate, 199
   nonlinearity of, 186
   observations of, 189
   relation to lambda gradient, 202

edge loss measurements, 220
efficiency of coaxial gun, 174
electrode materials, 239

electron inertia
  helicity conservation, 54
  single species helicity, 60
electron temperature measurement, 251
  Thomson scattering, 257
energy
  free for relaxation, 74
  inductive, 35
energy transport
  Bohm, 210
  charge exchange, 214
  classical/neoclassical, 211
  conduction, 213
  diffusive, 210
  line radiation, 213
  non-diffusive, 213
  via drift wave turbulence, 212
equilibrium field
  provided by coils, 145
  provided by wall, 145

field reversed configuration (FRC), 1
field-aligned current
  relation of uniformity of to MHD stability, 102
flipped spheromak, 167, 183
  relation to resonance of S functions, 167
flux amplification, 170
flux conserver
  skin depth and skin time, 239
flux coordinates, 42
flux surfaces
  closed, 207
  evolution with increasing lambda in driven configuration, 162
flux tube
  bifurcation, 298
  fragmentation, 298
  safety factor, 42
  twist, 42, 44
  twisted, 63
  voltage drop along, 217

flux tube linkage, 39
  ambiguous, 45
flux tubes
  splicing, 66
flux-core, 117
footpoints, 296
force-free equation
  derivation of in isolated relaxed configuration, 75
  isolated spheromak, 129
force-free states
  cylindrical, 77
frozen-in flux
  condition for, 33
  single species, 59
frozen-in magnetic fields
  failure to have, 60
frozen-in magnetic flux, 33
fuel injection into tokamaks, 274
fusion reactor, 263

gas puff valve
  operation of, 232
gas puffing and breakdown, 114
gas valves, 112, 114
gauge
  ambiguity, 45
  invariance, 45
    doubly connnected volume, 50
    relaxation of isolated configuration, 73
Gauss's theorem
  doubly-connected volumes, 37
gettering, 236
gettering experiments, 220
Grad-Shafranov equation, 197
ground loops, avoidance of, 242
gun flux definition, 157

Hall term, 54
helicity
  conducting wall as insulator for, 92, 143

helicity flow in Sun-Earth connection, 292
helicity injection current drive, 275
Helmholtz equation
  for driven coaxial system, 158
  in spherical geometry, 81
  solar prominence, 294
high temperature experiments, 221
hoop force, 20, 35, 117, 145, 147

impedance, 92, 155
  calculation of for driven coaxial gun, 177
  comparison with experiment, 180
implosion field, 121
impurities
  cleanup of, 234
  line radiation, 236
impurity radiation, 260
inductance
  budget, 240
  tendency of plasma to maximize, 35
insulator for helicity, wall as, 92
integrator circuit analysis, 244
interferometry, 252
interplanetary magnetic field, 288
ion heating
  anomalous, 222, 224
  measurement of, 223
  stochastic, 224
ion temperature
  Doppler measurement, 255
ion temperature measurement
  neutral particle analysis, 257
  via charge exchange, 223
  via Doppler shift, 224
isolated spheromak, 129
  flux, current, magnetic field, helicity, energy, 129
  safety factor, 136

Lagrange multiplier, 89
lambda

analogy to temperature, 126
as a Lagrange multiplier, 74
as a multivalued function of relative helicity, 176
as an eigenvalue, 75
gradient, 183, 196, 199, 202
in system with open field lines, 89
isolated spheromak, 131
lowest value and minimum energy state, 76
ratio of energy to helicity, 76
relation to gun flux and gun current in driven system, 157
relation to twist, 76
spheromak as a source, 184
various meanings of, 183
when independent parameter, 90
Langmuir probe, 249
Larmor radius
  poloidal, 26
Lawson criterion, 264
leaky system, 127
Legendre functions, associated, 81
line radiation, 213, 236
line-tying and relaxation, 300
lowest energy magnetic field, 17

magnetic axis
  development of as lambda increases, 165
  safety factor, 31
magnetic cloud, 284
magnetic energy
  different decay rates for toroidal, poloidal, 136
  isolated spheromak, 129, 135
  of an isolated state, 75
  relation to helicity in system with open field lines, 90
  relative, 90, 174
  relative, in driven coaxial system, 175
  transport, 214

INDEX

magnetic energy principle, 95
magnetic field
   curvature, 21
   differences between field, line, flux, 299
   errors, 26
   experimental measurements, 136
   field line trajectory, 295
   flux, 24
   force, 22
   isolated spheromak, 129
   lowest energy, 17
   poloidal, 18
   relation between field, field lines, flux, 26
   stress tensor, 20
   sudden filling of volume as lambda increases, 166
   symmetry, 24
   tension, 21
   toroidal, 18
   toroidal and poloidal in driven coaxial system, 133
   vacuum, 17
      in a coaxial plasma gun, 114
      relaxation of system with open field lines, 89
magnetic field line, 26
magnetic flux, 24
   annihilation during reconnection, 68
   conservation, 32
   frozen-in, 33
magnetic helicity, 37
   accounting rules, 62
   conservation
      compared to energy conservation, 57
      time-dependent volume, 56
   conservation during reconnection, 63
   conservation equation, 50
   conservation in doubly connected volume, 55
   conservation in fixed volume, 51
   conservation of during relaxation, 71
   definition, 39
   dissipation of, 53
   downhill flow of, 183, 184
   evolution of for driven coaxial system, 171
   flux of, 53
   geometric interpretation, 61
   injection, 56
   injection in a coaxial plasma gun, 112
   injection of in a coaxial plasma gun, 115
   injection rate, 92
   isolated spheromak, 129, 134
   relation to energy in system with open field lines, 90
   relative, 45
      in a driven system, 87
   resonances of in a driven system, 173
   single species, 58
   topological interpretation, 62
   transfer of, 66
   transport, 214
   writhe, linking, crossing, 61
magnetic probes, 243
magnetic reconnection, 60
   dissipation, 66
   observations of, 224
   studied by colliding spheromaks, 277
magnetohdrodynamic instability
   relaxation, 74
magnetohydrodyamic instability, 35
magnetohydrodyamics
   stability analysis of an equilibrium, 96
magnetohydrodynamic
   instabilities, 28
   instability

pressure driven and current driven, 101
instability, good and bad curvature, 103
stability
shear, 103
magnetohydrodynamics
critique of, 95
equations for ideal, 95
mechanical forces, 241
Mercier stability analysis, 108
merging spheromaks, 223
magnetic reconnection, 279
minimum energy magnetic field, 17
minimum energy state, 74
non-axisymmetric, in a long cylinder, 80
minimum energy states
in a long cylinder, 79
monolayer deposition, 235

neutral line, 293
noise, pulsed power
mitigation of, 241
non-analyticity, 38
non-axisymmetric behavior, 110

Ohmic ignition, 267
oil contamination, 236
opening switch, 281

parallel electric field
helicity dissipation, 53
Paschen curve, 217, 227
gas puff valve, 232
Paschen gas breakdown criterion, 114
Paschen minimum, 231
perfect conductor, 33
poloidal field energy
equality to toroidal field energy in a spheromak, 76
poloidal flux
eigenfunctions, 164

evolution of profile as lambda increases, 164
for driven coaxial spheromak, 163
poloidal magnetic field
magnetic field, 18
poloidal magnetic flux, 24
particle confinement, 25
power supply
load-line analysis, 179
power supply impedance, 156
prominence, 284
loss of equilibrium, 296
prominence simulation experiment, 300

q (see safety factor), 28

radiated power estimate, 238
radiative recombination, 236
regimes
classification of, 123
relative helicity
and Taylor relaxation in system with open field lines, 87
doubly connected volume, 49
evaluation of for driven coaxial spheromak, 171
relaxation, 71
in a system with open field lines, 87
relaxed state, 74
in spherical geometry, 80
reversed field pinch (RFP), 1, 71
Rogowski coils, 243

S function, definition of, 159
S-shape, 287, 297
safety factor, 28
in spherical geometry, 83
isolated cylindrical spheromak, 136
magnetic axis, 31
of a flux tube, 42
on magnetic axis in spherical geometry, 83
profile, 30

profile in a spherical spheromak, 83
tilt stability, 148
shear, 103
  in spherical geometry, 85
simply connected volume, 49
slingshot, 224
solar corona, 284
solar plasmas, 283
solar prominence, 284
solar prominence simulation, 293
space plasmas, 283
spheromak as a fusion reactor, 263
spheromak experiment
  operating point of driven system, 179
spheromak experiments
  gas pressure operation regime, 115
spheromak formation schemes, 109
  inductive method, 117
  non-axisymmetric method, 117
  Z-theta pinch, 120
spheromak fusion reactor
  design studies, 265
  Ohmic heating, 267
  parameters, 268
  power density, 268
spheromak gun voltage, 179
stability analysis, 99
Stark broadening, 256
stellarator, 1
Stokes's theorem
  doubly connected volumes, 37
Sun-Earth connection, 291
surface boundary condtions, 90
symmetry, 24, 208

Taylor relaxation, 71
  isolated configuration, 71
Taylor state
  deviation from, 196
  isolated configuration, 73
  stability of, 103
temperature measurement

Langmuir probe, 249
thermodynamics
  analogy to, 123
Thomson scattering, 257
tilt stability
  analysis, 148
  criterion for, 148
  relation to safety factor, 148
  role of wall, 146
tokamak, 1
  fusion reactor, 264
  refueling by spheromak injection, 274
topology
  doubly connected, 49
  evolution of in a coaxial spheromak, 157
  simply connnected, 49
  transformation as lambda increases, 110
toroidal confinement, comparison of methods, 2
toroidal current
  isolated spheromak, 133
toroidal field energy
  equality to poloidal field energy in a spheromak, 76
transport
  measurements of, 218
triple probe, 249
twist
  instability of localized, 66
  relation to lambda, 76

vacuum magnetic field, 17
variational problem, 74
  for configuration with open field lines, 89
vector identities, 315
vector potential
  in driven coaxial system, 134
vertical field, 145, 147
voltage collapse, 115

voltage measurements, 248

wall
  image currents, 146
  impurities, 234
  liquid, for fusion reactor, 265
  providing equilibrium, 143
  recycling, 235
  tilt stability, 146

x-ray sources, 281
x-rays, observation of, 222

Information on Use of Copyrighted Material

Copyrighted material has been used with the permission of the copyright owners who are listed below.

Figure	Reference	Copyright Owner (year of copyright)
1.2	[8]	Elsevier Science (1959)
1.3	[8]	Elsevier Science (1959)
1.4	[15]	American Institute of Physics (1964)
7.3	[60]	American Physical Society (1983)
7.5	[105]	American Institute of Physics (1989)
7.6	[106]	International Atomic Energy Agency (1985)
7.7	[25]	American Physical Society (1981)
7.8	[21]	American Physical Society (1980)
9.2	[23]	American Institute of Physics (1983)
9.3	[23]	American Institute of Physics (1983)
12.1	[115]	American Institute of Physics (1985)
12.2	[115]	American Institute of Physics (1985)
12.3	[115]	American Institute of Physics (1985)
12.4	[110]	American Institute of Physics (1991)
12.5	[69]	American Physical Society (1993)
13.1	[141]	International Atomic Energy Agency (1989)
16.1	[58]	International Atomic Energy Agency (1982)
16.4	[31]	American Physical Society (1988)
16.6	[35]	American Nuclear Society (1994)
16.7	[40]	American Physical Society (1990)
16.8	[42]	American Physical Society (1994)
16.9	[47]	American Institute of Physics (1993)